Lecture Notes in Computer Science 6125

Commenced Publication in 1973
Founding and Former Series Editors:
Gerhard Goos, Juris Hartmanis, and J.

Editorial Board

Agostino Dovier   Enrico Pontelli (Eds.)

A 25-Year Perspective on Logic Programming

Achievements of the Italian Association
for Logic Programming, GULP

 Springer

Volume Editors

Agostino Dovier
Università di Udine
Dip. di Matematica e Informatica
Via delle Scienze 206, 33100 Udine, Italy
E-mail: dovier@dimi.uniud.it

Enrico Pontelli
New Mexico State University
Department of Computer Science
P.O. Box 30001, MSC CS, Las Cruces, NM 88003, USA
E-mail: epontell@cs.nmsu.edu

Library of Congress Control Number: 2010929758

CR Subject Classification (1998): D.1.6, F.4.1, F.3, F.4, I.2.3, I.2

LNCS Sublibrary: SL 2 – Programming and Software Engineering

ISSN   0302-9743
ISBN-10   3-642-14308-3 Springer Berlin Heidelberg New York
ISBN-13   978-3-642-14308-3 Springer Berlin Heidelberg New York

springer.com

© Springer-Verlag Berlin Heidelberg 2010
Printed in Germany

Typesetting: Camera-ready by author, data conversion by Scientific Publishing Services, Chennai, India
Printed on acid-free paper   06/3180

Foreword

This book celebrates the 25th anniversary of GULP—the Italian Association for Logic Programming. Authored by Italian researchers at the leading edge of their fields, it presents an up-to-date survey of a broad collection of topics in logic programming, making it a useful reference for both researchers and students.

During its 25-year existence, GULP has organised a wide range of national and international activities, including both conferences and summer schools. It has been especially active in supporting and encouraging young researchers, by providing scholarships for GULP events and awarding distinguished dissertations.

We in the international logic programming community look upon GULP with a combination of envy, admiration and gratitude. We are pleased to attend its conferences and summer schools, where we can learn about scientific advances, catch up with old friends and meet young students. It is an honour for me to acknowledge our appreciation to GULP for its outstanding contributions to our field and to express our best wishes for its continuing prosperity in the future.

March 2010

Robert Kowalski
Imperial College London

Preface

On June 18, 1985, a group of pioneering researchers, including representatives from industry, national research labs, and academia, attended the constituent assembly of the *Group of researchers and Users of Logic Programming (GULP)* association. That was the starting point of a long adventure in science, that we are still experiencing 25 years later.[1] This volume celebrates this important event.

What about the editors of this volume? On that date, one of us was completing his secondary school studies, the other his mandatory military service. But only one year later, the two of us met in the introductory class of the computer science program at the University of Udine, and that was the beginning of a logic programming experience that spans the majority of our careers in academia.

With excitement, humbleness, and profound honor, and after a formal nomination as editors from the GULP assembly (during the CILC'08 meeting), we embarked upon the mission of developing this volume. The purpose of this effort is to celebrate an important milestone in the world of logic programming, the 25th anniversary of GULP. GULP is the oldest formal logic programming association (the international Association for Logic Programming, for example, was established in 1986), and, over the last 25 years, GULP has promoted research activities whose results and directions are at the core of the whole logic programming world.

Summarizing 25 years of research on logic programming in a single volume is a daunting and perhaps impossible task. We were forced to make difficult decisions in selecting the topical areas of logic programming to be analyzed in the various chapters; this task was particularly complex, due to the diversity of the research initiatives in logic programming that have developed over the years in Italy. In the end, we decided to concentrate on those areas that, historically, have been at the core of logic programming research in Italy; we wish to apologize to those researchers whose areas have been excluded from this volume.

Each chapter of this volume has been co-authored by several researchers. In particular, we have attempted to create a balance between historical developments and current state of the art by pairing, in each chapter, younger researchers with more established leaders in the field (but we will not explicitly identify who is who...). The response from the logic programming community to our invitations to author chapters was overwhelmingly positive; 35 researchers enthusiastically accepted to participate in this initiative. The effort resulted in 14 chapters, each providing a fresh and useful overview of a different area of logic programming. Thanks to the hard work of the authors, each chapter represents

---

[1] As a remark, Italy was the reigning football world champion in 1985, as they are today.

a great analysis of a specific research field, providing both historical perspectives as well as a precise discussion of the current state of the art. The authors also provide an interesting view of how the contributions of Italian researchers have shaped the field of logic programming over the years.

This volume represents the logical continuation of the volume edited by Maria I. Sessa in 1995, celebrating the 10th anniversary of GULP. While several of the chapters address analogous topics (i.e., theoretical foundations, program transformations, non-monotonic reasoning, constraint logic programming, concurrent logic programming, program verification), other chapters have either been replaced, due to the lack of intense research (e.g., metalogic programming) or expanded into more detailed chapters, to reflect the changes in directions within the field. For instance, the stable models chapter has evolved into the more mature answer set programming chapter, and the chapter on applications to software engineering has evolved into a wider scope applications chapter. We added other new chapters that represent very active fields, like databases and web, agents and multi-agent systems, two chapters on extensions of logic programming (functional logic programming and higher order programming), and a seminal paper on research in automated theorem proving.

The organization of this volume follows a structure that highlights what we perceived to be the historical dependencies among the various areas. These dependencies are summarized in the graph in Fig.1.

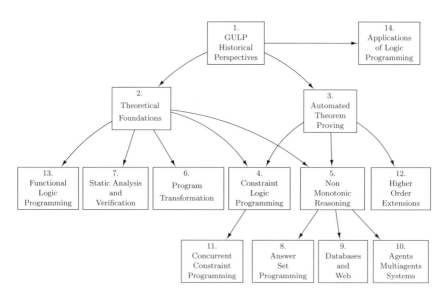

Fig. 1. The GULP tree and book structure

The volume opens with an historical perspective of the first 25 years of the association, written by the current GULP president, Gianfranco Rossi. Gianfranco has witnessed the evolution of GULP since its inception, and he reports a detailed history of the GULP association in his chapter. He also provides a personal view of the directions to be followed by GULP to avoid past mistakes and expand the success of logic programming in Italy, especially in terms of impact on the industrial world.

The roots of logic programming research in Italy can be traced back to the research efforts in the areas of automated theorem proving and theoretical computer science (e.g., programming languages semantics). Indeed, looking back at the areas of the various contributions reported in the first volumes of the proceedings of GULP, one can note how semantical foundations and theorem proving are at the backbone of many of the reported contributions. The editors decided to open the research overview of this volume with two chapters dedicated to these two foundational areas. These are identified at the top of the graph and reported in Chaps. 2 and 3.

A reason for the great initial success of logic programming was undoubtedly the elegance of its semantics. The set of ground atoms that can be inferred from a program P, using SLD resolution (operational semantics), can be proved to be equivalent to the minimum Herbrand model of P (logical semantics) and, in turn, equivalent to the least fixpoint of a continuous operator dependent on P (declarative semantics). Chapter 2, developed by Annalisa Bossi and Chiara Meo, gives an overview of the original roots of research in the theoretical foundations of logic programming. Work in this area was spearheaded by the group of Giorgio Levi (first president of GULP) and his colleagues in Pisa and Torino, and was instrumental in placing Italian logic programming research on the international map.

The clear ties of logic programming, since its inception, with mathematical logic and theorem proving, have provided ample opportunities for research in automated theorem, laying the foundations to the growth of logic programming. In particular, the completeness proof of SLD resolution as an inference method for first-order theories given as sets of definite clauses, and the Turing completeness of this fragment of first-order logic, are probably the two fundamental contributions of automated reasoning that allowed Kowalski to write the seminal contribution *Predicate Logic as Programming Language*. Chapter 3, developed by Andrea Formisano and Eugenio G. Omodeo revisits the original work in the area of theorem proving, highlighting the ties to logic programming.

A combination of the studies in theorem proving (based on theory-based resolution), and on a generalization of the semantics of logic programming (to the case of non-Herbrand domains) offers the foundations on which the area of constraint logic programming developed. Constraint logic programming enabled the first step towards enhancing the declarative nature of logic programming, often lost in the use of Prolog, and at the same time gaining a level of efficiency required by industrial-strength applications. The combination of declarativeness of logic programming and of efficiency of solvers in suitable theories allows one to solve efficiently real-life problems without the need of writing low-level code. A nice survey of this area is presented in Chap. 4, developed by Marco Gavanelli and Francesca Rossi.

The field of constraint logic programming, thanks also to the intense work conducted in the context of the Fifth Generation Computer Systems project, has evolved to create a revolutionary paradigm that combines logic programming, constraint programming, and concurrency; the challenging issues of concurrent constraint programming are reviewed in Chap. 11, written by Maurizio Gabbrielli, Catuscia Palamidessi, and Frank Valencia.

The original developments on the semantics of logic programming quickly moved towards the investigation of variants of the logic programming paradigm where the traditional elegant properties of logic programming semantics (e.g., uniqueness of the least Herbrand model) fail. This is particularly true in the case of extensions of logic programming developed to handle non-monotonicity, which are vital to the task of knowledge representation and commonsense reasoning. This volume dedicates two related chapters to the investigation of these aspects. The first is Chap. 5, by Laura Giordano and Francesca Toni, which explores the role of logic programming in the area of non-monotonic reasoning and knowledge representation. While techniques for non-monotonic reasoning moved originally in different directions, in recent years the field has witnessed a convergence of effort towards the use of stable model semantics proposed by Gelfond and Lifschitz. The embedding of stable model semantics in a concrete programming paradigm, originated from the concurrent work of Marek, Truszczyński, and Niemelä, led to what is now known as answer set programming. The field is now at the core of logic programming, thanks also to the development of highly competitive solvers. This area is surveyed in Chap. 8, by Piero Bonatti, Francesco Calimeri, Nicola Leone, and Francesco Ricca.

The field of logic-based intelligent agents also traces back its foundations to the area of logic programming and non-monotonic reasoning; this field has matured over the years and Italian research in this domain has gained reputation within the larger umbrella of the international artificial intelligence community. Chapter 10, by Matteo Baldoni, Andrea Omicini, Cristina Baroglio, Viviana Mascardi, and Paolo Torroni, provides an exciting review of work on logic-based methodologies for intelligent agents and multi-agent systems.

The foundations of non-monotonic reasoning have also been deeply tied with two other areas that have witnessed intense research with the support of GULP—databases and intelligent agents. The field of databases has been present at GULP since its beginning; work in the area of deductive databases has offered significant contributions not only to the database community, but also to the development of the foundations of modern logic programming (e.g., the work on DATALOG¬ contributed to answer set programming). Chapter 9, by Francesca Lisi and Sergio Greco, provides an overview of logic programming work in the area of databases and the web.

The work on semantics of logic programming has traditionally provided the foundations for enhancing the understanding of programs; this is essential in order to develop techniques for program transformation, to gain efficiency, and program verification, to guarantee correctness. These two aspects are analyzed in Chap. 6, by Alberto Pettorossi, Maurizio Proietti, and Valerio Senni—which covers the area of program transformations—and Chap. 7, by Giogio Delzanno, Roberto Giacobazzi, and Francesco Ranzato—which provides an overview of research in the areas of static analysis, abstract interpretation, and program verification.

From an automated reasoning point of view, traditional logic programming is just one particular instance of automated deduction with a given first-order language (with definite clauses) and with a particular proof engine (SLD resolution). One can enlarge this schema in several directions, for instance working on the proof structure (e.g., uniform proofs) or admitting higher-order predicates. This area is surveyed in Chap. 12, by Alberto Momigliano and Mario Ornaghi.

The overall area of declarative approaches to programming includes several other paradigms beyond logic programming. In particular, functional programming provides a number of features that are absent in logic programming and that are convenient in many programming tasks. The area of functional logic programming investigates attempts to combine logic programming and functional programming within a single paradigm, which provides the benefits of both logic programming (e.g., search, non-determinism) and functional programming (e.g., higher order constructs). The role of Italian research in functional logic programming has been predominant since its inception, and it is summarized in Chap. 13, by Maria Alpuente, Demis Ballis, and Moreno Falaschi.

Last but not least, Chap. 14, by Alessandro Dal Palù and Paolo Torroni, reviews in detail the main applications of logic programming developed in Italy and/or by Italian researchers in the last 25 years. This chapter represents an ideal closure to this volume—there is wide agreement that the continued success of the field of logic programming vitally depends on investigating the use of logic programming technology to solve concrete real-world problems. The chapter nicely illustrates successful work done and potential directions for future developments.

In closing this introduction, the editors would like to take the opportunity to extend their heartfelt thanks to a number of people who made this effort possible:

- The authors of the chapters, who have tirelessly worked on creating comprehensive overviews of research directions that have developed over 25 years of logic programming (most of them also acted as reviewers of other chapters):

Maria Alpuente	Matteo Baldoni	Demis Ballis
Cristina Baroglio	Piero Bonatti	Annalisa Bossi
Francesco Calimeri	Alessandro Dal Palù	Giogio Delzanno
Moreno Falaschi	Andrea Formisano	Maurizio Gabbrielli
Marco Gavanelli	Roberto Giacobazzi	Laura Giordano
Sergio Greco	Nicola Leone	Francesca Lisi
Viviana Mascardi	Chiara Meo	Alberto Momigliano
Andrea Omicini	Eugenio G. Omodeo	Mario Ornaghi
Catuscia Palamidessi	Alberto Pettorossi	Maurizio Proietti

Francesco Ranzato Francesco Ricca Francesca Rossi
Gianfranco Rossi   Valerio Senni   Francesca Toni
Paolo Torroni   Frank Valencia

- The additional reviewers who provided insightful comments on the various chapters under very strict time constraints:

Sergio Antoy   Nicoletta Cocco
Sandro Etalle   Camillo Fiorentini
Roberta Gori   Evelina Lamma
Michela Milano   Alessandra Mileo
Angelo Montanari   Carla Piazza
Germán Vidal   Alicia Villanueva García

- The Italian logic programming community and GULP, who have created in Italy a nurturing environment for countless young researchers to embrace and appreciate the beauty of logic programming. Personally, the editors would not have been coordinating this volume without the friendship and advice of Gianfranco Rossi and Eugenio Omodeo, who originally introduced us to logic programming.
- Bob Kowalski for his foreword and for what has done and is still doing for the logic programming community.

Finally, we would like to send a "thank you" to all those we love in this and other worlds.

March 2010   Agostino Dovier
                                                     Enrico Pontelli

Table of Contents

Twenty-Five Years of Logic Programming in Italy

Logic Programming in Italy:
A Historical Perspective

Gianfranco Rossi

Dipartimento di Matematica, Università di Parma
gianfranco.rossi@unipr.it

Abstract. The history of Logic Programming in Italy is largely that of
GULP, the Italian Association of Users and Researchers in Logic Programming. This paper provides a historical perspective on the birth and
development of GULP in the last 25 years. The paper is mainly concerned with what has been done in Italy, but it also points out the
many relationships and synergies that emerged—and still exist—in the
field of Logic Programming, between Italy and other countries all over
the world. I identify three main periods in the history of GULP, which
closely correspond to different seasons in the history of Logic Programming in general, and I try to characterize them in terms of activities the
GULP supported and of the achievements obtained by its members.

1 Introduction

The history of Logic Programming (LP) in Italy is largely that of GULP, the
Italian Group of Users and researchers in LP.

GULP was founded 25 years ago (1985) as a non-profit organization. It was
the first national LP association to be established in the world. Since its very
beginning, GULP has been constantly committed to keep the interest in LP and
related themes alive by promoting various initiatives both in research and in
education. Its main role—in my opinion—has been to provide an opportunity
for young researchers to be introduced into an active and challenging research
area in a very informal and friendly way.

LP in Italy, and its representative association GULP, have gone through several phases in the last 25 years. For the sake of the presentation, I will group
these phases in three main periods:

- The Early Years, approximately 1984—1993
- The APPIA-GULP-PRODE Years, approximately 1994—2003
- The CILC Years, approximately 2004—2010.

It turns out that these periods closely correspond to different seasons in the
history of LP in general.

This paper shows how the original activities of GULP has evolved in the last
25 years, mainly with reference to what has been done in Italy but with a wider,
world-wide perspective in mind. To this end, I will try to point out relationships

A. Dovier, E. Pontelli (Eds.): 25 Years of Logic Programming, LNCS 6125, pp. 1–14, 2010.

and synergies that have been established between Italian researchers in LP and the rest of the world. Needless to say, I will try to highlight not only positive but also negative aspects in the history of LP in general and particularly in Italy.

I will conclude by stressing that LP in Italy is still alive and there are many people who still believe in it, although it seems necessary to radically change the way of presenting LP and the LP community to the outside world.

2   The Early Years (ca. 1984—1993)

The idea of creating an association of Prolog users took place during a workshop organized by Luigi Marcolungo and other colleagues from the University of Padua in November 1984. The constituent assembly of GULP was held in Pisa on June 18th, 1985 (see Appendix A.4 for the first executive committee). This is the official starting date, that is widely accepted as the starting point for the history of LP in Italy.

Actually, various Italian researchers were interested in LP well before this date. In particular, as long ago as 1974, Enrico Pagello and some young colleagues from Milan and Padua installed a Prolog interpreter on their computers at Politecnico of Milan and at the University of Padua, using it for their applications in robotics (see also the paper by Dal Palù and Torroni in this volume). On the more theoretical side, various researchers from Pisa were already investigating Prolog programming in the seventies (e.g., [1]).

In the first year, GULP had more than 160 members. The interest in LP, from both the academic and the industrial sides, was constantly growing in Italy and around the world. Since 1986, GULP started to organize an annual national conference on LP (see Appendix A.1). These conferences represented—and still represent—the main occasion for all people (researchers, users, and developers) dealing with LP in Italy to meet and to exchange ideas and experiences.

Main topics of interests in LP in Italy in those years were:

- Transformation of logic programs (including partial evaluation)
- Metalogic programming
- Semantics of logic programs
- Non-monotonic reasoning
- Constraint Logic Programming
- Concurrent Logic Programming
- Program modularity and object-oriented in Prolog.

Italian contributions on all such topics were at the highest international level, as testified by the many contributions presented at LP international conferences and workshops between the 80s and the early 90s. Many connections with researchers all over the world were established in that period. To cite only a few of them, in strict alphabetic order: Maria Alpuente, Krzysztof Apt, Michael Codish, Philippe Codognet, Frank de Boer, Georg Gottlob, Manuel Hermenegildo, Pat Hill, Antonis Kakas, Bob Kowalski, Michael Maher, Germán Vidal, Carlo Zaniolo. All of them, and, of course, many others that I did not mention, had (and, in many cases, still have) strict collaborations with Italian researchers in LP.

A detailed report of the huge amount of work put forward in those years can be found in the book on ten years of Logic Programming in Italy, edited by Maria Sessa [8]. Developments on these and other topics are also discussed in more details in other companion papers in this volume.

Many universities and many centers of the National Research Council (CNR) were involved in these first years of LP in Italy. Among them, the University of Bologna, Calabria, Genoa, Padua, Pisa, Rome, Turin, and Udine and CNR centers of Genoa, Pisa, and Rome were the most active. The Pisa group, however, spurred by the restless efforts of Giorgio Levi, surely was the leading group in LP in Italy in those years.

Significant interests in LP came also from industries. Major Italian companies such as CSELT, DATAMONT, Digital, ELSAG, Enidata, and Olivetti were involved at some extent in research and development of LP. Also smaller companies, such as DS-Logics, ICON, and Systems and Management, widely used LP to develop concrete applications, in different fields. But also many other companies were interested in LP, even if not as a main tool. Looking at the list of participants at the constituent assembly of GULP in 1985, we can easily realize that almost half of the participants came from industries (see Appendix A.4). The interest of Italian companies in LP is also well testified by the many contributions presented at the first GULP conferences.

In the meantime, the LP paradigm was spreading around the world. These were the fabled heydays of LP with over 300 attendants at ICLP conferences. This was also the era of the Fifth Generation Computer Systems (FGCS) project, which launched the idea of (concurrent) logic programming as the key programming language of next generation computer systems. The project was launched in April 1982 with the opening of ICOT. The second FGCS conference held in Japan in 1984 was a very big event. The multi-billion yen budget of the FCGS project was carrying the LP field out of its narrow boundaries of the early days (see [4]).

Competing projects were set up in the U.S.A. and in Europe, such as the European Strategic Program of Research in Information Technology (ESPRIT). In 1984 ECRC (European Computer Industry Research Centre) was also founded in Munich, on the initiative of three major European manufacturers: Bull (France), ICL (UK), and Siemens (Germany). (Constraint) Logic Programming was one of the main research topics of ECRC since its foundations [5].

In Italy, a number of national projects, mostly founded by the Ministry of Education and by CNR, were devoted to LP and LP-related topics. Among them:

- "Languages and architectures for functional and logic programming", 1984–1987
- "Software Architectures for Intelligent Systems", 1985–1987
- "Automatic reasoning techniques in Intelligent Systems", 1987–1989
- "Intelligent Systems", 1990–1992
- "New Programming Languages", sub-project of the CNR project "Sistemi Informatici e Calcolo Parallelo", 1989—1994.

In particular, the last project involved, besides many universities, also a number of CNR research centers and Italian industries, such as DS-Logics, ICON, Italdata, and led to the development of some interesting applications using the LP paradigm. The main results of the project are summarized in [3].

Italian researchers were involved also in international projects focusing on LP. In particular, the ESPRIT project ALPES (P973) "Advanced Logic Programming Environments" started in June 1986 (actually preceded by a preliminary phase started in 1984). The objective of the project was to build the prototype of a high-level programming environment for logic programming and the Prolog language in particular. The consortium was formed of six partners, among which the Italian Software company Enidata, and five sub-contractors, including the Universities of Rome and Bologna, and an Institute of CNR in Rome.

An important event for the LP community in Europe, and in particular in Italy, was the launch, at the end of the 80s, of the ESPRIT Basic Research Action Compulog (3012) "Computational Logic", followed in years 1992—1995 by the Project Compulog 2 (6810). Furthermore, as a complement to the activity of the Compulog Project, in 1990 Bob Kowalski launched the idea of a Network of Excellence in the field of computational logic. Compulog Net officially started on April 15th, 1991. Luigia Carlucci Aiello, from the University of Rome, was appointed network coordinator and Consorzio Roma Ricerche began to take care of the coordination and administration of Compulog Net. The scientific objective of Compulog Net was to lay the foundations for an integrated software development environment for building knowledge-rich applications by extending the logic programming paradigm.

Each node in the network represented an institute, research laboratory or company active in the area of computational logic. The number of nodes in the network was initially 17 but after a few years the network consisted of more than 80 nodes.

The network had its First General Meeting in Rome in May 1991, jointly with a workshop of the Compulog I Project. In August 1994, the Italian nodes of the network were:

- IRST ("Istituto per la Ricerca Scientifica e Tecnologia"), Trento
- Università di Bologna
- Università di Genova
- Università di Milano
- Università di Padova
- Università di Pisa
- Università di Roma La Sapienza
- Università di Roma Tor Vergata
- Università di Torino.

In particular, the University of Pisa was the coordinating node of one of the five main research areas initially chosen for the network, namely Programming Languages. The first and second Compulog Net area meeting on Programming Languages were held in Pisa in April 1992 and May 1993, respectively.

A personal memory to conclude this section. At the beginning of the 90s, I met Eugenio Omodeo in Udine. From the synergy of his expertise in computable set theory and my skills in LP, and with the invaluable insight of two young (at that time :-)) students of the University of Udine–namely, Agostino Dovier and Enrico Pontelli–we concretized our idea of Logic Programming with Sets, which has been the leitmotiv of my research activity in the last fifteen years and one of the many research topics connected with LP.

3   The APPIA-GULP-PRODE Years (ca. 1994—2003)

The beginning of the 90s represents the period of maximum glory of LP in the world.

Besides the already well-established International Conference on LP (ICLP) and International Symposium on LP (ILPS), along with their joint editions (JICSLP), a number of new international conferences and workshops started in that period. Among them:

- PLILP - Int. Symposium on Programming Language Implementation and LP
- WELP - Int. Workshop on Extensions of Logic Programming
- PAP - Int. Conf. on the Practical Application of Prolog
- LOPSTR - Int. Workshop on Logic-based Program Synthesis and Transformation
- META - Workshop on Meta-Programming in Logic
- LP & NMR - Int. Workshop on LP and Non-Monotonic Reasoning
- LPAR - Int. Conf. on LP and Automated Reasoning
- CCL - Int. Conf. on Constraints in Computational Logic
- ILP - Int. Workshop on Inductive LP.

Moreover, various international schools were specifically devoted to LP, or they mentioned LP as a central topic of interest. In September 1992, Compulog Net supported a summer school on LP in Zurich (Switzerland), organized by Gerard Comyn (ECRC) and Norbert E. Fuchs (University of Zurich). The ESSLLI Summer School in Logic Language and Information was organized each year since 1989. Also, more general schools, such as the Int. School for Computer Science Researchers, organized each year by Alfredo Ferro (Università di Catania) and other colleagues on the island of Lipari (Italy) under the auspices of the European Association for Computer Science Logic (EACSL), saw a growing number of lectures devoted to LP.

Attention to applications and to the industrial transfer was very high in those years. In 1993 "Prolog 1000", a catalogue of Prolog applications edited by Chris D. S. Moss at Imperial College, contained about 500 entries. A first summary of the catalogue appeared in ALP Newsletter Vol. 6/2, February 1993, pages 3—7. Conferences such as "Prolog for Industry" and "INAP - Symp. and Exibition on Industrial Applications of Prolog", served to provide industrial attendees with examples of applications of LP in several industrial areas.

1994 is also the year of ICLP for the first time in Italy. The main LP Conference was organized by Maurizio Martelli in the magnificent surroundings of Genoa (Santa Margherita Ligure) in June 1994. In those years other important LP related events took place in Italy. Among them:

- WELP'92 - 3rd Int. Workshop on Extensions of Logic Programming, Bologna, 1992, organized by Evelina Lamma and Paola Mello
- ALP'92 - 3rd Int. Conf. on Algebraic and LP, Pisa, 1992, organized by Giorgio Levi and Helene Kirchner
- WSA'93 - 3rd Int. Workshop on Static Analysis, Padova, 1993, organized by Gilberto Filè.

The number of members (full, students, honorary) of the Association for Logic Programming (ALP) in June 1994 was quite high: 488. Many of them were also organized in local associations affiliated to ALP. In 1994 the affiliated societies were:

- AFCET (France) with 105 members
- ALP-UK (United Kingdom) with 131 members
- GLP (Austria, Germany, Switzerland) with 93 members
- GULP (Italy) with 113 members.

Furthermore, other related associations and special interest groups in Europe were more and more involved in LP. Many of their members had strong collaborations with members of GULP. Thus in 1993 the GULP executive committee decided to organize the next annual conference jointly with the Spanish conference on Declarative Programming PRODE ("Programación Declarativa"). The first joint conference on Declarative Programming GULP-PRODE'94 was held in Peñiscola (Spain) in 1994. Two years later, the conference was enlarged to another very active community in Europe, that of the Portuguese Association for Artificial Intelligence APPIA ("Associao Portuguesa para a Inteligência Artificial") founded in 1984 in Portugal. From 1996 to 2003, for eight years, the three communities met together at least once a year, alternatively in Italy, Portugal and Spain. A complete list of the GULP-PRODE and APPIA-GULP-PRODE Conferences is reported in Appendix A.1.

In the meantime, Compulog Net was fully operational. The interest in the network activity soon attracted new leading centers in addition to the initial ones: the number of nodes in the network quickly grew to over 80 units, involving several hundred members from more than 20 countries. In addition to Luigia Carlucci Aiello, the executive council of Compulog Net now included other two Italian representatives, Giorgio Levi and Paola Mello from the Universities of Pisa and Bologna, respectively.

On the other side of the Atlantic Ocean, at the end of 1996 the idea of Compulog Americas took shape, an organization of logic programming researchers in North America (but hoping to involve researchers from both North and South Americas as well). It was explicitly modelled after Compulog Europe from which it drew much of its inspiration. The activities of Compulog Americas were organized within several sub-areas, each with an area-coordinator. The initial chief coordinators of Compulog Americas were Gopal Gupta and I.V. Ramakrishnan.

Despite the growing number of initiatives concerning LP and the availability of more and more efficient implementations of Prolog, however, the interest for LP by the industrial world was progressively but inexorably decreasing. Prolog-based applications hardly were able to become real products.

This negative trend is particularly evident in Italy. Looking at the list of participants to the ICLP Conference in Genoa in 1994 it is evident that the industrial participation is almost absent. The same is true for GULP conferences: since the second half of the 90s, participants came only from universities and public research centers. More generally, participation of people from industries to the activities of GULP completely disappeared in those years. One at a time, industries were abandoning investments in LP.

The reasons for such a disappointing result were partly connected to the specific Italian weakness in advanced industrial research in those years (and, unfortunately, also nowadays), but they were surely connected with also more general world wide issues.

One reason for this is the general disappointment resulting from the perceived failure of the Japanese FGCS Project. It is widely accepted that the FGCS Project did not meet the expected success, though the discussion on this topic lasted long (see [4] for an account on results and possible developments of the Project). Since in the mind of many people LP was synonymous with the FGCS Project, LP was (and, unfortunately, often still is) perceived by many people as an experiment that was tried in the 80s and that did not work.

As a direct consequence, during the 90s, most industries stopped funding LP based research projects, and the research momentum developed by the FGCS Project disappeared.

Another phenomenon that occurred in that period is the birth, or simply the strengthening, of new associations and groups in neighboring areas, such as constraint programming, inductive logic programming, deductive databases, static analysis, knowledge representation. This caused a progressive migration of many researchers born and raised in the area of LP to these related areas, in which they still continued to use their background in LP but without considering themselves part of the LP community.

In Italy this is particularly true for the neighboring associations of Artificial Intelligence AI*IA ("Associazione Italiana per l'Intelligenza Artificiale") and the European Association for Theoretical Computer Science (EATCS). Many former GULP members moved to these associations and definitively abandoned GULP.

As a tangible result of this declining interest in LP, in particular in Italy, from the 60 papers presented at the GULP-PRODE Conference in Peñiscola in 1994 (with almost half by Italian authors) we arrived to only 20 papers presented at the APPIA-GULP-PRODE Conference in Madrid in 2002 (with only 7 Italian authors).

There are several articles and discussions about lights and shadows of LP in the literature and on the Web (see, e.g., [7] for Kowalski's personal opinion on why "LP has not made the impact in Computing that many of us once expected"). An analysis of the possible reasons for the lack of success of the LP paradigm and the subsequent loss of interest in it, especially from industries, is out of the scope of this paper. What I want to stress here is that this negative trend that characterized the history of LP in the world since the beginning of the 90s, characterized the history of LP in Italy and of its representative association GULP, as well.

Despite the widespread feeling of something not working in the right way, several efforts have been put forward in the second half of the 90s, both in the field of LP training and in research projects connected with LP.

From 1996 to 2002 GULP organized four very successful summer schools on LP in Sardinia and Calabria (see Appendix A.2). Italian researchers in LP still continued to propose national projects dealing, more or less explicitly, with LP. Among them:

- CNR special project "Logic Programming Languages" (coordinator M. Martelli), 1996–1997
- CNR coordinated project on "Logic Programming: program analysis and transformation tools, software engineering techniques, extensions with constraints, concurrency and objects" (coordinator M. Martelli), 1997–1998
- GNIM ("Gruppo Nazionale per l'Informatica Matematica") project "New computation paradigms: languages and models" (coordinator E. Omodeo), 1999–2000.

To the end of the 90s, however, the age of projects focusing on LP came at the end. The involvement of LP was rather on the inside of more general projects, where LP could play an important but, anyway, accessory role. One of them, to which I personally participated, is the M.U.R.S.T. Co-financed project "Automatic Program Certification by Abstract Interpretation", coordinated by Roberto Giacobazzi (1999—2001).

All these projects, however, involved only people from universities and CNR centers. At the end of the 90s, Italian industries had completely stopped to invest in LP research and development projects.

On the international side, an important achievement for the LP community was the opening of the new ACM journal "Transactions on Computational Logic (TOCL)", founded by K. Apt in 2000. Actually, as explicitly stated in the journal aims, TOCL is devoted to the research concerned with all the uses of logic in computer science; LP is one of the areas. This widening of horizons, from LP to the more general area of Computational Logic (CL), is a trend that characterizes the history of LP in Italy, as well as in the rest of the world, from the half of the 90s to nowadays.

A further example of this enlarged view of LP is the foundation of the Network of Excellence in Computational Logic CologNet. The network started in January 2002 and officially terminated in June 2005. It was an European-funded Network of Excellence which was intended to continue the role played by the Compulog Net network (ended in 2001). It published also an on-line newsletter which is still available at http://newsletter.colognet.org/.

Various Italian research centers participated in the network. In particular Francesca Rossi at the University of Padua coordinated the Constraint Logic

Programming site, while Enrico Franconi at the University of Bozen coordinated CologNet nodes working in Logic and Natural Logic Processing.

With the scientific sponsorship of CologNet and of many other European associations, since 2004 the University of Bozen offered (and still is offering) the European Masters Program in Computational Logic (EMCL). EMCL is an international distributed Master of Science course, in cooperation with Technische Universität Dresden, (Germany), Universidade Nova de Lisboa (Portugal), Technische Universität Wien (Austria), and Universidad Politécnica de Madrid (Spain).

4   The CILC Years (ca. 2004—2010)

As mentioned above, at the end of the 90s many LP researchers realized that it was absolutely necessary to widen their horizons, thinking LP not only *per se* but mainly as a key tool to understand problems and to support solutions in relation to other disciplines. Reporting on his period as the president of the ALP, Krzysztof Apt wrote in the ALP Newsletter issue of February 2001 "My main objective was to make logic programming more known outside of our own circle and to 'connect' it better with other areas of computer science. Fortunately, as it turned out, several of my colleagues independently shared this objective, as well . . . ".

In an attempt to meet this requirement, the GULP executive committee, in a meeting held in Venice in December 2003, decided to reorganize its annual conference. The conference changed its name to "Convegno Italiano di Logica Computazionale" (CILC), i.e., Italian Congress of Computational Logic, to open it to a larger audience. Moreover, in order to attract young people, the costs of participation were drastically cut down. Thus, the GULP annual conference moved towards a lighter organization and (unfortunately) we had to return to a national dimension, interrupting our collaboration with APPIA and PRODE. The list of CILC conferences held from 2004 to nowadays is shown in Appendix A.1.

In the meantime, new topics of interest for the LP community emerged, most of which were on the boundary with related disciplines, such as Artificial Intelligence and Deductive Databases. Among them I can mention:

- Multi-Agents systems
- Semantic Web
- Answer Set Programming
- Knowledge discovery and learning
- Static analysis
- Model checking
- Knowledge representation and automated reasoning
- Computational biology.

All of them were included in the topics of interest of CILC. This was in accordance with the new philosophy of GULP to enlarge its scope as much as possible.

Most of these topics represented, and still represent, important research areas within the Italian LP community. Consequently, several chapters of this book are devoted precisely to them.

Since the beginning of the CILC age, the LP community in Italy has been quite stable. During these years, the GULP association maintained a steady size of about 60 members, most of which were PhD students and young researchers. Members were spread over the whole country and abroad (about 30 universities and research centers are involved at present). The number of people attending the annual conference has been constantly more than 50, while the number of submitted papers has varied from 25 to 30. Unfortunately, also the presence of industries to all activities supported by GULP in the last years has been constantly very low (actually, almost nothing).

More or less the same situation occurred within the international LP community. A precise account of the past activities and what is going on in the LP community can be found in the ALP Newsletter, the official newsletter of ALP since 1987. The Newsletter is available on-line since May 2001. Sandro Etalle and, later, Enrico Pontelli have been the editors of the new electronic version of the Newsletter (by the way, Sandro and Enrico are two of the many Italian researchers in LP that approached LP within the GULP community and that now are living and working abroad). From May 2006, the Newsletter contains, among others, a very interesting column dedicated to presenting personal historical perspectives on the field of LP.

The main LP conference, ICLP, has been regularly held each year, many times co-located with other related conferences. In December 2008, in particular, it was held in Udine (Italy), organized by Agostino Dovier. In addition to ICLP, other conferences continued to be tightly connected with LP, such as the conferences on "Principles and Practice of Declarative Programming" (PPDP), "Practical Aspects of Declarative Languages" (PADL), and "Logic Programming and Non-Monotonic Reasoning" (LPNMR). Many LP contributions, however, have moved in those years to related conferences in classical neighboring areas such as Artificial Intelligence, Deductive Databases and Theoretical Computer Science, as well as in emerging new areas such as Semantic Web and Multi-Agent Systems.

Like in Italy, moreover, it is undeniable that attention to LP from industries was inexorably decreasing everywhere. These are also the years to reflect upon the problems (technical, social, ... ?) that was afflicting the LP community. The article by Tom Schrijvers "A Wake-Up Call for the Logic Programming Community", in the ALP Newsletter vol. 20 n. 3/4, is symptomatic to this respect. There was (and still there is) even the need to clarify the very notion of LP, as pointed out in the worried letter by Bob Kowalski entitled "Logic Programming in Wikipedia - A Call for Help", in the ALP Newsletter vol. 20 n. 1. The question is far from being closed as the recent article by Carl Hewitt [6] and the lively discussion that followed its publication (see, e.g., http://lambda-the-ultimate.org/node/2803) clearly demonstrate.

In the last years, a big effort has been devoted by the LP community all over the world to teaching LP and, more generally, to form young researchers with a correct LP background. Among the many initiatives that moved along these lines we can mention the international summer schools in "Constraint and LP" and in "LP and Computation Logic" that were held in Dallas, Texas (2004) and in Las Cruces, New Mexico (2008), following the highly successful 1st summer school in (C)LP held in Las Cruces, in 1999. The schools were especially directed to Ph.D. students who were just about to start research, post-doctoral students interested in entering a new area of research, and young researchers.

On the Italian side, an important initiative of GULP were the two editions of the Best Italian PhD Dissertation Award in Computational Logic that have been assigned in 2006 and in 2009 (see http://www-lia.deis.unibo.it/gulp/ Burocrazia/PhD-awards.html). Fifteen high quality thesis dealing with computational logic were submitted in the last edition of the award.

As teaching is concerned, an important fact that deserves to be noticed is the high number of courses dealing with LP in Italian universities. From a survey conducted by GULP in 2006 by sending a questionnaire by email to the mailing list of the association, it turned out that over 50 courses (or part of courses) in 20 universities were teaching LP, involving about 1500 students every year. Apart from a relatively small number of dedicated courses, LP was usually taught as part of more "classical" courses, such as courses on Artificial Intelligence, Knowledge representation and reasoning, Programming languages, Theoretical Computer Science, Logics.

Hence, despite of the little consideration that LP is receiving as a tool for real world applications, its educational value does not seem questioned.

5   The Future

The training of students and the interaction with other disciplines should be two major objectives of the LP community for the near future.

The community should emphasize, in every conceivable way, the role that LP has played, and still can play, in providing methods and tools to support ideas in related areas. As an example of the feasibility of this cross-fertilization, a forthcoming special issue of the Theory and Practice of Logic Programming journal, edited by Letizia Tanca and Giorgio Orsi, will be devoted to "Logic Programming in Databases: From Datalog to Semantic Web Rules". As a matter of fact, many people who were once logic programming researchers have moved into other areas of computer science and made major impacts.

In my experience, the LP ideas (and Prolog, in particular) played a fundamental role to open my mind and to stimulate me to face a large number of different topics. Knowledge representation, unification, search strategies, declarative programming, constraints, are all subjects that I met and appreciated through and thanks to LP.

I have been teaching a course on non-conventional programming languages for several years and I find Prolog an irreplaceable tool to prove to students that programming can be faced in a quite different way from what they are accustomed to. I do not think that Prolog should replace C++ or Java, but I think that it can be a unique vehicle to better understand programming, as well as many other problems and related disciplines.

The forthcoming years of LP in Italy will be probably (and hopefully) characterized by as many as possible efforts:

- to develop activities to improve LP teaching and training;
- to promote the participation of young researchers to these activities (e.g., through summer schools, incentives for students' participation to conferences, etc.);
- to improve the collaboration with other associations of researchers and practitioners in related areas;
- to improve visibility of our association outside of the LP community (e.g., through awards and workshops on specific topics of interest).

The opportunity to celebrate the 50 years of LP in Italy greatly depends on the success of these efforts!

References

1. Aiello, L., Attardi, G., Prini, G.: Towards a More Declarative Programming Style. In: Neuhold, E.J. (ed.) Formal Descriptions of Programming Concepts, pp. 121–137. North-Holland, Amsterdam (1978)
2. Apt, K.R., Marek, V.W., Truszczynski, M., Warren, D.S. (eds.): The Logic Programming Paradigm: A 25-Year Perspective, pp. 53–71. Springer, Heidelberg (1999)
3. Filè, G. (ed.): Ambienti per linguaggi di nuova concezione. Franco Angeli, Milano (1995)
4. Fuchi, K., Kowalski, R., Furukawa, K., Ueda, K., Kahn, K., Chikayama, T., Tick, E.: Launching the new era. Commun. ACM 36(3), 49–100 (1993)
5. Gallaire, H.: ECRC: a joint industrial research centre. Future Generation Computer Systems 3(4), 279–283 (1987)
6. Hewitt, C.: Middle History of Logic Programming Resolution, Planner, Prolog and the Japanese Fifth Generation Project (2009), http://arxiv.org/abs/0904.3036
7. Kowalski, R.: Logic Programming and the Real World. ALP Newsletter 14(1), 9–11 (2001)
8. Sessa, M.I. (ed.): 1985-1995: Ten years of Logic Programming in Italy. Palladio, Salerno (I) (1995)

A   Appendices

A.1   List of Conferences Organized by GULP

1. GULP (1986), Genova (Italy)
2. GULP (1987), Torino (Italy)
3. GULP (1988), Roma (Italy)
4. GULP (1989), Bologna (Italy)
5. GULP (1990), Padova (Italy)
6. GULP (1991), Pisa (Italy)

7. GULP (1992), Tremezzo (Italy)
8. GULP (1993), Gizzeria (Italy)

9. I GULP-PRODE (1994), Peñiscola (Spain)
10. II GULP-PRODE (1995), Vietri (Italy)
11. III APPIA-GULP-PRODE (1996), San Sebastian (Spain)
12. IV APPIA-GULP-PRODE (1997), Grado (Italy)
13. V APPIA-GULP-PRODE (1998), La Coruña (Spain)
14. VI APPIA-GULP-PRODE (1999), L'Aquila (Italy)
15. VII APPIA-GULP-PRODE (2000), La Habana (Cuba)
16. VIII APPIA-GULP-PRODE (2001), Évora (Portugal)
17. IX APPIA-GULP-PRODE (2002), Madrid (Spain)
18. X APPIA-GULP-PRODE (2003), Reggio Calabria (Italy)

19. I CILC (2004), Parma (Italy)
20. II CILC (2005), Roma (Italy)
21. III CILC (2006), Bari (Italy)
22. IV CILC (2007), Messina (Italy)
23. V CILC (2008), Perugia (Italy)
24. VI CILC (2009), Ferrara (Italy)
25. VII CILC (2010), Rende (Italy)

A.2   List of Doctoral Schools Organized by GULP

1. 1988 Advanced School on Foundations of Logic Programming, Alghero, Sardinia (organizers: Roberto Barbuti and Maurizio Martelli)
2. 1990 Advanced School on Foundations of Logic Programming, Alghero, Sardinia (organizers: Paolo Mancarella and Giuseppe Sardu)
3. 1996 Int'l Summer School on Advances in Logic Programming, Alghero, Sardinia (organizers: Nicoletta Cocco and Gianfranco Rossi)
4. 1998 Int'l Summer School on Logic Programming Perspectives in Hot Research Areas, Maratea, Basilicata (organizers: Patrizia Asirelli and Piero Bonatti)
5. 2000 First Int'l Summer School in Computational Logic ISCL 2000, Maratea, Basilicata (organizers: Sandro Etalle and Maurizio Gabbrielli)
6. 2002 Second Int'l Summer School in Computational Logic ISCL 2002, Maratea, Basilicata (organizers: Roberto Bagnara and Patricia Hill)

A.3   Past Presidents of GULP

- Giorgio Levi, Univ. di Pisa
- Roberto Barbuti, Univ. di Pisa
- Maurizio Martelli, Univ. di Genova
- Maurizio Gabbrielli, Univ. di Bologna
- Gianfranco Rossi, Univ. di Parma

A.4   The Formal Beginning

The constituent assembly of GULP was held in Pisa on June 18th, 1985. Here is the list of companies, universities and public research centers participating to the constituent assembly.

- CSELT, Torino
- SIPE Optimization, Roma
- Selenia, Roma
- CGD, Roma
- S&M, Pisa
- Digital, Milano
- Olivetti, Ivrea
- ELSAG, Genova
- LIST, Pisa
- INTECS, Pisa

- Univ. di Genova
- Univ. di Padova
- Univ. di Pisa
- Univ. di Salerno
- Univ. di Torino
- Univ. di Trento

- CNUCE, CNR - Pisa
- IEI, CNR - Pisa
- IMA, CNR - Genova
- ILC, CNR - Pisa
- Scuola Superiore G.Reiss Romoli, L'Aquila
- IDG, CNR - Firenze

The formal date of birth of the association was February 4th, 1986, in Pisa. Members of the first executive committee were:

- Giorgio Levi, Dip. Informatica, Univ. di Pisa (president)
- Giuliana Dettori, IMA, CNR, Genova (secretary)
- Luigi Marcolungo, ISI, Univ. di Padova (vice-president)
- Giovanni Adorni, DIST, Univ. di Genova
- Giovanna Ballaben, Selenia, Roma
- Pietro Jalamoff, Scuola superiore Reiss Romoli, L'Aquila
- Leonardo Roncarolo, ELSAG, Genova
- Gianfranco Rossi, Dip. di Informatica, Univ. di Torino
- Umberto Rugani, INTECS, Pisa
- Genoveffa Tortora, Dip. di Informatica e Applicazioni, Univ. di Salerno

Theoretical Foundations and Semantics
of Logic Programming

Annalisa Bossi[1] and Maria Chiara Meo[2]

[1] Dipartimento di Informatica, Università Ca' Foscari di Venezia, Italy
`bossi@dsi.unive.it`
[2] Dipartimento di Scienze, Università di Chieti-Pescara
Viale Pindaro 42, 65127 Pescara, Italy
`cmeo@unich.it`

Abstract. The paper provides an overview of an approach to the semantics of (constraint) logic programs, whose aim is providing suitable theoretical bases for modeling observable properties of logic programs in a compositional way. The approach is based on the idea of choosing (either equivalence classes or abstractions of) sets of clauses as semantic domain and provides an uniform framework for defining different compositional semantics for logic programs, parametrically with respect to a given notion of observability. Since some observable properties have a natural definition which is dependent on the selection rule, the framework has been adapted to cope also with a suitable class of rules, which includes the leftmost selection rule. This provides a formal description of most of the observable properties of Prolog derivations and can therefore be viewed as reference semantics for Prolog transformation and analysis systems.

1 Introduction

The paper provides an overview of an approach of the semantics of (constraint) logic programming which push forward the *s-semantic* approach [26] developed about twenty years ago. The aim of such an approach was that of providing a suitable base for program analysis by means of a semantics which really captures the operational semantics of logic programs and thus permits to model properties which can be observed in an SLD-tree (observables). For instance, in [26] two programs are equivalent if for any goal G they return the same (up to renaming) computed answers. That doesn't hold for the least Herbrand model semantics, namely, there exist programs which have the same least Herbrand model, yet compute different answer substitutions. Several ad-hoc semantics modeling various observables have been defined. These include correct answer substitutions, computed answer substitutions, partial answers [25], OR-compositional correct answers [9,8], call patterns [33], proof trees and resultants [30].

In addition there are several semantics specifically designed for static program analysis, which can handle various observables such as types and groundness dependencies. The idea of this approach is to define a framework which collects all the informations on SLD-derivations (for example in terms of resultants) and that permits to define denotations modeling various observables (thus inheriting basic constructions and results).

A. Dovier, E. Pontelli (Eds.): 25 Years of Logic Programming, LNCS 6125, pp. 15–36, 2010.

The relevant information for specific applications can be extracted from such a *collecting semantics* by suitable abstractions.

The paper is organized as follows. In the next section we recall the basic notions and introduce the terminology used in the paper. In Section 3 we describe the observables and their associated equivalence relations considered in the paper. Sections 4 and 5 describe a first general semantics schema and its principal instances. In Section 6 we discuss how the previous results can be specialized for a suitable class of selection rules. In Section 7 we introduce a framework for constraint logic programs. Finally, in Section 8 we describe a framework for bottom-up abstract interpretation.

2 Preliminaries

2.1 Logic Programming

The reader is assumed to be familiar with the terminology of and the main results on the semantics of logic programs [43,1]. We briefly recall here few basic notions.

Throughout the paper we assume programs and goals defined on a first order language given by a signature consisting of a finite set F of *data constructors*, a finite set Π of *predicate symbols*, a denumerable set V of *variable symbols*. T denotes the set of terms built on F and V. Variable-free terms are called ground. If E is any syntactic object, $Var(E)$ and $Pred(E)$ denote the set of (free) variables and of predicates occurring in E, respectively. A substitution is a mapping $\vartheta : V \to T$ such that the set $dom(\vartheta) = \{X \mid \vartheta(X) \neq X\}$ (domain of ϑ) is finite; ε is the empty substitution: $dom(\varepsilon) = \emptyset$. If ϑ is a substitution and E is a syntactic expression, we denote by $\vartheta_{|E}$ the restriction of ϑ to the variables in $Var(E)$.

The composition $\vartheta\sigma$ of the substitutions ϑ and σ is defined as the functional composition. A substitution ϑ is idempotent if $\vartheta\vartheta = \vartheta$. A renaming is a (nonidempotent) substitution ρ for which there exists the inverse ρ^{-1} such that $\rho\rho^{-1} = \rho^{-1}\rho = \varepsilon$. The result of the application of the substitution ϑ to a term t is an *instance* of t denoted by $t\vartheta$. We define $t \preceq t'$ (t is more general than t') iff there exists ϑ such that $t\vartheta = t'$. A substitution ϑ is a grounding for t if $t\vartheta$ is ground and $Ground(t)$ denotes the set of ground instances of t. The relation $\preceq$ is a preorder and $\approx$ denotes the associated equivalence relation (variance). A substitution θ is a unifier of terms t_1 and t_2 if $t_1\theta = t_2\theta$ (where $=$ denotes syntactic equality). If two terms are unifiable then they have an idempotent most general unifier which is unique up to renaming. Therefore $\mathrm{mgu}(t_1, t_2)$ denotes any such an idempotent most general unifier of t_1 and t_2. All the above definitions can be extended to other syntactic objects in the obvious way.

We restrict our attention to idempotent substitutions, unless differently stated.

An *atom* A is an object of the form $p(t_1, \ldots, t_n)$, where $p \in \Pi$ and $t_1, \ldots t_n \in T$. A (definite) *clause* is a formula of the form $H :- A_1, \ldots, A_n$ with $n \geq 0$, where H (the *head*) and $A_1, \ldots, A_n$ (the *body*) are atoms. $: -$ and $,$ denote logic implication and conjunction respectively, and all variables are universally quantified. If the body is empty the clause is a *unit clause*. A (positive) *program* is a finite set of definite clauses and a (positive) *goal* is a conjunction of atoms $A_1, \ldots, A_m$. The empty goal is denoted by $\square$. $\mathcal{A}$ and $\mathcal{C}$ denote the sets of atoms and of clauses, respectively, while $\wp(S)$ denotes the powerset of a set S.

In the following $\mathbf{t}, \mathbf{X}$ denote tuples of terms and of *distinct* variables respectively, while $\mathbf{B}$ denotes a (possibly empty) conjunction of atoms.

The ordinal powers of a generic monotonic operator f on a complete lattice $(D, \leq)$ with bottom $\perp$ are defined as usual, namely $f \uparrow 0 = \perp$, $f \uparrow (\alpha + 1) = f(f \uparrow \alpha)$, for α successor ordinal and $f \uparrow \alpha = lub(\{f \uparrow \beta \mid \beta \leq \alpha\})$ if α is a limit ordinal.

The *Herbrand base* $\mathcal{B}_P$ of a program P is the set of all ground atoms whose predicate symbols are in $Pred(P)$. An *Herbrand interpretation* I for a program P is any subset of the Herbrand base $\mathcal{B}_P$. An Herbrand model for a program P is any Herbrand interpretation M which satisfies all the clauses of P. The intersection $\mathcal{M}(P)$ of all the Herbrand models of a (positive) program P is a model (least Herbrand model).

Definite clauses have a natural computational reading based on the resolution procedure. The specific resolution strategy called SLD can be described as follows. Let $\mathbf{G} = A_1, \ldots, A_m$ be a goal and $c = H : -\mathbf{B}$ be a (definite) clause. $\mathbf{G}'$ is derived from $\mathbf{G}$ and c by using ϑ iff there exists an atom A_j, $1 \leq j \leq m$ such that $\vartheta = \mathtt{mgu}(A_j, H)$ and $\mathbf{G}' = (A_1, \ldots, A_{j-1}, \mathbf{B}, A_{j+1}, \ldots, A_m)$. Given a goal $\mathbf{G}$ and a program P, an SLD-derivation (or simply a derivation) of $P \cup \mathbf{G}$ (of $\mathbf{G}$ in P) consists of a (possibly infinite) sequence of goals $\mathbf{G}_0, \mathbf{G}_1, \mathbf{G}_2, \ldots$ called resolvents, together with a sequence $c_1, c_2, \ldots$ of variants of clauses in P which are *renamed apart* (i.e. such that c_i does not share any variable with $\mathbf{G}_0, c_1, \ldots, c_{i-1}$) and a sequence $\vartheta_1, \vartheta_2, \ldots$ of idempotent mgu's such that $\mathbf{G} = \mathbf{G}_0$ and, for $i \geq 1$, each $\mathbf{G}_i$ is derived from $\mathbf{G}_{i-1}$ and c_i by using ϑ_i. An SLD-refutation of $P \cup \mathbf{G}$ is a finite SLD-derivation of $P \cup \mathbf{G}$ which has the empty clause $\square$ as the last goal in the derivation.

Following [1], a *selection rule* R is a function which when applied to a "history" containing all the clauses and the mgu's used in the derivation $\mathbf{G}_0, \mathbf{G}_1, \ldots, \mathbf{G}_i$, returns an atom in $\mathbf{G}_i$ (the selected atom in $\mathbf{G}_i$). Given a selection rule R, an SLD-derivation is called via R if all the selections of atoms in the resolvents are performed according to R.

In the following $\mathbf{G} \overset{\vartheta}{\leadsto}_{P,R}^* \mathbf{B}$ denotes a finite SLD-derivation of $P \cup \mathbf{G}$ via selection rule R, which has length ≥ 0, where ϑ is the composition of the mgu's introduced and $\mathbf{B}$ is the last resolvent in the derivation. If R is omitted, we mean that any selection rule can be used (and the definition is independent from the selection rule). Moreover, when the length of the derivation is 0, we assume that $\vartheta = \varepsilon$ and $\mathbf{B} = \mathbf{G}$.

The computed answer substitution of a refutation $\mathbf{G} \overset{\vartheta}{\leadsto}_P^* \square$ is the substitution obtained by the restriction of ϑ to the variables occurring in $\mathbf{G}$. $\mathbf{G} \overset{\vartheta}{\rightarrow}_P \square$ will denote explicitly the refutation of $\mathbf{G}$ in P with computed answer substitution ϑ.

2.2   Galois Insertions and Abstract Interpretation

Abstract interpretation [19,20] is a theory developed to reason about the abstraction relation between two different semantics. The theory requires the two semantics to be defined on domains which are complete lattices. $(C, \preceq)$ (the concrete domain) is the domain of the concrete semantics, while $(A, \leq)$ (the abstract domain) is the domain of the abstract semantics. The partial order relations reflect an approximation relation. The two domains are related by a pair of functions α (abstraction) and γ (concretization), which form a Galois insertion.

(Galois insertion). Let $(C, \preceq)$ be the concrete domain and $(A, \leq)$ be the abstract domain. A Galois insertion $(\alpha, \gamma) : (C, \preceq) \to (A, \leq)$ is a pair of maps $\alpha : C \to A$ and $\gamma : A \to C$ such that α and γ are monotonic, for each $x \in C$, $x \preceq \gamma(\alpha(x))$ and for each $y \in A$, $\alpha(\gamma(y)) = y$.

Given a concrete semantics and a Galois insertion between the concrete and the abstract domain, we want to define an abstract semantics. The concrete semantics is the least fixpoint of a semantic function $F : C \to C$. The abstract semantic function $\tilde{F} : A \to A$ is correct if for all $x \in C$, $F(x) \preceq \gamma(\tilde{F}(\alpha(x)))$. F can be defined as composition of "primitive" operators. Let $f : C^n \to C$ be one such an operator and assume that $\tilde{f}$ is its abstract counterpart. Then $\tilde{f}$ is (locally) correct w.r.t. f if for all $x_1, ..., x_n \in C$, $f(x_1, ..., x_n) \preceq \gamma(\tilde{f}(\alpha(x_1), ..., \alpha(x_n)))$. The local correctness of all the primitive operators implies the global correctness. According to the theory, for each operator f, there exists an optimal (most precise) locally correct abstract operator $\tilde{f}$ defined as $\tilde{f}(y_1, ..., y_n) = \alpha(f(\gamma(y_1), ..., \gamma(y_n)))$. However the composition of optimal operators is not necessarily optimal. The abstract operator $\tilde{f}$ is precise if $\tilde{f}(\alpha(x_1), ..., \alpha(x_n)) = \alpha(f(x_1, ..., x_n))$. The above definitions are naturally extended to "primitive" semantic operators from $\wp(C)$ to C.

3 Observables and Composition Operators

The concrete operational semantics of (logic) programs can be specified by means of a set of inference rules which specify how derivations are made and by defining which are the "observables" we are interested in. In pure logic programming, we can be interested in different observable properties such as successful derivations, finite failures, (partial) computed answer substitutions, etc. Therefore a program can have different concrete operational semantics depending on which properties are observed.

A given choice of the observable x induces an "observational" equivalence on programs. Namely $P \approx_x Q$ iff P and Q are observationally indistinguishable according to x. When also composition of programs is taken into account, for a given observable property we can obtain different equivalences depending on which kind of program composition we consider. Given an observable x and a syntactic program composition operator $\circ$, the induced equivalence $\approx_{(x,\circ)}$ is defined as follows. $P \approx_{(x,\circ)} Q$ iff for any program R, $P \circ R$ and $Q \circ R$ are observationally indistinguishable according to x (i.e. P and Q are observationally indistinguishable under any possible context allowed by the composition operator). A semantics $\mathcal{S}$ is correct wrt $(x, \circ)$, if $\mathcal{S}(P) = \mathcal{S}(Q)$ implies $P \approx_{(x,\circ)} Q$, for each logic programs P and Q. $\mathcal{S}(P)$ is fully abstract wrt $(x, \circ)$ when also the converse of the previous implication holds.

A semantic $\mathcal{S}$ is compositional wrt the program composition operator $\circ$, if the semantics of the composition of programs P and Q can be obtained from the semantics of P and the semantics of Q, i.e. if for a suitable composition operator f, $\mathcal{S}(P \circ Q) = f(\mathcal{S}(P), \mathcal{S}(Q))$.

If $\mathcal{S}$ is correct wrt x and is compositional wrt $\circ$, then $\mathcal{S}$ is also correct wrt $(x, \circ)$.

If we are concerned with the input/output behavior of programs we should just observe computed answers. However some semantic based techniques (such as program analysis, debugging and transformation), require to observe and take into account other features of the derivation, which make visible internal computation details. In principle, one could be interested in the complete information about the SLD-derivation, namely the sequences of goals, most general unifiers and variants of clauses. The resultants, introduced in [44] in the framework of partial evaluation, are a compact representation of the relation between the initial goal **G** and the current $\langle goal, substitution \rangle$ pair in a SLD-derivation of **G**, where the *substitution* is the (restriction to $Var(\mathbf{G})$ of the) composition of the mgu's computed in the SLD-derivation from **G** to the current goal.

Definition 1. *Let* P *be a program and let* R *be a selection rule.* $\mathbf{G}\vartheta :-\mathbf{B} \in \mathcal{C}$ *is an* R-*computed resultant for* **G** *in* P *iff there exists a SLD-derivation via* R *such that* $\mathbf{G} \overset{\vartheta}{\rightsquigarrow}_{P,R}{}^{*} \mathbf{B}$. *Moreover* Φ *is a computed resultant of* **G** *in* P *if there exists a selection rule* R *such that* Φ *is an* R-*computed resultant for* **G** *in* P.

In the following, given the (R-)computed resultant $\mathbf{G}\vartheta :-\mathbf{B}$ for the goal **G**, we will say that $\vartheta_{|\mathbf{G}}$ is the substitution associated to the resolvent **B**.

Resultants are a logical representation, which is quite convenient to study transformation techniques of logic programs such as partial evaluation and Fold/Unfold [41,52]. In fact, since these transformations are based on unfolding, i.e. on the application of some SLD-derivation steps to the program clauses, their intermediate and final results and also their basic properties can be naturally expressed in terms of resultants. For example, in addition to the above mentioned use, resultants have been used in [4] to study loop checking mechanisms and in [24] to prove the correctness of a modular Unfold/Fold transformation system.

The resultants are the basic observables to introduce a semantic scheme in Section 4 which collects informations on SLD-derivations. We will then derive as instances of the scheme other semantics which model (in some cases compositionally) more abstract observables, formally defined in Definition 2. These observables are:

partial answers. (denoted by pa), which are the substitutions associated to a resolvent in any SLD-derivation, and correct partial answers (denoted by cpa), which are the substitutions associated to a resolvent in any SLD-refutation. The knowledge about partial answers is important for program analysis [11], to characterize the semantics of concurrent languages [25] and to characterize universal termination, which in turn is useful for the semantics of PROLOG [2,5]

call patterns. (denoted by pt), which are the atoms (procedure calls) selected in any SLD-derivation, and correct call patterns (denoted by cpt), which are the atoms (procedure calls) selected in any SLD-refutation. Call patterns make it possible to derive properties of procedure calls, which are clearly relevant to program optimization and play an important role in most program analysis frameworks based on abstract interpretation (see [22] for a broad overview).

computed answers. (denoted by ca), which are the substitutions associated to the last resolvent ($\square$) in an SLD-refutation, and

successful derivations. (denoted by s), where we just observe successful termination.

In the following sections we will show, as instances of the general scheme, a semantics (in some cases compositional) for each one of the previous observables. Each semantics $\mathcal{F}_x$ is obtained by setting a parameter in the scheme in Section 4, according to the corresponding observational equivalence $\approx_x$. Moreover each $\mathcal{F}_x$ is correct wrt the corresponding $\approx_x$. In several cases also full abstraction is obtained.

We formally define now the observational equivalences that we will consider.

Computed answers and successful derivations are known to be independent from the selection rule. This property is based on the switching lemma [1] and on the fact that these observables are obtained from successful derivations, where all the atoms have been evaluated. This is not the case for partial answers and call patterns which therefore depend on the selection rule. We first consider only notions which are independent from the selection rule. Therefore, in the case partial answers and call patterns, we introduce the independence in the definition by considering any selection rule.

Definition 2. *Let P be a program, R be a selection rule and let $\mathbf{G}$ be a goal such that there exists a derivation $\mathbf{G} \overset{\gamma}{\leadsto}_{P,R}^* \mathbf{B}$.*

1. *ϑ is a R-partial answer for $\mathbf{G}$ in P iff $\vartheta = \gamma_{|\mathbf{G}}$,*
2. *ϑ is a correct R-partial answer for $\mathbf{G}$ in P iff $\vartheta = \gamma_{|\mathbf{G}}$ and $\mathbf{B}$ has a refutation in P,*
3. *A is a R-call pattern for $\mathbf{G}$ in P iff A is the atom selected by R in $\mathbf{B}$,*
4. *A is a correct R-call pattern for $\mathbf{G}$ in P iff A is the atom selected by R in $\mathbf{B}$ and $\mathbf{B}$ has a refutation in P.*

Moreover ϑ is a (correct) partial answer for $\mathbf{G}$ in P iff there exists a selection rule R such that ϑ is a (correct) R-partial answer for $\mathbf{G}$ in P. Analogously for (correct) call patterns.

Note that computed answers are a special case of (correct) partial answers.

The only notion of program composition (the *OR-composition*) we will consider in the following is a generalization of program union $\cup_\Omega$ defined in [8]. First an Ω-open program P is a (positive) program in which the predicate symbols belonging to the set Ω are considered partially defined in P. P can be composed with another program Q which may further specify the predicates in Ω. Such a composition is denoted by $\cup_\Omega$ and $P \cup_\Omega Q$ is defined only if the predicate symbols occurring in both P and Q are contained in Ω. When Ω contains all the predicate symbols of P and Q we get the standard $\cup$-composition, while if $\Omega = \emptyset$ the composition is allowed only on programs which do not share predicate symbols.

The combination of the above defined six observables with the composition operator gives six observational equivalences. We list below their definitions.

Definition 3. *Let P and Q be Ω-open programs, $\mathbf{G}$ be a goal and let W denote a program such that $P' = P \cup_\Omega W$ and $Q' = Q \cup_\Omega W$ are defined. Assume that $x \in \{s, ca, pa, cpa, pt, cpt\}$. Then $P \approx_{(\Omega,x)} Q$ iff i_x holds for any $\mathbf{G}$ and for any W, where the conditions i_x are defined as follows*

i_s: $\mathbf{G}$ has a refutation in P' iff $\mathbf{G}$ has a refutation in Q',
i_{ca}: $\mathbf{G}$ has the same set of computed answers in P' and in Q',
i_{pa} (i_{cpa}): $\mathbf{G}$ has the same set of (correct) partial answers in P' and in Q',
i_{pt} (i_{cpt}): $\mathbf{G}$ has the same set of (correct) call patterns in P' and in Q'.

The case $\Omega = \emptyset$ is equivalent to considering no composition at all and therefore in order to simplify the notation we will denote $\approx_{(\emptyset,x)}$ by $\approx_x$.

4   A General Semantic Scheme

The scheme which has been proposed in [30,31] is a generalization of the open semantics introduced in [9,8] to obtain compositionality wrt program union. The standard
semantics based on atoms is not compositional wrt union of programs. Consider for
instance the programs $P = \{q(a), p(X) : -r(X)\}$, $Q = \{q(a)\}$ and $R = \{r(a)\}$. The
least Herbrand model semantics $\mathcal{M}(P)$ identifies P and Q, since $\mathcal{M}(P) = \mathcal{M}(Q) = \{q(a)\}$. However $\mathcal{M}(P \cup R) \neq \mathcal{M}(Q \cup R)$. In order to obtain the semantics of the
union $P \cup R$ from those of the components, the semantics of P should contain also the
information given by the clause $p(X) : -r(X)$. For this reason, the open semantics
was then defined on domains containing equivalence classes of sets of clauses (called
π-interpretations).

If we abstract from the specific equivalences in [9,8], the open semantics can be
viewed as a semantic framework for correctly modeling $\approx_{(\circ,x)}$ equivalences. Similarly
to what happens for least Herbrand model semantics [23] the semantics built on π-inter-
pretations is a mathematical object which is defined in model-theoretic terms and which
can be computed both by a top-down and a bottom-up construction. The link between
the top-down and the bottom-up constructions is given by an unfolding operator [42],
denoted by Γ.

In the following a π-interpretation is a $\sim$-equivalence class $[I]$ where $I \subseteq \mathcal{C}$. $\mathcal{I}$ is
the set of all the π-interpretations and we define $\iota(I) = a$ where a is the renamed apart
version of any element in $I \in \mathcal{I}$. All the definitions which use elements from $\mathcal{I}$ are
parametric wrt an equivalence $\sim$. However, in the remaining of this section, we omit
the $\sim$ index in order to simplify the notation.

The general semantics scheme in [31,30] is defined in terms of π-interpretations and
hence parametrically wrt $\sim$. We give two equivalent characterizations. The top-down
one has a definition in the style of an operational semantics, while the bottom-up one is
based on the fixpoint of a general immediate consequences operator. Let us first define
the top-down semantics $\mathcal{O}(P)$.

Definition 4 (Operational Semantic Scheme). *Let P be a program.* $\mathcal{O}(P) = [\{\Phi \in \mathcal{C} \mid \Phi$ *is a resultant for a goal of the form* $p(\mathbf{X})$ *in* $P\}] \in \mathcal{I}$.

Note that $\mathcal{O}(P)$ is a π-interpretation and it is the (equivalence class of the) set of all the
resultants obtained from goals of the form $p(\mathbf{X})$ in P for any possible selection rule.
In [7] the resultants are extended by collecting also sequences of clause identifiers in
order to obtain the maximum amount of information on computations so to observe all
the internal details of SLD-derivations. Moreover, by modifying $\mathcal{O}(P)$, it is possible
to obtain semantics compositional w.r.t. other composition operators, as for example
inheritance mechanisms [6].

The semantics $\mathcal{O}(P)$ can be obtained also by a fixpoint construction. The suitable
immediate consequences operator can be defined in terms of an unfolding operator. To
this aim, first it is necessary organize the set of π-interpretations in a lattice $(\mathcal{I}, \sqsubseteq)$ based
on a suitable partial order relation $\sqsubseteq$. Second, an immediate consequences operator T_P
is defined and proved monotonic and continuous on $(\mathcal{I}, \sqsubseteq)$. This allows us to define the
fixpoint semantics $\mathcal{F}(P)$ for P as $\mathcal{F}(P) = T_P \uparrow \omega$, which is proved equivalent to the
operational semantics.

We require $\sim$ to be a congruence wrt infinite unions, i.e. if, for all $n \in N$, $I_n, J_n \subseteq C$ and $I_n \sim J_n$, then $\bigcup_{n \in N} I_n \sim \bigcup_{n \in N} J_n$. Since $\sim$ is a congruence wrt infinite unions, given $X \subseteq \mathcal{I}$ we can define $\bigsqcup X = [\bigcup_{I \in X} \iota(I)]$ and for $I, J \in \mathcal{I}$, $I \sqsubseteq J$ if and only if $I \sqcup J = J$. The relation $\sqsubseteq$ is an ordering on $\mathcal{I}$ and $(\mathcal{I}, \sqsubseteq)$ is a complete lattice (with $\sqcup$ as lub and $[\emptyset]$ as the bottom element).

Let us introduce the basic syntactic operator Γ which will be used to construct the general immediate consequence operator $\mathcal{T}$. Given a program P and a set of clauses I, $\Gamma_P(I)$ generates all the clauses obtained by "partially" unfolding P wrt I, i.e. it generates also those clauses obtained by rewriting a (possibly empty) subset of the atoms in the bodies of clauses in P.

In the following Id_Ω be the set of clauses $\{p(\mathbf{X}) \colon\! -p(\mathbf{X}) \mid p \in \Omega\}$.

$$\Gamma_P(I) = \{(A \colon\! -\mathbf{D_1}, \ldots, \mathbf{D_n})\vartheta \mid \exists \text{ a clause } A \colon\! -B_1, \ldots, B_n \in P,$$
$$\exists \, n \text{ renamed apart clauses in } I \cup Id_\Pi :$$
$$H_1 \colon\! -\mathbf{D_1}, \ldots, H_n \colon\! -\mathbf{D_n},$$
$$\exists \vartheta = \mathtt{mgu}((B_1, \ldots, B_n), (H_1, \ldots, H_n))\}.$$

Now, in order to define the fixpoint semantics we require that $\sim$ is a congruence wrt the Γ operator, i.e. if $I \sim J$, then for any program P, $\Gamma_P(I) \sim \Gamma_P(J)$. This restriction will guarantee the correctness of the definition of the general fixpoint semantics. $\mathcal{T}_P$ is defined simply as the semantic counterpart of the syntactic operator Γ_P.

Definition 5. *Let P be a program. Then* $\mathcal{T}_P : \mathcal{I} \to \mathcal{I}$ *is the function*

$$\mathcal{T}_P(I) = [\Gamma_P(\iota(I))].$$

$\mathcal{T}_P(I)$ is well defined, i.e. its definition is independent from the element chosen in the equivalence class I, because Γ is a congruence wrt $\sim$. Moreover $\mathcal{T}_P$ is continuous on $(\mathcal{I}, \sqsubseteq)$ and $\mathcal{T}_P \uparrow \omega$ is the least fixpoint of $\mathcal{T}_P$.

Definition 6 (Fixpoint Semantic Scheme). *Let P be a program.*

$$\mathcal{F}(P) = \mathcal{T}_P \uparrow \omega \in \mathcal{I}.$$

Because of the previously mentioned ability of Γ_P (and therefore of $\mathcal{T}_P$) to produce also the result of partial unfoldings, $\mathcal{F}(P)$ gives a bottom-up description of partial derivations, i.e. it contains also the intermediate results of non-terminated (and possibly non-terminating) computations. Indeed, no matter which specific $\sim$ equivalence is used, the equality of the top-down and the bottom-up constructions holds [30]. This general result simplifies the treatment in specific cases since it is usually easier proving the congruence requirements on $\sim$ rather than proving the stated equality.

Lemma 1 (Equivalence). *Let P be a program,* $\sim$ *be an equivalence on* $\wp(\mathcal{C})$ *which is a congruence wrt infinite unions and wrt the Γ operator. Then* $\mathcal{F}(P) = \mathcal{O}(P)$.

By instantiating $\sim$ to a specific equivalence $\sim_{(\circ,x)}$, which depends on the composition operator $(\circ)$ and the observable (x), we can obtain suitable $\mathcal{T}_P$ operators and (equivalent operational and fixpoint) semantics for the corresponding $\approx_{(\circ,x)}$ equivalences.

When considering as $\sim$ the identity on $\wp(\mathcal{C})$ we obtain a kind of "collecting semantics" which correctly models resultants. The semantics modeling resultants is clearly correct wrt the equivalence induced by any notion of observability considered in the previous section. However, we are interested in defining, for specific observables, coarser $\sim$ equivalences in order to obtain a more (possibly fully) abstract semantics, while preserving the correctness.

In the following we will then introduce a suitable $\sim$-equivalence to obtain a correct (in some cases fully abstract) semantics for any $\approx$-equivalence considered in the previous section. The instances of the generic constructions $\mathcal{I}$, $\mathcal{T}$, $\mathcal{O}$ and $\mathcal{F}$, obtained by using a specific $\sim_i$-equivalence, will be denoted by $\mathcal{I}_i$, $\mathcal{T}^i$, $\mathcal{O}_i$ and $\mathcal{F}_i$, respectively. When the subscripts are omitted we mean that $\sim$ is the identity on $\wp(\mathcal{C})$.

5 Getting Instances from the General Schema

5.1 Computed Answers Substitutions and Successful Derivations

In this section we consider first the composition of programs which do not share predicates (i.e. $\Omega = \emptyset$). As previously discussed, this is the same as the case of no composition at all. Here the observables we are concerned with are computed answer substitutions and successful derivations. The induced equivalences on programs have been previously denoted by $\approx_{ca}$ and $\approx_s$. We first show that suitable definitions of $\sim_{ca}$ and $\sim_s$ allow us to obtain the s-semantics [26] and the least Herbrand model as instances of the scheme. Then we consider the relation of these semantics to $\approx_{ca}$ and $\approx_s$. Since here we are not concerned with compositions, it is sufficient to extract from each set of clauses I only the information given by the unit clauses contained in I. Two sets of clauses can then be considered equivalent if they contain the same unit clauses (up to variance). Moreover, in the case of successful derivations, we only need the information given by the ground instances of the clauses. We define then $\sim_{ca}$ and $\sim_s$ as follows.

Definition 7. Let $I, J \subseteq \mathcal{C}$. $I \sim_{ca} J$ iff $I \cap \mathcal{A} = J \cap \mathcal{A}$. Moreover $I \sim_s J$ iff $Ground(I \cap \mathcal{A}) = Ground(J \cap \mathcal{A})$.

$\sim_{ca}$ and $\sim_s$ are congruences wrt infinite unions and wrt the Γ operator and therefore, we obtain automatically from the scheme for any program P, $\mathcal{I}_{ca}$, $\mathcal{T}^{ca}$, $\mathcal{O}_{ca}$ and $\mathcal{F}_{ca}$, (analogously for $\sim_s$).

Let us first consider the instances of the general definitions obtained by using $\sim_{ca}$. For any $I \in \mathcal{I}_{ca}$, the set of unit clauses (modulo variance) of any element $\iota(I)$ can be considered the canonical representative of the equivalence class I. $\mathcal{T}_P^{ca}$ defined in terms of canonical representatives is essentially the immediate consequence operator T_P^{s-sem} originally defined in [26]. The s-semantics is the least fixpoint $T_P^{s-sem} \uparrow \omega$ of such an operator. As an obvious consequence, the s-semantics as originally defined is the canonical representative of $\mathcal{F}_{ca}(P)$ [31].

The strong completeness theorem in [26] shows that the s-semantics is fully abstract wrt $\approx_{ca}$. The mentioned correspondence with $\mathcal{F}_{ca}$ implies that $\mathcal{F}_{ca}(P)$ is fully abstract wrt $\approx_{ca}$ [31]. The same result was obtained in [35] using a proof theoretic approach.

Lemma 2. Let P and Q be programs. Then $P \approx_{ca} Q$ iff $\mathcal{F}_{ca}(P) = \mathcal{F}_{ca}Q)$.

Analogously, in the case of $\sim_s$, the canonical representative $\iota_s(J)$ of $J \in \mathcal{I}_s$ can be obtained by taking the ground instances of the unit clauses in $\iota(J)$. T_P^s defined in terms of canonical representatives is essentially the standard immediate consequence operator T_P [23]. Also in this case, the two formulations are equivalent and the least fixpoint of T_P (the least Herbrand model $\mathcal{M}(P)$) is the canonical representative of $\mathcal{F}_s(P)$ [31]. The mentioned correspondence between $\mathcal{M}(P)$ and $\mathcal{F}_s(P)$ implies that the latter semantics is fully abstract wrt $\approx_s$. More precisely the following holds [31].

Lemma 3. *Let P and Q be programs defined on a signature Σ which contains infinitely many constant symbols. Then $P \approx_s Q$ iff $\mathcal{F}_s(P) = \mathcal{F}_s(Q)$.*

5.2   Compositional Equivalences

We consider now equivalences obtained by considering $\cup_\Omega$ as composition operator. We first focus on computed answers as observable to obtain from the scheme the semantics which is correct wrt $\approx_{(\Omega,ca)}$. Finally we take into account successful derivations: by using an equivalence $\sim_{(\Omega,s)}$ based on weak subsumption equivalence [45], we obtain the semantics $\mathcal{F}_{(\Omega,s)}(P)$ which is fully abstract wrt $\approx_{(\Omega,s)}$.

A semantics correct wrt $\approx_{(\Omega,ca)}$. We show now the instance of the scheme $\mathcal{F}_{(\Omega,ca)}(P)$, which is compositional wrt $\cup_\Omega$ and correctly models computed answers, i.e. it is correct wrt $\approx_{(\Omega,ca)}$. A semantics with these features was already defined in [8] by using sets of clauses as interpretations. [31,30] show how such a semantics can be obtained from the general scheme.

   We first define a syntactic equivalence $\simeq$ on (sets of) clauses which is correct wrt $\approx_{(\Omega,ca)}$ (for any Ω) and hence can be used to define π-interpretations for the compositional case when considering computed answers. A distinction can be made among the atoms in the body of a clause, by identifying those *relevant* atoms which can share variables with the head in a derivation, and those which cannot. Clearly, only the atoms of the first type can contribute to the answer computed in a derivation. The others can only be *tested* for their successful derivation, but their derivation cannot give any useful binding for the computed answer, since such an answer is always restricted to the variables in the goal. Hence the following.

Definition 8. *An atom B in the body of a clause c is called relevant if either it shares variables with the head of c or, inductively, it shares variables with another atom B' in the body of c which is relevant. The multiset of relevant atoms in c is denoted by $Rel(c)$.*

In the following $Set(M)$ denotes the set of the elements which appear in the multiset (or sequence) M. Moreover, when applied to multisets, $\subseteq$ denotes multiset inclusion.

   Note that, in the following definitions relevant atoms in clause bodies are considered as multisets rather than sets. This is because in general a relevant atom in the body $\mathbf{B}$ of a clause cannot be deleted (even if a copy of the atom appear in $\mathbf{B}$) without changing the operational meaning of the clause in terms of computed answers. Recall that a clause $c_1 = H_1 :- \mathbf{A}$ subsumes a clause $c_2 = H_2 :- \mathbf{B}$ if there exists a substitution ϑ such that $H_1 \vartheta = H_2$ and $Set(\mathbf{A})\vartheta \subseteq Set(\mathbf{B})$. Now, let c_1 and c_2 be two clauses which do not share variables and whose heads are H_1 and H_2, respectively. We say that $c_1 \leq_c c_2$ iff c_1 subsumes c_2 and there exists a renaming ρ such that $H_1 \rho = H_2$, $Rel(c_2)\rho \subseteq Rel(c_1)$ and $Set(Rel(c_2)\rho) = Set(Rel(c_1))$.

The equivalence $\simeq$ is then defined as the symmetric closure of the Smith preordering induced on sets of clauses by $\leq_c$. It can be proved (see [31,30]) that $\simeq$ equivalent sets of clauses can be interchanged in any context while preserving the computed answer substitutions semantics. In fact, given $I, J \subseteq C$, if $I \simeq J$ then the two sets of clauses are indistinguishable by $\approx_{(\Pi,ca)}$. We can then use $\simeq$ to define the equivalence $\approx_{(\Omega,ca)}$. Moreover, since $\cup_\Omega$ allows us to compose programs which share predicate symbols in Ω only, we only need the information given by clauses in C^Ω, where C^Ω denotes the set of clauses $H :\!-\mathbf{A}$ such that $Pred(\mathbf{A}) \subseteq \Omega$.

Definition 9. *Let $I, J \subseteq C$. We define $I \simeq J$ iff for any $c \in I$ there exists $c' \in J$ such that $c' \leq_c c$ and vice versa. Moreover $I \sim_{(\Omega,ca)} J$ iff $I \cap C^\Omega \simeq J \cap C^\Omega$.*

It can be shown that $\sim_{(\Omega,ca)}$ is finer than (and hence correct wrt) $\approx_{(\Omega,ca)}$. $\sim_{(\Omega,ca)}$ is a congruence wrt infinite unions and wrt the Γ operator and therefore, we obtain automatically from the scheme for any program P, $\mathcal{I}_{(\Omega,ca)}$, $\mathcal{T}^{(\Omega,ca)}$, $\mathcal{O}_{(\Omega,ca)}$ and $\mathcal{F}_{(\Omega,ca)}$ by using $\sim_{(\Omega,ca)}$ as $\sim$.

Essentially the same results have been given in [9,8] by using the identity on $\wp(C)$ as $\sim_{(\Omega,ca)}$ equivalence.

Lemma 4. *Let P and Q be programs. If $\mathcal{F}_{(\Omega,ca)}(P) = \mathcal{F}_{(\Omega,ca)}Q$ then $P \approx_{(\Omega,ca)} Q$.*

The converse of the previous statement does not hold, i.e. the semantics $\mathcal{F}_{(\Omega,ca)}(P)$ is not fully abstract wrt $\approx_{(\Omega,ca)}$. The difficulty here is related to the use of clauses in the semantic domain (the full abstraction result in [34] was obtained using a domain not containing clauses).

A semantics correct and fully abstract wrt $\approx_{(\Omega,s)}$. Now we consider the usual program composition $\cup_\Omega$ but we will focus on successful derivations as observable. We will obtain from the general scheme a semantics $\mathcal{F}_{(\Omega,s)}(P)$ is fully abstract wrt $\approx_{(\Omega,s)}$.

According to the general construction, we have only to define a suitable equivalence $\sim_{(\Omega,s)}$ on clauses. First, note that the clause c is a tautology iff the body of c contains a copy of the head. Given $I, J \in C$, we say that I and J are subsumption equivalent iff for any $c \in I$ there exists $c' \in J$ such that c' subsumes c and vice versa. I and J are weakly subsumption equivalent iff $I \backslash Taut(I)$ is subsumption equivalent to $J \backslash Taut(J)$, where $Taut(I)$ denotes the set of tautologies in I. Since here we are concerned only with successful derivations, $\sim_{(\Omega,s)}$ can simply be defined in terms of weak subsumption equivalence. Indeed, if c_1 subsumes c_2 then each successful derivation of a goal $\mathbf{G}$ can be performed by using c_1 instead of c_2. Moreover, if $\mathbf{G}$ has a successful derivation which uses the tautology c, $\mathbf{G}$ has also a derivation which does not use c. In other words, tautological clauses can be deleted. These remarks can be formalized as follows.

Definition 10. *Let $I, J \subseteq C$. $I \sim_{(\Omega,s)} J$ iff $I \cap C^\Omega$ is weakly subsumption equivalent to $J \cap C^\Omega$.*

$\sim_{(\Omega,s)}$ is a congruence wrt infinite unions and wrt the Γ operator and therefore, we obtain automatically from the scheme for any program P, $\mathcal{F}_{(\Omega,s)}$ by using $\sim_{(\Omega,s)}$ as $\sim$. We have the following result.

Lemma 5. *Let P, Q be (finite) programs. $P \approx_{(\Omega,s)} Q$ iff $\mathcal{F}_{(\Omega,s)}(P) = \mathcal{F}_{(\Omega,s)}(Q)$.*

Note that the previous result holds also for infinite programs which contain only finitely many function symbols. It does not hold for generic infinite programs (for a counterexample consider the programs P and $Ground(P)$).

5.3 A Semantics for Partial Answers and Call Patterns

A fixpoint semantics for partial answers has been defined in [25]. [31,30] extend such a characterization by obtaining, from the general scheme, a fully abstract semantics for partial answers and a correct semantics for correct partial answers. Semantics for call patterns is also given.

We give just the intuition on how these semantics are obtained. More details can be found in the cited literature. For the sake of simplicity, we consider only the case $\Omega = \emptyset$. The compositional case can be obtained by using techniques similar to those used in the above section.

From the clauses in $\mathcal{F}(P)$ it is possible to extract the information needed to model partial answers and call patterns for any goal $\mathbf{G}$. For example, since each clause $H :- \mathbf{B}$ in $\mathcal{F}(P)$ corresponds to a derivation $p(\mathbf{X}) \overset{\beta}{\leadsto}_{P,R}{}^* \mathbf{B}$ (where $H = p(\mathbf{X})\beta$) ϑ is a partial answer for the goal $p(\mathbf{X})$ if there exists a clause $H :- \mathbf{B}$ in $\mathcal{F}(P)$ such that $\gamma = \mathrm{mgu}(p(\mathbf{X}), H)$ and $\vartheta = \gamma_{|p(\mathbf{X})}$. Moreover ϑ is a correct partial answer for $p(\mathbf{X})$ if there exists also a conjunction $\mathbf{C}$ containing atoms from $\mathcal{F}(P)$ such that $\mathbf{B}$ and $\mathbf{C}$ unify. This example can be extended to the general case in a obvious way.

Note that, when considering partial answers, we only need the information in the heads of the clauses in $\mathcal{F}(P)$, while for correct partial answers clearly we have to consider also bodies. In fact bodies contain the information needed to check if the partial derivation is part of a refutation. First of all, given $J \subseteq \mathcal{C}$, we define $Heads(J) = \{H \in \mathcal{A} \mid H :- \mathbf{B} \in J\}$ and therefore, according to the previous considerations the equivalences $\sim_{pa}$ and $\sim_{cpa}$ are defined as follows.

Definition 11. Let $I, J \subseteq \mathcal{C}$. $I \sim_{pa} J$ iff $Heads(I \cup Id_{\Pi}) = Heads(J \cup Id_{\Pi})$. Moreover $I \sim_{cpa} J$ iff $I \sim_{(\Pi, ca)} J$.

$\sim_{pa}$ and $\sim_{cpa}$ are congruences wrt infinite unions and wrt the Γ operator and therefore, the semantics for partial answers and correct partial answers can be automatically obtained as usual from the general scheme for any program P, by using $\sim_{pa}$ and $\sim_{cpa}$ as $\sim$, respectively. Moreover $\mathcal{F}_{pa}(P)$ is fully abstract wrt $\approx_{pa}$.

Lemma 6. Let P, Q be programs. Then $P \approx_{pa} Q$ iff $\mathcal{F}_{pa}(P) = \mathcal{F}_{pa}(Q)$.

For $\mathcal{F}_{cpa}(P)$ we have only the following correctness result. The problems for obtaining full abstraction here are the same as those mentioned for compositional computed answers.

Lemma 7. Let P, Q be programs. If $\mathcal{F}_{cpa}(P) = \mathcal{F}_{cpa}(Q)$ then $P \approx_{cpa} Q$.

The information needed to model call patterns can be obtained from the clauses in $\mathcal{F}(P)$ as well. For example, if $H :- B_1, \dots, B_n \in \mathcal{F}(P)$ and $\vartheta = \mathrm{mgu}(A, H)$ then $B_i\vartheta$ is a call pattern for the goal A. Since we are not considering a specific selection rule, we only need the information on the relation between the head and the various atoms in the body. In other words, the clause $H \mathrel{:-} B_1, \ldots, B_n$ is equivalent to the set of clauses $\{H \mathrel{:-} B_1, \ldots, H \mathrel{:-} B_n\}$. Therefore the following.

Definition 12. *Let* $c = H \mathrel{:-} B_1, \ldots, B_n \in \mathcal{C}$. $Krom(c) = \{H \mathrel{:-} B_1, \ldots, H \mathrel{:-} B_n\}$.

The Krom operator, which transforms (equivalence classes of) clauses into sets of binary clauses, is extended in the obvious way to subsets of $\mathcal{C}$.

Definition 13. *Let* $I, J \subseteq \mathcal{C}$. $I \sim_{pt} J$ *iff* $Krom(I) = Krom(J)$.

$\sim_{pt}$ is a congruence wrt infinite unions and the operator Γ, therefore we have the usual definition of the semantics as instance of the scheme.

Definition 14. *(Call patterns semantics) Let* P *be a program. The semantics* $\mathcal{F}_{pt}(P)$ *for call pattern is defined as the instance of* $\mathcal{F}(P)$ *obtained by using* $\sim_{pt}$.

From the previous observations, we have the correctness results for the call pattern semantics.

Lemma 8. *Let* P, Q *be programs. If* $\mathcal{F}_{pt}(P) = \mathcal{F}_{pt}(Q)$ *then* $P \approx_{pt} Q$.

6   Introducing the Selection Rule

[32] shows how all the previous results can be specialized for a suitable class of selection rules. We discuss the idea of the specialization and give as an example the definition of the R-partial answer semantics. For the sake of simplicity, we consider only the case $\Omega = \emptyset$. The compositional case can be obtained by using techniques similar to those used in the Section 5.2.

First we focus on R-computed resultants, i.e. on those resultants which describe derivations which use the selection rule R. This provides a sort of collecting semantics which describes most of the observable properties of R-derivations. As in Section 4 a π-interpretation is a $\sim$-equivalence class $[I]$ where $I \subseteq \mathcal{C}$. $\mathcal{I}$ is the set of all the π-interpretations and we define $\iota(I) = a$ where a is the renamed apart version of any element in $I \in \mathcal{I}$. All the definitions which use elements from $\mathcal{I}$ are parametric wrt an equivalence $\sim$. However, in the remaining of this section, we omit the $\sim$ index in order to simplify the notation.

Definition 15 (Operational Semantic Scheme). *Let* P *be a program.* $\mathcal{O}_R(P) = [\{\Phi \in \mathcal{C} \mid \Phi \text{ is a } R\text{-computed resultant for a goal of the form } p(\mathbf{X}) \text{ in } P\}] \in \mathcal{I}$.

The problems arise with the fixpoint definition. If we consider a generic selection rule, we cannot obtain a fixpoint (bottom-up) semantics equivalent to the operational one [32]. Therefore, in order to be able to reconstruct exactly the derivation from the bottom, [32] introduces the local rules, as specified by the following definition.

Definition 16 (Local rule). *Let* ϕ *be a given bijection on the set of integer numbers. A selection rule* R *is local, if it satisfies the following conditions:*

1. *if* $\mathbf{G} = A_1, \ldots, A_n$ *is the initial goal, then the atom selected by* R *in* $\mathbf{G}$ *is the atom* A_s, *such that* $\phi(s) < \phi(i)$ *for any* $i \in [1, n]$, $i \neq s$,

2. *if* **G** *is a generic resolvent, assume that* $A_1, \ldots, A_n$ *is the sequence of atoms in* **G** *introduced by the last derivation step. Then, as before, the atom selected is* A_s, *such that* $\phi(s) < \phi(i)$ *for any* $i \in [1, n]$, $i \neq s$.

Rules which select one of the most recently introduced atoms were called local in [40] and were studied since they produce SLD-trees with a simple structure, suitable for efficient searching techniques. Clearly the rules that we consider are also local in the sense of [40]. Note also that the PROLOG leftmost rule is local by defining ϕ as follows: $\phi(i) = i$.

It is possible to define a fixpoint semantics for R-computed resultants, where R is a local rule. Moreover, since the leftmost selection rule is a local rule, this semantics can therefore be viewed as a reference semantics for Prolog transformation and analysis systems, by setting R equal to the leftmost selection rule. Suitable abstractions of this semantics allow the characterization of observables useful for specific applications. We will consider explicitly the abstraction which gives a (fully abstract) semantics for partial answers.

The intuition behind the definition of the bottom-up semantics is the following. According to the previous definition, if A_j is the atom selected by a local rule R in the resolvent $A_1, \ldots, A_n$, then all the atoms derived from A_j are fully evaluated before the selection of the atoms A_i, $i \neq j$. Moreover a function ϕ is used to establish an ordering on the atoms of the query and of the clauses used in the derivation.

The ordering ϕ can then be used *locally* on the bodies of clauses in P, to establish how to rewrite the bodies (by using clauses in I in $\Gamma_{P,R}(I)$, see Definition 17). Namely, when considering a clause $H \mathbin{:-} B_1, \ldots, B_n \in P$ in the definition of $\Gamma_{P,R}(I)$, we take any partition K, J of the indexes $\{1, \ldots, n\}$ such that $\phi(k) < \phi(j)$ for any $k \in K$ and $j \in J$. This means that any atom B_k, with $k \in K$, is fully evaluated before any B_j with $j \in J$, in any derivation which uses the clause $H \mathbin{:-} B_1, \ldots, B_n$. Accordingly, the B_k's are unified with atoms in I. Moreover we consider an atom B_s such that $s \in J$ and the value of $\phi(s)$ is the minimum among the $\phi(j)$'s for $j \in J$. This means that B_s is the first atom selected after the evaluation of the B_k's has been completed. Since the evaluation of (the atoms derived by) B_s can also be not completed, B_s is unified with the head of a generic clause in I.

In order to simplify the notation, given a query $\mathbf{G} = A_1, \ldots, A_n$ and a set of indexes $K = \{k_1, \ldots, k_m\} \subseteq \{1, \ldots, n\}$, in the following we denote by $\mathbf{G}_K$ the query $A_{k_1}, \ldots, A_{k_m}$ and by $\mathbf{G}_{-K}$ the query obtained from $\mathbf{G}$ by deleting A_k for any $k \in K$.

Definition 17. *Let* P *be a program,* R *be a local selection rule and let* I *be a set of clauses.*

$$\begin{aligned}
\Gamma_{P,R}(I) = \{ (A \mathbin{:-} \mathbf{D})\vartheta \mid &\exists\, A \mathbin{:-} \mathbf{B} \in P \text{ with } \mathbf{B} = B_1, \ldots, B_n, \\
&\exists\, K \subseteq \{1, \ldots, n\}, J = \{1, \ldots, n\} \setminus K, \\
&\exists\, s \in J, \text{ such that for any } k \in K \text{ and for any } j \in J \\
&\phi(k) < \phi(s) \le \phi(j) \\
&\exists\, \text{a sequence } \mathbf{H} \text{ of atoms in } I \text{ and} \\
&\exists\, \text{a clause } H' \mathbin{:-} \mathbf{B}' \text{ in } I \cup Id_\Pi \text{ such that} \\
&\vartheta = \mathtt{mgu}((\mathbf{B}_K, B_s), (\mathbf{H}, H')) \text{ and} \\
&\mathbf{D} \text{ is obtained from } \mathbf{B}_{-K} \text{ by replacing } B_s \text{ with } \mathbf{B}' \}.
\end{aligned}$$

All the results shown in Section 4 hold also for this specialized version of the immediate consequence operator. In particular if $\sim$ is a congruence wrt infinite unions and the Γ operator, then $\mathcal{T}_{P,R} = [\Gamma_{P,R}(\iota(I))]$ (the semantic counterpart of the syntactic operator $\Gamma_{P,R}$) is well defined. Moreover $\mathcal{T}_{P,R}$ is continuous on $(\mathcal{I}, \sqsubseteq)$ and $\mathcal{F}_R(P) = \mathcal{T}_{P,R} \uparrow \omega$ (the least fixpoint of $\mathcal{T}_{P,R}$) is equal to the operational semantics $\mathcal{O}_R(P)$.

Now, we show as it is possible to model the R-partial answer semantics. For all the other observables it is possible to follows a similar construction.

Definition 18. *Let P and Q be programs. $P \approx_{pa,R} Q$ iff for any goal $\mathbf{G}$, $\mathbf{G}$ has the same set of R-partial answers in P and in Q.*

From $\mathcal{F}_R(P)$ it is possible to extract the R-partial answers as follows. Analogously to the case of partial answers (without considering the selection rule) in Section 5.3, since each clause $H := \mathbf{B}$ in $\mathcal{F}_R(P)$ corresponds to a derivation $p(\mathbf{X}) \overset{\beta}{\leadsto}_{P,R}{}^* \mathbf{B}$ (where $H = p(\mathbf{X})\beta$) in order to model R-partial answer we only need keep the heads of the resultants and therefore, in the definition of $\sim_{pa,R}$, we can abstract from the bodies. However, we need to distinguish among partial answers those which are also computed answers, i.e. we need to distinguish between heads of non unit clauses and heads of unit clauses in $\mathcal{F}_R(P)$. Consider for example the goal $q(X), r(Y)$ and assume that R is the leftmost selection rule. If $X = a$ is a computed answer for $q(X)$ in the program P (i.e. if $\mathcal{F}_R(P)$ contains the unit clause $q(a)$) and $Y = b$ is a leftmost partial answer for $r(Y)$ in P, then $\{X = a, Y = b\}$ is a leftmost partial answer for $q(X), r(Y)$ in P. This in general is not the case if $X = a$ is a leftmost partial answer (and not a computed answer) for $q(X)$ (i.e. if $\mathcal{F}_R(P)$ contains a non unit clause $q(a) := \mathbf{B}'$ and does not contain a unit clause $q(a)$).

According to the above considerations, the equivalences $\sim_{pa,R}$ is defined as follows.

Definition 19. *Let $I, J \subseteq \mathcal{C}$. $I \sim_{pa,R} J$ iff $Heads(I) = Heads(J)$ and $I \cap \mathcal{A} = J \cap \mathcal{A}$.*

$\sim_{pa,R}$ is a congruence wrt infinite unions and wrt the $\Gamma_{P,R}$ operator and therefore, we obtain automatically from the scheme $\mathcal{F}_{pa,R}$ by using $\sim_{pa,R}$ as $\sim$. We have the following result.

Lemma 9. *Let P, Q be programs and let R be a local selection rule. $P \approx_{pa,R} Q$ iff $\mathcal{F}_{pa,R}(P) = \mathcal{F}_{pa,R}(Q)$.*

7 A Semantic Scheme for Constraint Logic Programs

The *Constraint Logic Programming* paradigm CLP($\mathcal{X}$) (CLP for short) has been proposed by Jaffar and Lassez [38,37] in order to integrate a generic computational mechanism based on constraints with the logic programming framework. The benefits of such an integration are several. From a pragmatic point of view, CLP($\mathcal{X}$) allows one to use a specific constraints domain $\mathcal{X}$ and a related constraint solver within the declarative paradigm of logic programming. From the theoretical viewpoint, CLP provides a unified view of several extensions to pure logic programming (arithmetics, equational programming, object-oriented features, taxonomies) within a framework which preserves the unique semantic properties of logic programs, in particular the existence of equivalent operational, model theoretic and fixpoint semantics [38]. Moreover, since the computation is performed over the specific domain of computation $\mathcal{X}$, CLP($\mathcal{X}$) programs have an equivalent "algebraic" semantics [38] directly defined on the algebraic structure of $\mathcal{X}$.

[28] introduces a framework for defining various semantics, each corresponding to a specific observable property of computations, thus applying to the CLP case the methodology proposed in [7,31]. Analogously to the case of (standard) Logic Programming in Section 4, each semantics can be equivalently defined either operationally (top-down) or declaratively (bottom-up) as the least fixpoint of a suitable operator. The construction is based on a new notion of interpretation (which is a modified version of that given in Section 4), on a natural extension of the standard notion of truth and on the definition of various immediate consequences operators, whose least fixpoints on the lattice of interpretations are models corresponding to various observable properties. All the semantics defined in [38] can be reconstructed within the framework proposed in [28]. The main issue however is the definition of some new semantics and the investigation of their relation, in terms of correctness and full abstraction, wrt the program equivalences induced by various observable properties.

Some of the semantics considered in [28] are the generalization to the CLP case of the non-ground semantics for (positive) logic programs in [26] and of the compositional semantics in [8]. Indeed, most semantic constructions and results lift directly from logic programming to CLP. Moving to a non-ground semantics is even more natural in the case of CLP, since the computation structure may not even include constants so that there might be no "ground" objects.

In particular, [28] first defines a fully abstract semantics which characterizes computed answer constraints for constraint logic programs and then a semantics which models answer constraints and which is compositional wrt programs union. Such a semantics is the natural extension of the previous one obtained by using a semantic domain based on clauses.

Since the compositional semantics contains the "maximum" amount of information on computations, it can also be used to model other non-standard observable properties. Indeed suitable abstractions of this compositional semantics allow us to obtain a correct (in one case fully abstract) semantics for partial answer constraints and call patterns for constraint logic programs.

The definitions of the semantics are mainly interesting for their applications. Thus, the answer constraint semantics can be taken as the basis of a correct notion of program equivalence to be preserved by program transformation techniques. Suitable abstract versions of the immediate consequence operators introduced in [28] can be used for bottom-up abstract interpretation (i.e. fixpoint computation of the abstract model). More interestingly, the compositional semantics was used in [24] to develop a framework for the modular analysis of CLP programs. This is particularly relevant for practical applications where modularity can help to reduce the size and the complexity of the analysis. The semantics for partial answers and call patterns was used for the analysis of constraint logic programs too. For example, informations on partially computed constraints can be used to detect "independence" of (sub)goals [21], thus providing the conditions for optimizations of CLP programs based on AND-parallelism and intelligent backtracking.

8   A Semantic Scheme for Static Program Analysis

Static program analysis aims at determining properties of the behavior of a program without actually executing it. Static analysis is founded on the theory of abstract interpretation ([18]) for showing the correctness of analysis with respect to a given semantics. Thus, it is essentially a semantic-based technique and different semantic definition styles lead to different approaches to program analysis. In the field of logic programs we find two main approaches which correspond to the two main possible constructions of the semantics: top-down and bottom-up. The main difference between them is related to goal dependency. In particular, a top-down analysis starts with an abstract goal (see [10,39]), while the bottom-up approach (see [46,47]) determines an approximation of the success set which is goal independent. It propagates the information "bottom-up" as in the computation of the least fixpoint of the immediate consequences operator T_P.

Thanks to the equivalence between top-down and bottom-up constructions of the concrete semantics, by using an approach analogous to that given in Section 4, it is possible to get a goal independent top-down and bottom-up construction of the abstract model. This was the leading principle in the development of the framework for bottom-up abstract interpretation proposed in [3]. An instance of the framework consists in the specialization of a set of basic abstract operators like abstract unification, abstract substitution application and abstract union. By means of these abstract operators, [3] gives a bottom-up definition of an abstract model, i.e. a goal independent approximation of the concrete denotation. Different instances produce different analysis.

The concrete semantics considered in [3] is the semantics of computed answer substitutions. It is worth noticing that previous attempts [46,47], based on concrete semantics which do not contain enough information on the program behavior, failed on non-trivial analysis (like mode analysis). The problem was that they were too abstract to be useful to capture program properties like variable sharing or ground dependencies.

The ability to determine call patterns was also usually associated to goal dependent top-down methods. [11,29] showed that the choice of an adequate (concrete) semantics allows us to determine goal independent information on both partial answer substitutions and call patterns and that this information can be computed both top-down and bottom-up. This facilitates the analysis of concurrent logic programs (ignoring synchronization) and provides a collecting semantics which characterizes both successes and call patterns. Many other analysis had been defined based on a "non-ground Tp" semantics like groundness dependency analysis, depth-k analysis, and a "pattern" analysis to establish most specific generalizations of calls and success sets (see [12]). A similar methodology has been applied also to CLP programs [36], leading to a framework where abstraction simply means abstraction of the constraint system.

[14] builds upon the idea in [13] of providing an algebraic characterization of the observables. [14] extends the approach, by taking two basic semantics: a denotational semantics and a transition system which define SLD-derivations. In addition, the semantic properties of the observables are expressed as compositionality properties. This

leads to a more flexible classification of the observables, where it is possible reason about properties such as OR-compositionality and existence of abstract transition systems. Using abstract interpretation techniques to model abstraction allows us to state very simple conditions on the observables which guarantee the validity of several general theorems.

The idea is to define the denotational semantics and the transition system for SLD-derivations in terms of four semantic operators, directly related to the syntactic structure of the language. The observables are defined as Galois insertions and it is possible to characterize various classes of observables in terms of simple properties of the Galois insertion and of the basic semantic operators.

The reconstruction of an existing semantics or the construction of a new semantics in the framework requires just a few very simple steps.

1. First of all, we define an observable property domain, namely, a set of properties of derivations with an ordering relation which can be viewed as an approximation structure. An observation consists of looking at an SLD-derivation and extracting some property (abstraction). The formalization of the property o we want to model is a Galois insertion $\langle \alpha_o, \gamma_o \rangle$ between SLD-derivations and the property domain.

2. Once we have an observable o, we want to systematically derive the abstract semantics. The idea is to define the optimal abstract versions of the various semantic operators and then check under which conditions (on $\langle \alpha_o, \gamma_o \rangle$) we obtain the optimal abstract semantics. This will allow us to identify some interesting classes of observables and to assign the observable property to the right class of observables.

3. Depending on the class, we automatically obtain the new denotational semantics, transition system, top-down ($\mathcal{O}_{\alpha_o}(P)$) and bottom-up ($\mathcal{F}_{\alpha_o}(P)$) denotations (simply replacing the concrete semantic operators by their optimal abstract versions), together with several interesting theorems (equivalence, compositionality w.r.t. the various syntactic operators, correctness and minimality of the denotations).

Since it is based on standard operational and denotational semantic definitions, the framework can be adapted to other programming languages.

Finally [14] considers two classes of observables, *complete* and *approximate*. For every complete or approximate observable, the abstract operational semantics and the abstract denotational semantics are equivalent. This will allow us to define equivalent top-down and bottom-up analysis algorithms. The above equivalence property requires the observable to be condensing. Condensing is a compositionality property which tells that the abstract semantics of a procedure call can be derived (without losing precision) from the abstract semantics of the procedure declaration. This property is needed in abstract diagnosis [17,15,16] where the specification is a post-condition describing a (goal-independent) property of a set of procedure declarations. It is worth noting that the observables corresponding to the declarative semantics are condensing and that the declarative semantics do indeed characterize procedure declarations. Note also that several observables used in program analysis (for mode, type and groundness analysis) are also condensing and that a non-condensing observable can systematically be transformed into a (more concrete) condensing observable, by using domain refinement operators (see, for example, how the condensing domain $\mathcal{POS}$, for groundness analysis can be derived from the non-condensing domain $\mathcal{DEF}$ [50]). The results of the diagnosis for approximate observables are also valid for non-condensing domains, which are sometimes convenient to use in practice for efficiency reasons.

As expected from abstract interpretation theory, the difference between complete and approximate observables is related to precision. Namely, the abstract semantics coincides with the abstraction of the collecting semantics, in the case of complete observables, while it is just a correct approximation, in the case of approximate observables. On the other side, approximate observables correspond to noetherian domains. Hence their abstract semantics is finite, while (in general) it is infinite for complete observables. The class of complete observables includes the observables (ground instances of) computed answers and correct answers which allow us to reconstruct the declarative semantics used in declarative debugging, i.e., the least Herbrand model used in [51] and the least term model (atomic logical consequences or c-semantics) used in [27]. Moreover includes all the observable introduced in Section 3. On the other hand, the class of approximate observables includes $depth(k)$ [49] and several domains proposed for type, mode and groundness analysis (for example the domain $\mathcal{POS}$ [48] for groundness analysis).

Note that the AND-compositionality property (i.e., the compositionality with respect to the conjunction of atoms) of all the semantics defined by this approach, including their abstract versions, allows us to proceed in a goal independent way since we can obtain the result for any specific goal $\mathbf{G}$ just by executing $\mathbf{G}$ in $\mathcal{O}_{\alpha(o)}(P)$.

9   Conclusions

In the last twenty years, several semantics for logic programs had been developed according to an approach which push forward the s-semantics introduced by Moreno Falaschi, Giorgio Levi, Maurizio Martelli and Catuscia Palamidessi in [26]. The common aim was that of providing suitable theoretical bases for program analysis of different operational behaviors of logic programs. Each semantics captures properties which can be observed in an SLD-tree and is correct (in some cases fully abstract) wrt an equivalence relation induced by the considered property. We provided an overview of these semantics emphasizing their mutual relations and characteristics.

Acknowledgment

First of all we wish to thank Giorgio Levi, the main promoter of the s-semantics approach and together with him we wish to thank all those who have contributed to the development of this approach.

References

1. Apt, K.R.: Logic programming. In: Handbook of Theoretical Computer Science, Volume B: Formal Models and Sematics (B), pp. 493–574 (1990)
2. Barbuti, R., Codish, M., Giacobazzi, R., Maher, M.J.: Oracle semantics for Prolog. Information and Computation 122(2), 178–200 (1995)

3. Barbuti, R., Giacobazzi, R., Levi, G.: A general framework for semantics-based bottom-up abstract interpretation of logic programs. ACM Transactions on Programming Languages and Systems (TOPLAS) 15(1), 133–181 (1993)
4. Bol, R.N., Apt, K.R., Klop, J.W.: An analysis of loop checking mechanisms for logic programs. Theor. Comput. Sci. 86(1), 35–79 (1991)
5. Bossi, A., Bugliesi, M., Fabris, M.: A new fixpoint semantics for Prolog. In: ICLP 1993: Proceedings of the Tenth Int'l Conference on Logic Programming, pp. 374–389. MIT Press, Cambridge (1993)
6. Bossi, A., Bugliesi, M., Gabbrielli, M., Levi, G., Meo, M.C.: Differential logic programming. In: POPL 1993: Proceedings of the 20th ACM SIGPLAN-SIGACT symposium on Principles of programming languages, pp. 359–370 (1993)
7. Bossi, A., Gabbrielli, M., Levi, G., Martelli, M.: The s-semantics approach: theory and applications. Journal of Logic Programming 19(20), 149–197 (1994)
8. Bossi, A., Gabbrielli, M., Levi, G., Meo, M.C.: A compositional semantics for logic programs. Theoretical Computer Science 122(1-2), 3–47 (1994)
9. Bossi, A., Menegus, M.: Una semantica composizionale per programmi logici aperti. In: Sesto convegno sulla programmazione logica, pp. 95–109 (1991)
10. Bruynooghe, M.: A practical framework for the abstract interpretation of logic programs. Journal of Logic Programming 10(2), 91–124 (1991)
11. Codish, M., Dams, D., Yardeni, E.: Bottom-up abstract interpretation of logic programs. Theoretical Computer Science 124(1), 93–125 (1994)
12. Codish, M., Søndergaard, H.: Meta-circular abstract interpretation in Prolog, pp. 109–134 (2002)
13. Comini, M., Levi, G.: An algebraic theory of observables. In: SLP, pp. 172–186 (1994)
14. Comini, M., Levi, G., Meo, M.C.: Compositionality in sld-derivations and their abstractions. In: ILPS, pp. 561–575 (1995)
15. Comini, M., Levi, G., Meo, M.C., Vitiello, G.: Proving properties of logic programs by abstract diagnosis. In: Dam, M. (ed.) LOMAPS-WS 1996. LNCS, vol. 1192, pp. 22–50. Springer, Heidelberg (1997)
16. Comini, M., Levi, G., Meo, M.C., Vitiello, G.: Abstract diagnosis. Journal of Logic Programming 39(1-3), 43–93 (1999)
17. Comini, M., Levi, G., Vitiello, G.: Abstract debugging of logic program. In: Fribourg, L., Turini, F. (eds.) LOPSTR 1994 and META 1994. LNCS, vol. 883, pp. 440–450. Springer, Heidelberg (1994)
18. Cousot, P.: Program analysis: the abstract interpretation perspective. ACM Computing Surveys 28(4es), 165 (1996)
19. Cousot, P., Cousot, R.: Abstract interpretation: A unified lattice model for static analysis of programs by construction or approximation of fixpoints. In: POPL, pp. 238–252 (1977)
20. Cousot, P., Cousot, R.: Systematic design of program analysis frameworks. In: POPL, pp. 269–282 (1979)
21. García de la Banda, M.J., Hermenegildo, M.V., Marriott, K.: Independence in constraint logic programs. In: ILPS, pp. 130–146 (1993)
22. Debray, S.K.: Formal bases for dataflow analysis of logic programs, pp. 115–182 (1994)
23. Van Emden, M.H., Kowalski, R.A.: The semantics of predicate logic as a programming language. Journal of the ACM 23(4), 733–742 (1976)
24. Etalle, S., Gabbrielli, M.: Transformations of clp modules. Theor. Comput. Sci. 166(1&2), 101–146 (1996)
25. Falaschi, M., Levi, G.: Finite failures and partial computations in concurrent logic languages. Theor. Comput. Sci. 75(1&2), 45–66 (1990)

26. Falaschi, M., Levi, G., Martelli, M., Palamidessi, C.: Declarative Modeling of the Operational Behaviour of Logic Languages. Theoretical Computer Science 69, 289–318 (1989)
27. Ferrand, G.: Error diagnosis in logic programming, an adaptation of E.Y. Shapiro's method. Journal of Logic Programming 4(3), 177–198 (1987)
28. Gabbrielli, M., Dore, G.M., Levi, G.: Observable semantics for constraint logic programs. J. Log. Comput. 5(2), 133–171 (1995)
29. Gabbrielli, M., Giacobazzi, R.: Goal independency and call patterns in the analysis of logic programs. In: SAC, pp. 394–399 (1994)
30. Gabbrielli, M., Levi, G., Meo, M.C.: Observational equivalences for logic programs. In: Proceedings of the Joint Int'l Conference and Symposium on Logic Programming, pp. 131–145 (1992)
31. Gabbrielli, M., Levi, G., Meo, M.C.: Observable behaviors and equivalences of logic programs. Inf. Comput. 122(1), 1–29 (1995)
32. Gabbrielli, M., Levi, G., Meo, M.C.: Resultants semantics for prolog. J. Log. Comput. 6(4), 491–521 (1996)
33. Gabbrielli, M., Meo, M.C.: Fixpoint semantics for partial computed answer substitutions and call patterns. In: Kirchner, H., Levi, G. (eds.) ALP 1992. LNCS, vol. 632, pp. 84–99. Springer, Heidelberg (1992)
34. Gaifman, H., Shapiro, E.: Fully abstract compositional semantics for logic programs. In: POPL 1989: Proceedings of the 16th ACM SIGPLAN-SIGACT symposium on Principles of programming languages, pp. 134–142. ACM Press, New York (1989)
35. Gaifman, H., Shapiro, E.: Proof theory and semantics of logic programs. In: Proceedings of the Fourth Annual Symposium on Logic in computer science, pp. 50–62. IEEE Press, Los Alamitos (1989)
36. Giacobazzi, R., Debray, S.K., Levi, G.: A generalized semantics for constraint logic programs. In: Proceedings of the Int'l Conference on Fifth Generation Computer Systems, pp. 581–591. ACM Press, New York (1992)
37. Jaffar, J., Lassez, J.-L.: Constraint logic programming. Technical report, Department of Computer Science, Monash University (June 1986)
38. Jaffar, J., Lassez, J.-L.: Constraint logic programming. In: POPL, pp. 111–119 (1987)
39. Janssens, G., Bruynooghe, M.: Deriving descriptions of possible values of program variables by means of abstract interpretation. Journal of Logic Programming 13(2-3), 205–258 (1992)
40. Kawamura, T., Kanamori, T.: Preservation of stronger equivalence in unfold/fold logic program transformation. Theor. Comput. Sci. 75(1&2), 139–156 (1990)
41. Komorowski, H.J.: A specification of an Abstract Prolog Machine and Its Applications to Partial Evaluation. Phd thesis, Linköping University (1981)
42. Levi, G.: Models, unfolding rules and fixpoint semantics. In: Proc. of the Fifth Int'l Conference and Symposium on Logic Programming, vol. 2, pp. 1649–1665. MIT Press, Cambridge (1991)
43. Lloyd, J.W.: Foundations of logic programming. Springer, New York (1984)
44. Lloyd, J.W., Shepherdson, J.C.: Partial evaluation in logic programming. J. Log. Program. 11(3&4), 217–242 (1991)
45. Maher, M.J.: Equivalences of logic programs. In: Foundations of Deductive Databases and Logic Programming, pp. 627–658 (1988)
46. Marriott, K., Søndergaard, H.: Bottom-up abstract interpretation of logic programs. In: Proc. Fifth Int'l Conf. on Logic Programming, pp. 733–748. MIT Press, Cambridge (1988)
47. Marriott, K., Søndergaard, H.: Semantics-based dataflow analysis of logic programs. In: IFIP Congress, pp. 601–606. North-Holland, Amsterdam (1989)
48. Marriott, K., Søndergaard, H.: Precise and efficient groundness analysis for logic programs. LOPLAS 2(1-4), 181–196 (1993)

49. Sato, T., Tamaki, H.: Enumeration of success patterns in logic programs. Theor. Comput. Sci. 34, 227–240 (1984)
50. Scozzari, F.: Logical optimality of groundness analysis. Theor. Comput. Sci. 277(1-2), 149–184 (2002)
51. Shapiro, E.Y.: Algorithmic Program Debugging. MIT Press, Cambridge (1983)
52. Tamaki, H., Sato, T.: Unfold/fold transformation of logic programs. In: ICLP, pp. 127–138 (1984)

Theory-Specific Automated Reasoning

Andrea Formisano[1] and Eugenio G. Omodeo[2]

[1] Università di Perugia, Italy
formis@dipmat.unipg.it
[2] Università di Trieste, Italy
eomodeo@units.it

Abstract. In designing a large-scale computerized proof system, one is often confronted with issues of two kinds: issues regarding an underlying logical calculus, and issues that refer to theories, either specified axiomatically or characterized by indication of either a privileged model or a family of intended models. Proof services related to the theories most often take the form of satisfiability decision or semi-decision procedures (in a sense, polyadic inference rules), while some of the services offered by the calculus (e.g., the Davis-Putnam propositional satisfiability checker) provide low-level mechanisms for integrating services of the former kind. Integration among services can ensure speed-up (i.e., lower number of steps) in the proofs, but it must always be legitimatized by a conservativeness result. Interoperability among proof checkers and autonomous theorem provers is another key point of integration.

In discussing these and related issues, this paper refers to Set Theory as the unifying background, and to a specific proof-checker based on a slightly unorthodox formalization of it as an arena for experimentation.

Keywords: proof assistant, decision algorithm, inference mechanism, Set Theory.

Introduction

Computer-aided verification of formal proofs can be applied extensively in mathematics, and one can likewise check for correctness sophisticated algorithms and computer programs, as well as critical hardware designs. Evidence of this has been achieved many times through *proof assistants* such as Mizar, Coq, HOL light, and Isabelle, to mention only a few [67, 5]. These are leading to the creation of big repositories of formalized mathematical knowledge.

After the discovery, around the mid 1960s [41], of easily implementable powerful inference methods (the linked conjunct method [40, 85], the resolution principle [104], and the Knuth-Bendix algorithm [73, 46]), the quest for automatic strategies that would effectively drive the exploration of the infinite search space of all proofs towards a specific goal remained, for many years, the main focus of the research on theorem-proving. Gradually, the expectation that theorem-provers would rapidly gain enough autonomy to even outperform human capabilities in many situations, was replaced by the conception that automated deduction systems must primarily assist, like reliable and fast technicians,

A. Dovier, E. Pontelli (Eds.): 25 Years of Logic Programming, LNCS 6125, pp. 37–63, 2010.

the working mathematician and computer scientist, either by interacting with her/him in the construction of detailed proofs, or by proof-checking script files which we call *proof-scenarios*, often written by hand and consisting at times of many thousand lines.

This explains why, in the pages that follow, we will insist more on proof-checking than on theorem-proving, and, as regards investigation topics relevant to the present, more on proof-engineering issues than on issues pertaining to computational logic *per se*. In particular, it seems to us that *integration* among the various systems (both at the level of formal proof language and at the level of proof-exchange across proof assistants) must be tackled as a crucially important issue in order that proof technology can be raised above its present limitations.

We will first trace a fundamental distinction between issues that regard the effectiveness of the inferential engine—which must be clever mainly in the discovery of small proofs—, and issues that have to do with the development of large-scale proof scenarios. We ascribe the former to *computational logic* (and, to some extent, to the broader field of artificial intelligence) and the latter to the relatively new area of *proof-engineering*, some of whose history we will trace back to publications of the 1970s.

Then we will contend that a set-oriented proof-language (mainly first-order, but with a second-order feature primarily aimed at easing proof-reuse) is the ideal support for a proof-checking system which upholds a style of proof akin to common, semi-formal mathematical language. We will suggest that a proof-checker of this nature—based on a broad-gauge theory, rather than on a logical calculus—offers natural hooks for the integration of proofs achieved with heterogeneous proof assistants, including among such "assistants" the implementations of specialized decision algorithms.

Within this eclectic view of proof assistants, where set theory serves as a sort of cement, decision algorithms can play various roles. In some cases, especially when they handle basic fragments of set theory (but this is also the case of a type-finding method expounded in [87, Sec. 4]), these algorithms can be exploited very straightforwardly, just as inference rules whose deductive power reflects how much of the proving effort can be delegated in full to automatic tools. Likewise, in declarative programming, specialized solvers encompass standard problem-solving techniques from whose details programmers want to be alleviated. When decision algorithms are intended this way, namely as basic inference mechanisms acting behind the scenes, their study still belongs to computational logic.

There is another dimension, in the study of decision algorithms, which poses its own challenges to the automated reasoning field. How can these methods be integrated with one another? To what extent their integration can boost the capabilities of an automated assistant? In which cases it is not viable due to complexity limitations or because integration would disrupt decidability?

Because of the significant involvement of Italian researchers in the study of decision algorithms from both angles—existence and complexity on the one hand, implementation and cross-combination on the other—, we devote an ample section of this paper to surveying achievements and trends in this area.

1   Proof-Engineering and Automated Reasoning in Logic

Leibniz's classic bipartition of logic, into a *calculus* for reasoning and an *ideography* of concepts, still today retains some influence on the organization of the field of automated deductive reasoning. On the one hand, many contributions to this field aim at providing mechanical rules (e.g., the resolution principle) for reliable reasoning, and these are most often rooted in a calculus: typically, in a version of first-order predicate logic. On the other hand, many contributions tend, through a cluster of axiomatic theories, to form the framework within which the notions needed for applications (the notion of real number, say, or the ones basic to general topology) can be tersely defined.

Two additional areas are essential for the deployment of a technology of formal proofs: if the challenges in front of us [1] are to be tamed effectively, we need a *heuristic*, which means an art of quick proof-discovery; and skills and tools for *proof-engineering*.

As for quick proof-discovery, we cannot rely on fully generic means to speed up explorations of the (usually infinite) search space where proofs are buried. Long ago Bledsoe warned [13] that automated proof-techniques would be confronted with limited success unless they embodied specific knowledge of the various mathematical disciplines. The great majority of heuristic techniques which have been devised to date enhance proof methods which are directly based on first-order predicate calculus; however, it would be quite an amazing coincidence if techniques of this sort proved to be clever at exploiting the peculiarities of an axiomatic theory without their design having been oriented to the specific purpose: when they do, it is likely that considerable human effort went into finding a well-conditioned formulation of the axioms [9]. In particular, as we view—in agreement with [9]—Set Theory as the main arena where automatic theorem-provers and automated proof-assistants should be put to work [64, 65], we must look for *ad hoc* proof-methods and heuristic search techniques.

Checking, by means of a proof-verifier, any elaborate argument—the explanation, say, of a sophisticated algorithm, or the proof of some profound mathematical theorem—requires that a large number of logical statements be fed into the system. These statements must formalize a line of reasoning that leads from bare rudiments of logic and mathematics to the specialized topic of interest—for instance, graph theory, or mathematical analysis—and then to a target conclusion. Such an enterprise can only be managed effectively if suitable constructs ensuring "modularity", which means the possibility to subdivide a long argument into wieldy chunks, are available. The obvious goal of proof modularization is to avoid repeating similar steps when the proofs of two theorems are closely analogous. Modularization must also conceal the details of a proof once they have been fed into the system and successfully certified. These considerations underlie some recent research trends in proof-engineering, whose highlight we postpone to Sec. 2, contenting ourselves for the time being to display in Fig. 1 the typical structure of a "chunk of mathematical knowledge" which can be invoked repeatedly during a proof-development session very much like a procedure during a program execution.

$$\mathsf{Splits}(P, S) \ \leftrightarrow_{\mathrm{Def}} \ \big\langle \forall\, b \in P,\ \forall\, b' \in P \mid (b = b' \ \leftrightarrow \ b \cap b' \neq \emptyset)\ \&\ b \subseteq S \big\rangle$$

$\text{THEORY } \ \mathsf{eq\_classes}(\mathsf{s_0}, \mathsf{Eq}(\mathsf{X}, \mathsf{Y}))$
$$\big\langle \forall\, v, w, z \mid \{v, w, z\} \subseteq \mathsf{s_0} \ \to \ \mathsf{Eq}(v, v)\ \&\ \big(\mathsf{Eq}(v, w) \& \mathsf{Eq}(z, w) \ \to \ \mathsf{Eq}(v, z)\big)\big\rangle$$
$\Rightarrow (\mathsf{quot}_\ominus, \mathsf{cl\_of}_\ominus) \quad \textit{-- quotient-set and canonical embedding}$
$\qquad \mathsf{Splits}(\mathsf{quot}_\ominus, \mathsf{s_0})$
$$\big\langle \forall\, x \mid \big(x \in \mathsf{s_0} \ \to \ x \in \mathsf{cl\_of}_\ominus(x)\big)\ \&\ \mathsf{cl\_of}_\ominus(x) \in \mathsf{quot}_\ominus \cup \{\{\mathsf{s_0}\}\}\big\rangle$$
$$\big\langle \forall\, x \in \mathsf{s_0},\ \forall\, y \in \mathsf{s_0} \mid \mathsf{Eq}(x, y) \ \leftrightarrow \ \mathsf{cl_of}_\ominus(x) = \mathsf{cl_of}_\ominus(y)\big\rangle$
$\text{END } \ \mathsf{eq\_classes}$

Fig. 1. Partitioning of a set into equivalence classes

Modularization alone does not suffice, of course, to answer the host of proof-engineering problems which arise in the development of very extensive proofs and in the creation, testing, and maintenance of automated proof-systems. Nevertheless, it contributes in many ways to various aspects of proof-engineering, ranging from the readability of proofs to soundness-preserving extensibility (see [43, 44]) of the verifiers.

2 Set Theory as a Background for Discussion

Historically, Set Theory grew out of efforts aimed at providing a single foundation and a sort of *lingua franca* for the diverse areas of mathematics; consequently, when constructions and proofs of classical mathematics are developed in full within the framework of a theory of sets such as Zermelo-Fraenkel (ZF), they will resemble the corresponding specifications as found in an ordinary textbook (were it not for the extra amount of formal detail needed to make them digestible to a proof verifier). E.g., the set-theoretic jargon can be successfully exploited to carry out any of the classical constructions of the field of real numbers. To pick another example, the Stone representation theorem for Boolean algebras is quite naturally stated and proved in set-theoretic terms.

Specification of algorithms, and algorithm correctness verification, can also benefit from a set-theoretic language [98], as one can judge from the very existence of set-based programming languages such as SETL [108] and {log} [55] (for another proposal, see [114]). To support formal reasoning in the realm of algorithms, a theory of hereditarily finite sets [117] would suffice, but a full-fledged set theory such a ZF is even better, as it enables one to treat in a uniform framework [94] algorithmic issues such as, e.g., the correctness of the Davis-Putnam-Logemann-Loveland satisfiability test DPLL, and non-constructive ones such as, e.g., the compactness of propositional logic, proved via the Zorn lemma.

A proof verifier based on set theory (thought of as a "big theory" [61] by means of which all reasoning is performed within a single, powerful and highly expressive language) has been described in [87].[1] Decidable fragments of set theory which we will discuss in Sec. 5 play, in this system, roles comparable to those of resolution and paramodulation in autonomous theorem provers. In order to support proof reuse and various ways of extending the inferential armory, this verifier relies on a version of ZF which offers a second-order construct named 'Theory' [93], inspired by the mechanism for parameterized specifications of the Clear language [16]. As the tiny example in Fig. 1 shows, these theories, like procedures in a programming language, have lists of formal input parameters (s_0 and $\mathsf{Eq}(\_,\_)$, in the example at hand). Each Theory requires its parameters to meet a set of assumptions. When "applied" to a list of actual parameters that have been shown to meet the assumptions, a theory will instantiate several additional "output" symbols (quot_Θ and $\mathsf{cl\_of}_\Theta$ in our example)[2] standing for sets (e.g. quot_Θ), functions (e.g. $\mathsf{cl\_of}_\Theta$), and predicates, and then supply a list of claims initially proved explicitly by the user inside the theory itself. These are theorems generally involving the new symbols.

A convenient format for Theory invocation is the one exemplified here, where $\mathsf{EqVenn}(\mathsf{X},\mathsf{Y}) \leftrightarrow_{\mathrm{Def}} \{\, t \in \mathsf{s} : \mathsf{X} \in t \,\} = \{\, t \in \mathsf{s} : \mathsf{Y} \in t \,\}$:

APPLY $\langle \mathsf{quot}_\Theta : \text{regions} \rangle$ eq_classes$\big(\, \mathsf{s}_0 \mapsto \bigcup\mathsf{s}, \ \ \mathsf{Eq}(\mathsf{X},\mathsf{Y}) \mapsto \mathsf{EqVenn}(\mathsf{X},\mathsf{Y}) \,\big) \Rightarrow$
     Theorem venn $\cdot$ 1: [Venn's partition] Splits$(\text{regions}, \bigcup\mathsf{s})$.

Acceptance of this single-line proof of Theorem venn $\cdot$ 1 on the part of a proof-checker presupposes verification of the assumption $\langle \forall v, w, z \mid \{v, w, z\} \subseteq \mathsf{s}_0 \rightarrow$ $\mathsf{Eq}(v, v) \,\&\, \cdots \rangle$ of the invoked theory, inside which s_0 and Eq are replaced by $\bigcup\mathsf{s}$ and by EqVenn, respectively, where the following definition of the *unionset* operator applies:

$$\bigcup S =_{\mathrm{Def}} \{\, u : v \in S,\, u \in v \,\}.$$

To see how the Theory construct can be exploited to enhance the inferential armory, consider the first example in Fig. 2: this theory, which provides a mechanism constructing a key entity for a refutation, implements an *induction principle*, seen here, rather than as a new inference rule, as a tactic for instantiating cleverly an existential variable. The principle under consideration enables us to prove that some property $\varphi(F)$ holds for all finite sets F, via an argument organized as follows: (1) Assume that a counterexample f_0 exists, i.e., suppose that $\mathsf{Is\_finite}(f_0) \ \& \ \neg\varphi(f_0)$. (2) By binding f_0 and $\neg\varphi(\_)$ as actual parameters to the formal parameters f and $\mathsf{P}(\_)$ of finite_induction, i.e. by invoking

APPLY $\langle f_\Theta : f_1 \rangle$ finite_induction$\big(\mathsf{f} \mapsto f_0, \ \ \mathsf{P}(\mathsf{X}) \mapsto \neg\varphi(\mathsf{X}) \big)$,

---

[1] This system, conceived by Jacob T. Schwartz, is sometimes called Referee (or 'Ref' for brevity), and sometimes called ÆtnaNova. An on-line tutorial for it is available at the URL http://setl.dyndns.org/EtnaNova/login/Ref_user_manual.html, while the fragments of Ref scenarios occurring in this paper are often drawn from http://setl.dyndns.org/EtnaNova/login/search_folder/scenario.pdf

[2] Such output symbols, whose meanings are specified inside the Theory, carry the Θ subscript.

THEORY finite_induction $(f, P(X))$
    Is_finite(f) & $P(f)$
$\Rightarrow$ (f_Θ)
    $f_\Theta \subseteq f$ & $P(f_\Theta)$ & $\langle \forall t \subseteq f_\Theta \mid t \neq f_\Theta \rightarrow \neg P(t) \rangle$
END finite_induction

THEORY membership_induction $(s, P(X))$
    $P(s)$
$\Rightarrow$ (s_Θ)
    $s_\Theta \in$ ult_membs$(\{s\})$ & $P(s_\Theta)$ & $\langle \forall k \in s_\Theta \mid \neg P(k) \rangle$
END membership_induction

Fig. 2. Two inference mechanisms introduced through THEORYes

get a *smallest* finite set f_1 such that $\neg\varphi(f_1)$ holds. (3) Through details that depend on the peculiarities of $\varphi(F)$, strive to derive a contradiction from the alleged minimality of f_1, so as to get the desired conclusion $\langle \forall f \mid$ Is_finite$(f) \rightarrow \varphi(f) \rangle$.

A proof-strategy analogous to this one can be associated with any well-founded relation. In the example just seen, this is the inclusion relationship over the class of *finite* sets;[3] in the other example of Fig. 2 this is the membership relation over *all* sets, which is well-founded according to von Neumann's *regularity axiom*—a built-in postulate in our set-based verifier (cf. Sec. 4).

While examining the THEORY membership_induction, we take the opportunity to illustrate another benefit arising from the axioms of set theory: sometimes, definitions serve just as a syntactic device enabling one to introduce shorthand notation, such as e.g.

$$\text{next}(X) =_{\text{Def}} X \cup \{X\};$$

but our proof-scenarios can contain recursive definitions justified by the regularity axiom, on which semantics [17] has a much heavier bearing. Examples of this nature are the definitions

$$\text{ult\_membs}(X) =_{\text{Def}} X \cup \bigcup \{ \text{ult\_membs}(y) : y \in X \},$$
$$\text{rk}(X) =_{\text{Def}} \bigcup \{ \text{next}(\text{rk}(y)) : y \in X \},$$

of which:[4]

– the former specifies the set of all *ultimate members* of any given set X, namely the set consisting of all those y from which a membership chain $y = y_0 \in y_1 \in \cdots \in y_n = X$ leads to X;

---

[3] When set-inclusion gets restricted to the natural numbers, the THEORY finite_induction specializes into the most familiar *arithmetic induction* principle.

[4] These definitions yield, among others, that ult_membs$(\{\{\{\emptyset\}\}\}) = \{\emptyset, \{\emptyset\}, \{\{\emptyset\}\}\} = \{0, 1, \{1\}\}$, rk$(\{\{\{\emptyset\}\}\}) = \{\emptyset, \{\emptyset\}, \{\emptyset, \{\emptyset\}\}\} = \{0, 1, 2\} = 3$, ult_membs$(\{0, 2\}) = $ ult_membs$(\{\emptyset, \{\emptyset, \{\emptyset\}\}\}) = 3 = $ rk$(\{0, 2\})$, and ult_membs$(\mathbb{N} \setminus f) = $ rk$(\mathbb{N} \setminus f) = \mathbb{N}$ if $\mathbb{N}$ designates the natural numbers and f is a finite set.

− the latter specifies the *rank* of X: intuitively speaking, an ordinal measure of how deeply the ultimate members of X are nested inside X.

The THEORY membership_induction tells us that when a set s_0 violating some property $\varphi(S)$ exists, so that $\neg\varphi(s_0)$ holds, then a rank-minimal such set s_1 exists: more specifically, there is an s_1 either coinciding with s_0 or appearing among its ultimate members, which meets the condition $\neg\varphi(s_1)$ whereas $\varphi(k)$ holds for all $k \in s_1$. This THEORY hints, again, at a strategy for proving claims of the form $\langle \forall x \mid \varphi(x) \rangle$: (1) suppose that an s_0 exists such that $\neg\varphi(s_0)$ holds; (2) let s_1 be a membership-minimal such set, so that $\neg\varphi(s_1)$ & $\langle \forall k \in s_1 \mid \varphi(k) \rangle$ holds; (3) strive to get a contradiction from the alleged minimality of s_1.

It should be clear that when a formal parameter, like P(_) in the two THE-ORYes just examined, refers to a general property of sets, one can assign to it as an actual parameter a first-order formula φ with one free variable: our set-language, in fact, to follow ZF closely, does not provide explicit means to speak about proper classes (in particular, its individual logical variables can only take set values).[5] Analogously, an output parameter like the cl_of$_\Theta$ of Fig. 1, as it stands for a function defined over the universe of all sets, designates a proper class of ordered pairs. As we are seeing, the THEORY construct, which we have introduced mainly motivated by proof-engineering considerations, lifts the expressiveness of our formal language well above the usual limitations of ZF: as a matter of fact, we can indefinitely extend the signature of our (essentially first-order) set-theoretic language thanks to a second-order Skolemization mechanism implicit in the functioning of THEORYes. We can, at times, raise considerably the import of a THEORY: see, for example, in Fig. 3, the much enhanced version of the THE-ORY of Fig. 1. While implementing the internals of this new THEORY, one can define ch$_\Theta$(X) to be an $\in$-minimal element of $\{ w \in \text{ult\_membs}(x_0) \mid \text{Eq}(w, X) \}$, where $x_0 = \{ u \in \text{c}(X) \mid \text{Eq}(u, X) \ \& \ \text{P}(u) \}$. Thus, even when the Eq-class of X is not a set (e.g., this class might consist of all ordinals which can be put in one-one correspondence with one another), ch$_\Theta$(X) will be an $\in$-minimal element of this class, depending solely on the class and not on X.

An invocation of this theory could be as follows (where EqVenn is as before):

APPLY $\langle$ch$_\Theta$: repr$\rangle$ circumscribed_eq_classes(Eq(X, Y) $\mapsto$ EqVenn(X, Y),

   P(X) $\mapsto$ X $\in$ {s} $\cup \bigcup$ s,   c(X) $\mapsto$ **if** X $\in \bigcup$ s **then** $\bigcup$ s **else** {s} **fi**) $\Rightarrow$

   THEOREM venn · 2: [Venn's representatives]

   $\langle \forall v, w \mid \text{EqVenn}(v, w) \ \leftrightarrow \ \text{repr}(v) = \text{repr}(w) \rangle$

   $\langle \forall v, w \mid \text{EqVenn}(v, w) \ \longrightarrow \ v \notin \text{repr}(w) \rangle.$

(This invocation presupposes, of course, that all three assumptions of the invoked theory, suitably instantiated, have been verified.)

A reason why set theory can do well as "glue" for the integration of proof assistants is that one can shallowly amalgamate into it the semantics of a formal deductive system: an illustration of this, referring to first-order predicate calculus

---

[5] The reader who finds the set-class distinction unfamiliar to him/her, can skim through this paragraph superficially.

THEORY circumscribed_eq_classes(Eq(X, Y), P(X), c(X))
$\quad\Big\langle \forall v, w, z \mid \mathsf{Eq}(v, v) \,\&\, \big(\mathsf{Eq}(v, w) \,\&\, \mathsf{Eq}(z, w) \;\rightarrow\; \mathsf{Eq}(v, z)\big)\Big\rangle$
$\quad\Big\langle \forall v \mid \big\langle \exists u \mid \mathsf{Eq}(u, v) \,\&\, \mathsf{P}(u)\big\rangle\Big\rangle$
$\quad\Big\langle \forall v, u \mid \mathsf{Eq}(u, v) \,\&\, \mathsf{P}(u) \;\rightarrow\; u \in \mathsf{c}(v)\Big\rangle$
$\Rightarrow (\mathsf{ch}_\Theta)$ -- *choice of an $\in$-minimal representative from each* Eq-*class*
$\quad\Big\langle \forall v \mid \mathsf{Eq}(\mathsf{ch}_\Theta(v), v)\Big\rangle$
$\quad\Big\langle \forall v, w \mid \mathsf{Eq}(v, w) \;\leftrightarrow\; \mathsf{ch}_\Theta(v) = \mathsf{ch}_\Theta(w)\Big\rangle$
$\quad\Big\langle \forall v, w \mid \mathsf{Eq}(v, w) \;\rightarrow\; v \notin \mathsf{ch}_\Theta(w)\Big\rangle$
END circumscribed_eq_classes

Fig. 3. Enhanced version of the selection of class representatives

will be provided in Sec. 3—analogously one could treat a more sophisticated, e.g. a strongly typed, logical calculus.

When needed (but more rarely, as this approach is more laborious), one can tackle interoperability among logical systems at a deeper level: one can proceed to "arithmetize" a logical system, i.e., to encode in set-theoretic terms both its syntax and its deductive apparatus. On a very small scale, this is illustrated by the today standard representation of CNF formulae as sets of sets of literals. Speaking in general, set theory can be very naturally used to support meta-level reasoning: this emerges vividly, for example, from the relative ease with which limitative results such as the celebrated Gödel theorems can be proved within a theory dealing with aggregates explicitly [96, 80], compared to an arithmetic of numbers (where the treatment of sets, lists, derivations, etc., sometimes calls for unwieldy encodings).

3   Interoperability among Reasoners

Besides being useful for the avoidance of repeated proofs in closely analogous contexts, for information-hiding, and for sound extensions of the logical armory, the THEORY construct gives us a mechanism for gluing together results obtained with different proof-assistants. To illustrate the point, we choose the mathematical theory of ordered Abelian groups as our example of an *outer* (or "external") theory. This can be very naturally stated as a first-order theory (cf. Fig. 4), but its integration in a scenario developed with our set-based proof-verifier presupposes various slight changes. For instance, unrestricted quantifiers must be restricted (i.e., $\langle \forall x \, \psi \rangle$ and $\langle \exists x \, \psi \rangle$ become $\langle \forall x \mid \mathsf{In\_dom}(x) \rightarrow \psi \rangle$ and $\langle \exists x \mid \mathsf{In\_dom}(x) \,\&\, \psi \rangle$ respectively), to reflect the fact that the primary domain of discourse remains the universe of all sets, even though one is momentarily focusing—while reasoning inside the outer theory—on the support domain of an ordered commutative group. Restricting quantifiers becomes necessary both in the statements of postulates and in the theorem claims. As another issue, let us mention the fact that a theorem-prover may offer no special means to separate axioms from the definitions of symbols which are not indispensable in the signature of the outer theory (e.g., in the case at hand, a symbol designating the

```
-- Abelian group axioms
⟨∀x, y, z | (x⊕y)⊕z=x⊕(y⊕z)⟩              -- associativity
⟨∀x | x⊕e=x⟩                              -- right unit
⟨∀x | x⊕⊖x=e⟩                             -- right inverse
⟨∀x, y | x⊕y=y⊕x⟩                         -- commutativity
-- ordering axioms (axioms concerning non-negativeness)
⟨∀x, y | Nneg(x) & Nneg(y) → Nneg(x⊕y)⟩
⟨∀x | Nneg(x) ∨ Nneg(⊖x)⟩
⟨∀x | Nneg(x) & Nneg(⊖x) → x=e⟩
-- definitional extensions
⟨∀x | Nneg(x) → ‖x‖=x⟩                    -- definition of the absolute value ...
⟨∀x | ¬Nneg(x) → ‖x‖=⊖x⟩                  -- ... definition of the absolute value
⟨∀x, y | x≼y ↔ Nneg(y⊕⊖x)⟩               -- definition of comparison
```

Fig. 4. Outer theory of ordered Abelian groups (postulates and definitional extensions)

absolute value operation); nonetheless, when one interfaces the theory, one wants to stress the different roles of the assumptions regarding the symbols (postulates on one side, definitions on the other).

Fig. 5 shows what form a THEORY interface with the theory of ordered Abelian groups may take.[6] Observe the suffix to the THEORY keyword, indicating a specific syntax to be adopted in external files (resulting from the interaction between a user and an external prover). A standardization of the syntaxes adopted by the different provers seems to be necessary to favor the integration between theorem assistants. In this example, observe that the domain to which the quantified variables are restricted is treated as a property $\mathsf{In\_dom}(\_)$ of sets, even though most typically it will satisfy the biimplication $\mathsf{In\_dom}(X) \leftrightarrow X \in \mathsf{theory\_dom}$ for a suitable set $\mathsf{theory\_dom}$. This design choice makes it possible for the user to, e.g., invoke the THEORY orderedGroups, after defining globally

$$Z^{\smallsmile} \;=_{\mathrm{Def}} \; \textbf{if } \emptyset \in Z \textbf{ then } Z \setminus \{\emptyset\} \textbf{ else } Z \cup \{\emptyset\} \textbf{ fi} \,,$$

with actual parameters of such generality as

$$\begin{aligned}
X \oplus Y &= \{z \in X \mid z^{\smallsmile} \notin Y\} \cup \{z \in Y \mid z^{\smallsmile} \notin X\}\,, \\
\ominus X &= \{z^{\smallsmile} : z \in X\}\,, \\
\mathsf{In\_dom}(X) &\leftrightarrow X \cap \ominus X = \emptyset\,, \\
\mathsf{Nneg}(X) &\leftrightarrow X = \emptyset \vee (\min\{z \cup \{\emptyset\} : z \in X\} \in X)\,,
\end{aligned}$$

where the min operation refers to a fixed well-ordering of the entire universe of sets. (Without too much effort the support-domain of this group could be so extended to encompass all sets.)

---

[6] This is a decidable theory—see below. One might hence consider introducing a decider for it as an inference rule, but the usefulness of an *ad hoc* inference rule is debatable, since the lemmas appearing in Fig. 5—instantiated to the case of rational numbers—are already adequate for the construction of the reals.

THEORY_*outer* orderedGroups$(\mathsf{In\_dom}(x), x \oplus y, e, \ominus x, \mathsf{Nneg}(x), x \preccurlyeq y)$
   -- closure laws
 $\big\langle \forall x, y \mid \mathsf{In\_dom}(x)\ \&\ \mathsf{In\_dom}(y) \rightarrow \mathsf{In\_dom}(x \oplus y) \big\rangle\ \&\ \mathsf{In\_dom}(e)$
 $\big\langle \forall x \mid \mathsf{In\_dom}(x) \rightarrow \mathsf{In\_dom}(\ominus x) \big\rangle$
   -- axioms proper
 $\big\langle \forall x, y, z \mid \mathsf{In\_dom}(x)\ \&\ \mathsf{In\_dom}(y)\ \&\ \mathsf{In\_dom}(z) \rightarrow (x \oplus y) \oplus z = x \oplus (y \oplus z) \big\rangle$

     $\vdots$ $\vdots$ $\vdots$

 $\big\langle \forall x \mid \mathsf{In\_dom}(x) \rightarrow \mathsf{Nneg}(x)\ \&\ \mathsf{Nneg}(\ominus x) \rightarrow x = e \big\rangle$
   -- shorthand notation
 $\big\langle \forall x, y \mid \mathsf{In\_dom}(x)\ \&\ \mathsf{In\_dom}(y) \rightarrow x \preccurlyeq y \leftrightarrow \mathsf{Nneg}(y \oplus \ominus x) \big\rangle$
 extdfn $\Rightarrow \|X\|_{\ominus} =_{\mathrm{Def}}$ **if** $\mathsf{Nneg}(X)$ **then** X **else** $\ominus X$ **fi**
$\Rightarrow$

 $\big\langle \forall x, y \mid \mathsf{In\_dom}(x)\ \&\ \mathsf{In\_dom}(y) \rightarrow \ominus(x \oplus \ominus y) = y \oplus \ominus x \big\rangle$
   -- cancellation laws
 $\big\langle \forall x, y, z \mid \mathsf{In\_dom}(x)\ \&\ \mathsf{In\_dom}(y)\ \&\ \mathsf{In\_dom}(z) \rightarrow x \oplus y = x \oplus z \rightarrow y = z \big\rangle$
 $\big\langle \forall x, y, z \mid \mathsf{In\_dom}(x)\ \&\ \mathsf{In\_dom}(y)\ \&\ \mathsf{In\_dom}(z) \rightarrow x \oplus z = y \oplus z \rightarrow x = y \big\rangle$
   -- totality, reflexivity, and three transitivity laws
 $\big\langle \forall x, y \mid \mathsf{In\_dom}(x)\ \&\ \mathsf{In\_dom}(y) \rightarrow (x \preccurlyeq y \vee y \preccurlyeq x)\ \&\ x \preccurlyeq x \big\rangle$
 $\big\langle \forall x, y, z \mid \mathsf{In\_dom}(x)\ \&\ \mathsf{In\_dom}(y)\ \&\ \mathsf{In\_dom}(z) \rightarrow x \preccurlyeq y\ \&\ y \preccurlyeq z \rightarrow x \preccurlyeq z \big\rangle$
 $\big\langle \forall x, y, z \mid \mathsf{In\_dom}(x)\ \&\ \mathsf{In\_dom}(y)\ \&\ \mathsf{In\_dom}(z) \rightarrow x \preccurlyeq y\ \&\ x \neq y\ \&\ y \preccurlyeq z \rightarrow x \neq z \big\rangle$
 $\big\langle \forall x, y, z \mid \mathsf{In\_dom}(x)\ \&\ \mathsf{In\_dom}(y)\ \&\ \mathsf{In\_dom}(z) \rightarrow x \preccurlyeq y\ \&\ y \preccurlyeq z\ \&\ y \neq z \rightarrow x \neq z \big\rangle$
   -- two isotony laws
 $\big\langle \forall x, y, z \mid \mathsf{In\_dom}(x)\ \&\ \mathsf{In\_dom}(y)\ \&\ \mathsf{In\_dom}(z) \rightarrow x \preccurlyeq y \rightarrow x \oplus z \preccurlyeq y \oplus z \big\rangle$
 $\big\langle \forall x, y, z \mid \mathsf{In\_dom}(x)\ \&\ \mathsf{In\_dom}(y)\ \&\ \mathsf{In\_dom}(z) \rightarrow x \preccurlyeq y\ \&\ x \neq y \rightarrow x \oplus z \neq y \oplus z \big\rangle$
   -- laws concerning the absolute value
 $\big\langle \forall x \mid \mathsf{In\_dom}(x) \rightarrow \mathsf{In\_dom}(\|x\|_{\ominus})\ \&\ \|x \ominus \ominus x\|_{\ominus} = e\ \&\ x \preccurlyeq \|x\|_{\ominus}\ \&\ e \preccurlyeq \|x\|_{\ominus} \big\rangle$
 $\big\langle \forall x \mid \mathsf{In\_dom}(x) \rightarrow \big\| \|x\|_{\ominus} \big\|_{\ominus} = \|x\|_{\ominus}\ \&\ \big(\|x\|_{\ominus} = e \leftrightarrow x = e\big)\ \&\ \|\ominus x\|_{\ominus} = \|x\|_{\ominus} \big\rangle$
 $\big\langle \forall x, y \mid \mathsf{In\_dom}(x)\ \&\ \mathsf{In\_dom}(y) \rightarrow x \oplus y \preccurlyeq \|x\|_{\ominus} \oplus \|y\|_{\ominus}\ \&\ \|x \oplus y\|_{\ominus} \preccurlyeq \|x\|_{\ominus} \oplus \|y\|_{\ominus} \big\rangle$
 $\big\langle \forall x, y \mid \mathsf{In\_dom}(x)\ \&\ \mathsf{In\_dom}(y) \rightarrow \neg \mathsf{Nneg}(x) \rightarrow x \preccurlyeq \|y\|_{\ominus}\ \&\ x \neq \|y\|_{\ominus} \big\rangle$
 $\big\langle \forall x, y, z \mid \mathsf{In\_dom}(x)\ \&\ \mathsf{In\_dom}(y)\ \&\ \mathsf{In\_dom}(z) \rightarrow \|x \ominus y\|_{\ominus} \preccurlyeq z \rightarrow y \preccurlyeq x \oplus z \big\rangle$
 $\big\langle \forall x, y, z \mid \mathsf{In\_dom}(x)\ \&\ \mathsf{In\_dom}(y)\ \&\ \mathsf{In\_dom}(z) \rightarrow \|x \ominus z\|_{\ominus} \preccurlyeq \|x \ominus y\|_{\ominus} \oplus \|y \ominus z\|_{\ominus} \big\rangle$
 $\big\langle \forall x, y \mid \mathsf{In\_dom}(x)\ \&\ \mathsf{In\_dom}(y) \rightarrow \mathsf{Nneg}(y) \rightarrow x \oplus \ominus y \preccurlyeq x \oplus y \big\rangle$
 $\big\langle \forall x, y \mid \mathsf{In\_dom}(x)\ \&\ \mathsf{In\_dom}(y) \rightarrow \big\| \|x\|_{\ominus} \oplus \ominus \|y\|_{\ominus} \big\|_{\ominus} \preccurlyeq \|x \ominus \ominus y\|_{\ominus} \big\rangle$
 $\big\langle \forall x, y \mid \mathsf{In\_dom}(x)\ \&\ \mathsf{In\_dom}(y) \rightarrow \big\| x \big\|_{\ominus} \oplus \ominus \big\| \|y\|_{\ominus} \oplus \ominus \|x\|_{\ominus} \big\|_{\ominus} \preccurlyeq \|y\|_{\ominus} \big\rangle$
 $\big\langle \forall x \mid \mathsf{In\_dom}(x) \rightarrow \|x\|_{\ominus} =$ **if** $\mathsf{Nneg}(\ominus x)$ **then** $\ominus x$ **else** x **fi** $\big\rangle$
END orderedGroups

Fig. 5. Interface THEORY for ordered Abelian groups (assumptions, various lemmas)

 Notice also that in some cases we cannot adopt this policy of restricting the quantified variables of a theory to a *possibly proper* class when translating its axioms into set-theoretic assumptions of a THEORY: in fact, one is frequently confronted with cases when there is an infinite axiom scheme (think, e.g., of the continuity postulate of elementary geometry [119], or of the induction postulate of Peano arithmetic) which admits a much more straightforward set-theoretic translation if one takes the domain of discourse to be a set.

Often, even when a theory is not meant to refer to a proper class, one may ease reasoning within it by extending its domain of discourse to all sets, so that any restriction of quantifiers becomes superfluous. An example of this has been given in [38], where a toggling function representing negation over the set of propositional literals gets plainly extended to the entire universe of sets; likewise, proving the correctness of various decidable extensions of multi-level syllogistic [26] relies on global extensions of functions or relations enjoying particular properties. It is not entirely clear to us—but we deem it useful to investigate this point—when the set-theoretic rendering of a first-order theory can dispense with the restriction of quantifiers.

At present, our set-based verifier implements only a form of loose coupling with outer provers. Stronger, more dynamic, forms of interaction with outer proof-assistants should be devised.

4   Bringing Algorithmic Specifications into Play

The legitimacy, in a set-based verifier, of the built-in form of recursion illustrated by the definitions of ult_membs(_) and rk(_) in Sec. 2, rests on the global well-foundedness of membership (the regularity axiom), statable as

$$\langle \forall x \mid \mathbf{arb}(x) \in x \cup \{x\} \ \& \ \mathbf{arb}(x) \cap x = \emptyset \rangle.$$

Likewise, as explained in [93], one can resort to a more tortuous recursive definition to introduce a function whose domain is a set s, whenever a binary relation has been shown to be well-founded on s. The THEORY wellfounded_recursive_fcn of Fig. 6, or some specialized variant of it, such as the THEORY finite_recursive_fcn of the same figure, can be exploited to do this. E.g., one can define summation over a monoid so as to meet the specification of Fig. 8, inside sigma_theory, by putting $\Sigma_\Theta(G) =_{\mathrm{Def}} \Sigma'(G, \emptyset)$ after invoking

APPLY $\langle \mathrm{rec}_\Theta : \Sigma' \rangle$ finite_recursive_fcn $\big(\mathsf{f}(\mathsf{B}, \mathsf{X}, \mathsf{T}) \mapsto \mathbf{arb}(\mathsf{B}),$
   $\mathsf{g}(\mathsf{R}, \mathsf{Y}, \mathsf{X}, \mathsf{T}) \mapsto \mathsf{R} \oplus \mathbf{arb}(\mathsf{X})^{[2]}, \ \mathsf{P}(\mathsf{R}, \mathsf{Y}, \mathsf{X}, \mathsf{T}) \mapsto \mathsf{Y} = \mathsf{X} \setminus \{\mathbf{arb}(\mathsf{X})\} \big).$

(This example involves various notions related to mappings, i.e. functions represented as sets of pairs; hence we are providing a quick prospect of those notions in Fig. 7.)

By means of wellfounded_recursive_fcn one can, occasionally, specify a terminating algorithm of which one wants to show the correctness relative to the specification of a problem which the algorithm is intended to solve. A substantial exercise of this kind was carried out in [94], to check correctness of the DPLL algorithm. This approach to algorithm-correctness verification is formally impeccable but not expedient: not only it calls for technical ingenuity in the actualization of the recursion parameters, but it often ends in scarcely readable specifications (as we have just seen with the sigma_theory example).

THEORY wellfounded_recursive_fcn$\big(s, Y \lhd X, f(B, X, T), G(A, Y, X, T), P(A, Y, X, T)\big)$

$\big\langle \forall t \subseteq s \mid t \neq \emptyset \rightarrow \big\langle \exists x \in t, \forall y \in t \mid \neg y \lhd x \big\rangle\big\rangle$

-- $\lhd$ *is thereby assumed to be irreflexive and well-founded on* s

$\Rightarrow (\mathsf{rec}_\Theta, \mathsf{rk}_\Theta)$

$\big\langle \forall x, t \mid x \in s \rightarrow \mathsf{rec}_\Theta(x, t) =$
$f\big(\{g(\mathsf{rec}_\Theta(y, t), y, x, t) : y \in s \mid y \lhd x \,\&\, P(\mathsf{rec}_\Theta(y, t), y, x, t)\}, x, t\big)\big\rangle$

$\big\langle \forall x, t \mid x \in s \rightarrow \mathsf{rk}_\Theta(x, t) = \bigcup \{\mathsf{next}(\mathsf{rk}_\Theta(y, t)) : y \in s \mid y \lhd x \,\&\, [y, x] \in t\}\big\rangle$

END wellfounded_recursive_fcn

THEORY finite_recursive_fcn$\big(f(B, X, T), g(R, Y, X, T), P(R, Y, X, T)\big)$

$\Rightarrow (\mathsf{rec}_\Theta)$

$\big\langle \forall x, t \mid \mathsf{Finite}(x) \rightarrow \mathsf{rec}_\Theta(x, t) =$
$f\big(\{g(\mathsf{rec}_\Theta(y, t), y, x, t) : y \subseteq x \mid y \neq x \,\&\, P(\mathsf{rec}_\Theta(y, t), y, x, t)\}, x, t\big)\big\rangle$

END finite_recursive_fcn

Fig. 6. Two versatile schemes of recursive definition

$[L, R] =_{\mathrm{Def}} \big\{\{L\}, \{\{L\}, \{\{R\}, R\}\}\big\}$ -- ordered pair and its projections

$P^{[1]} =_{\mathrm{Def}} \mathbf{arb}(\mathbf{arb}(P))$ $P^{[2]} =_{\mathrm{Def}} \big(\mathbf{arb}(P \setminus \{\mathbf{arb}(P)\}) \setminus \{\mathbf{arb}(P)\}\big)^{[1]}$

$\mathbf{domain}(F) =_{\mathrm{Def}} \big\{p^{[1]} : p \in F\big\}$ $\mathbf{range}(F) =_{\mathrm{Def}} \big\{p^{[2]} : p \in F\big\}$

$F_{|A} =_{\mathrm{Def}} \big\{p \in F, x \in A \mid p = [x, p^{[2]}]\big\}$ $F{\upharpoonright}X =_{\mathrm{Def}} \mathbf{arb}\big(F_{|\{X\}}\big)^{[2]}$

$\mathsf{Is\_map}(F) \leftrightarrow_{\mathrm{Def}} F = F_{|\mathbf{domain}(F)}$ $\mathsf{Svm}(F) \leftrightarrow_{\mathrm{Def}} F = \{[p^{[1]}, F{\upharpoonright}p^{[1]}] \mid p \in F\}$

$F^{\leftarrow} =_{\mathrm{Def}} \big\{[p^{[2]}, p^{[1]}] : p \in F\big\}$ $F \curvearrowright B \leftrightarrow_{\mathrm{Def}} \mathbf{range}(F^{\leftarrow}_{|B})$

$\mathsf{1{-}1}(F) \leftrightarrow_{\mathrm{Def}} \mathsf{Svm}(F) \,\&\, \mathsf{Svm}(F^{\leftarrow})$

Fig. 7. Map-related notions

In its current implementation, our set-based proof-checker does not support any genuine algorithmic specification language; but carrying out a hybridization between such a language and the proof-specification language available inside our verifier seems worth the effort: on the one hand, it would enable the user to annotate her/his algorithms with logical statements; on the other hand, it would make the inductive arguments underlying many mathematical proofs much more transparent. Consider, for example, the following proposition [91, Sec. 1]:

Discrimination lemma: Every finite nonnull set $\mathcal{F}$ has a set $\mathcal{D} \subseteq \bigcup \mathcal{F}$ of lower cardinality than its own cardinality $|\mathcal{F}|$, satisfying the equality $|\mathcal{F}| = |\{v \cap \mathcal{D} : v \in \mathcal{F}\}|$.

As discussed earlier, the THEORY finite_induction of Fig. 2 provides a mighty tool for handling the proofs of claims of this nature; however, the essential of the proof would be much better conveyed by the explicit construction, shown in Fig. 9, of a $\mathcal{D}$ meeting the claim of the discrimination lemma. Let us note in

THEORY sigma_theory$(s, X \oplus Y, e)$
$\quad \big\langle \forall x \in s, y \in s \mid x \oplus y \in s \big\rangle$ & $e \in s$
$\quad \big\langle \forall x \in s, y \in s, z \in s \mid (x \oplus y) \oplus z = x \oplus (y \oplus z) \big\rangle$
$\quad \big\langle \forall x \in s \mid x \oplus e = x \big\rangle$
$\quad \big\langle \forall x \in s, y \in s \mid x \oplus y = y \oplus x \big\rangle$
$\Rightarrow (\Sigma_\Theta)$ -- $\Sigma_\Theta(f)$ will be defined for any single-valued mapping f with values in s
$\quad \Sigma_\Theta(\emptyset) = e$
$\quad \big\langle \forall c \mid c^{[2]} \in s \rightarrow \Sigma_\Theta(\{c\}) = c^{[2]} \big\rangle$
$\quad \big\langle \forall f \mid \mathbf{Finite}(f) \ \& \ \mathbf{range}(f) \subseteq s \rightarrow \Sigma_\Theta(f) \in s \big\rangle$
$\quad \big\langle \forall c, f \mid c \in f \ \& \ \mathbf{Finite}(f) \ \& \ \mathbf{range}(f) \subseteq s \rightarrow \Sigma_\Theta(f) = \Sigma_\Theta(f \setminus \{c\}) \oplus c^{[2]} \big\rangle$
$\quad \big\langle \forall f, t \mid \mathbf{Finite}(f) \ \& \ \mathbf{Is\_map}(f) \ \& \ \mathbf{range}(f) \subseteq s \rightarrow$
$$\Sigma_\Theta(f) = \Sigma_\Theta(f_{|\mathbf{domain}(f) \cap t}) \oplus \Sigma_\Theta(f_{|\mathbf{domain}(f) \setminus t}) \big\rangle$$
$\quad \big\langle \forall f, g \mid \mathbf{Finite}(f) \ \& \ \mathbf{Svm}(f) \ \& \ \mathbf{Svm}(g) \ \& \ \mathbf{domain}(f) = \mathbf{domain}(g) \ \& \ \mathbf{range}(f) \subseteq s \rightarrow$
$$\Sigma_\Theta(f) = \Sigma_\Theta\Big(\Big\{ \Big[y, \Sigma_\Theta(f_{|g^{\cap}\{y\}}) \Big] : y \in \mathbf{range}(g) \Big\} \Big) \big\rangle$$
$\quad \big\langle \forall f, g \mid \mathbf{Finite}(f) \ \& \ \mathbf{Svm}(f) \ \& \ \mathbf{1\text{-}1}(g) \ \& \ \mathbf{domain}(f) = \mathbf{domain}(g) \ \& \ \mathbf{range}(f) \subseteq s \rightarrow$
$$\Sigma_\Theta(f) = \Sigma_\Theta\big(\{[y, f\!\restriction\!(g^{\leftarrow}\!\restriction\! y)] : y \in \mathbf{range}(g)\} \big) \big\rangle$$
END sigma_theory

Fig. 8. Interface of a THEORY of finite summation over an Abelian monoid

```
procedure discriminant (F);
    claim Is_finite(F) & F ≠ ∅;
    a := sel(F);     -- i.e., draw an element a from F without removing it
    if F = {a} then D := ∅;
    else             -- construct D recursively, as follows
        Δ := discriminant (F \ {a});
        claim |{v ∈ F \ {a} | v ∩ Δ = a ∩ Δ}| < 2;
        D := Δ ∪ {sel((v \ a) ∪ (a \ v)) : v ∈ F \ {a} | v ∩ Δ = a ∩ Δ};
    end if;
    claim D ⊆ ⋃ F & |D| < |F| = |{v ∩ D : v ∈ F}| &
        (∀x ∈ D | |{v ∩ (D \ {x}) : v ∈ F}| < |F|);
    return D;
end discriminant;
```

Fig. 9. Algorithmic specification of a proof of the discrimination lemma

passing that this construction is not really executable, in spite of its algorithmic appearance, when infinite sets occur among the elements of $\mathcal{F}$, as infinite sets can be of an utterly unmanageable nature.

A big variety of (pseudo-)algorithmic languages can be proposed for the specification of logically annotated constructions (see, e.g., [74] for a proposal pertaining to number theory and analysis). In our opinion, a promising start, to go hand-in-hand with the set-theoretic foundation of our proof-checker, could draw inspiration from the already cited programming language SETL.

5 The Role of Decision Algorithms

The history of proof assistants begins with an implementation of the decision algorithm for the additive Presburger arithmetic [101, 39, 41]. A considerably more versatile approach to theorem-proving gained ground in the 1960s, when Prawitz first [100], then Davis and Putnam [42], and finally Robinson with his celebrated resolution principle [104], proposed semidecision methods exploitable for *any* finitely axiomatized first-order theory—including, therefore, even the Gödel-Bernays class theory [123, 102]. The resolution-based approach to theorem-proving dominated the scene so much and for so long [76]—in spite of the already cited warning [13]—that, over the years, a host of refinements to resolution were proposed; resolution also evolved into a machinery underlying various systems for theory-based reasoning [115, 99, 66], and even into a method exploitable in order to deal with Church's typed lambda calculus [2, 3].

Due to the long-lasting popularity of resolution, the research on decision algorithms—sometimes referring to *fragments* of mathematical theories (or of logical calculi [59]), sometimes to theories in their full extent (cf., e.g., [60])—had only sporadically an impact on the automated deduction field, save for a few happy exceptions such as the papers by Nelson-Oppen and Shostak [81, 82, 111]. Concerning the works just cited, it should be noted that rather than offering a specific contribution to the inventory of decidable theories, they address an issue of integration between decision algorithms: in this sense, their significance and long-term influence [109, 121] can be compared to the ones of DPLL, whose role as a ubiquitous inference mechanism is much more relevant than its direct usability as a test for propositional logic.

5.1 Decidable Theories

Progressively, the attention bestowed to decision algorithms by researchers in the Automated Deduction community has increased significantly: decision algorithms related to different mathematical disciplines, and general inference methods into which they can be built, have become a main thread of research. In what follows we select, from among many decidability results, some which are likely to improve the quality of support provided by proof-assistants. Not all of these results are recent: actually, only a few have been obtained directly inside the Automated Deduction field, whereas others have simply migrated into it.[7]

If implemented in full, the already mentioned decision method for Presburger arithmetic—as well as a few variants of it relying, like it, on quantifier elimination techniques—has little practical applicability: as a matter of fact, as shown by [63], any decision algorithm for this theory suffers from doubly exponential worst-case complexity. In spite of this general limiting result, lower complexity can be achieved by restraining consideration to specific fragments of this theory. A number of possibilities have been explored: for instance, when only formulae

---

[7] Due to space limitations, we will pass under silence many contributions to the general unification field [113] and on rewriting systems [46], although several of these, e.g. [57], are likely to be quite relevant for the development of our set-based proof-checker.

devoid of quantifier alternations are treated, the decision problem acquires single exponential complexity (actually, it becomes NP-complete [95]). Particular fragments of Presburger arithmetic are of great interest in the field of automated verification. This is the case of the fragment named *UTVPI*, in which only formulae of the form $ax \leqslant by + c$ with $a, b \in \{0, 1, -1\}$ are admitted. In this case, polynomial algorithms are available. A similar result holds for the easier collection of Boolean combinations of atoms of the form $x \leqslant y + c$, forming the so-called *difference logic*. This fragment of Presburger arithmetic has recently received greater attention, because of its connection with stable model semantics: [84], among others, shows how decision methods for difference logic can be the basis for efficient mechanization of answer set semantics and proficient integration of decision methods into answer set solvers.

The decidability of universal Presburger arithmetic in presence of uninterpreted function and predicate symbols has been assessed in [110]. This result represents a first step towards the integration of solvers, as well as [21] that describes a decision method for unquantified formulae of Presburger arithmetic extended with sets.

Another stream of research investigated the decision properties of the elementary geometry and of the algebra of real numbers. The seminal paper [118] provides decidability results and basic decision techniques for these theories (see also [119]). A first upper bound on the complexity of the decision problem for real closed fields is provided by [37], which proposes an approach based on *cylindrical algebraic decomposition*. The resulting decision method has doubly exponential complexity in the number of variables that occur in the input formula; however, if a fixed number of variables are allowed to appear in the formulae, then the complexity becomes simply exponential. A refined decision algorithm which is doubly exponential in the number of quantifier alternations appears in [69]: this enables efficient implementations of deciders under strong limitations on quantifier nesting.

Lowering the complexity is possible, as ever, by restraining the collection of formulae which can be treated. A very interesting, and useful, collection of formulae of real algebra consists of the purely existentially quantified linear constraints. In this case, since the decision problem becomes essentially a linear programming problem, polynomial methods exist. Notice that, in general, for purely existentially quantified formulae, the complexity is expected to be at least exponential [63]. Notwithstanding, polynomial methods for existential formulae with a fixed number of variables are available [103]. In view of these results, deciders for various specific collections of formulae have been proficiently built into various computer algebra systems. The reader is referred to [7, 35] for a survey of classical results and contributions relevant to the field.

As regards elementary geometry, decidability follows from the decidability of real closed fields [118]. However, a direct reduction to the quantifier elimination techniques developed for the theory of real fields does not yield efficient decision methods. More viable approaches, such as Gröbner bases [15], have been proposed, and integrated in automated systems. See e.g. [36, 124, 105] for more details.

Decidability issues for classes of formulae in general topology and real analysis have been investigated too [18]. For example, [28] describes a decision method for formulae involving continuous functions. The result is obtained through reduction to the decidability problem for *two-level syllogistic*, which by itself is NP-complete. Another paper in this context is [19]. In this case the authors address the decision problem for a fragment of real analysis, consisting of unquantified formulae which, in addition to the operators of Tarski's theory of reals, involve predicates of comparison, monotonicity, concavity, and convexity of continuous real functions, over possibly unbounded intervals. The result is obtained via a reduction to Tarski's existential theory of reals.

The decidability of many algebraic theories was assessed long ago (for a comprehensive survey, endowed with a rich bibliography, see [60]). Among others, we mention the decidability results for the theories of Abelian groups [116], Boolean algebras, linearly ordered sets, free groups [72]. These have rarely had a direct impact on the design of proof-assistants, but in preparation for its embedding into our set-based inferential framework, the decidability result about ordered Abelian groups [70], originally referring to the first-order theory in its entirety, was downsized into a practical decision algorithm [30, Sec. 3] for a fragment of that theory.

5.2   Computable Set Theory

The discovery and classification of many decidable fragments of Set Theory constitutes a prolific stream of research begun with [107], more directly related to the conception of set-based proof-verification advocated for in this paper (for a remote historical antecedent, see [8]). Despite the decision algorithms in this area often being prohibitively time-consuming, singling them out seemed to be an unavoidable labor (cf. [23, 29]) before any sensible proof-search method could be implemented for Set Theory. It would have been silly [122] to hope that the full collection of set-theoretic sentences would eventually be brought under the jurisdiction of some mighty decision algorithm; rather, there was hope that the decision algorithms discovered for diverse fragments of set theory could be integrated in some broad-spectrum inferential armory which, properly driven by human experts, would then offer some flexible support for a good deal of proof-verification work. As we will mention in a short while, this expectation was not deceived.

Sometimes a decidable class of formulae is circumscribed by means of syntactical restrictions placed on the form of its quantificational prefix. This way of proceeding is parallel to the one adopted in the study of decision problems for predicate calculus [59], but in the set-theoretic context an underlying weak set-theory is assumed (in the form of a kernel of proper axioms). Results obtained in this frame of mind can be found in [10, 11, 12, 14, 47, 48, 51, 89, 88, 90, 91].

A somehow "orthogonal" approach uses of collections of set-constructors chosen from among the usual $\_\cap\_$, $\_\cup\_$, $\_\setminus\_$, $\{\_\}$, $\mathcal{P}(\_)$, $\bigcup\_$, etc., whose intended meanings are characterized through suitable axioms added to the common weak kernel set-theory. Valuable decidability results were obtained along this stream in a long series of papers, [20, 22, 24, 25, 26, 27, 131, 32, 97] to mention a few.

Most of the methods proposed in these papers, while ensuring that certain collections of formulae (sometimes very challenging) have a decidable satisfiability problem, do not appear to be polished enough to support the design of efficient decision procedures. The most promising approach, to the aim of getting effective decision algorithms, involves the synthesis of a tableau-based procedure. This approach is adopted, for instance, in [34, 33], and revealed, at least in principle, viable in a wide range of cases. Actually, the implementation of the most central of all inference primitives of our set-based verifier, called ELEM, is based on this approach. This rule ELEM, often used implicitly by other inference primitives, e.g. 'Suppose_not', 'Discharge', 'Use_def', 'Assump', EQUAL, ALGEBRA, and others [87], embodies an extended form of *multi-level syllogistic* [62]. It determines whether a given unquantified set-theoretic formula involving individual set variables, the set operators $_-\cap_-$, $_-\cup_-$, $_-\setminus_-$, $\{_-\}$, the pair assembly and decomposition operators $[_-,_-]$, $_-^{[1]}$, $_-^{[2]}$ (cf. Fig. 7), and a global selection operator $\mathbf{arb}(_-)$, is satisfiable. By using this decision algorithm, the verifier can identify many cases in which a conjunction constructed by negating one statement S of a proof and conjoining a selection of earlier steps is unsatisfiable, which implies that S follows from the preceding context.

When not all the constructs appearing in a context (e.g. quantifiers and set-formers) are accessible to multilevel syllogistic, a preprocessing step must precede its application. This replaces all parts of the current context whose lead operators are not recognized by the decision algorithm by 'blobs', i.e. by new variables designating either sets (when they occur as terms) or propositions (when they occur as subformulae). *Blobbing*, as we call this operation, replaces syntactically identical (or recognizably equal) parts of a formula by the same variable. It is also able to treat as equal well-formed parts which only differ by the renaming of bound variables in quantifiers or set-formers. Blobbing also treats existential quantifiers as negated universal quantifiers.

5.3   Integration of Decision Algorithms

Once the decidability of a theory—or of fragments of it—has been assessed and decision procedures have been designed for manageable portions of it, we are only half the way through. Indeed, it is a common situation that the decision procedures must be integrated into a pre-existing framework, be it an automatic theorem-prover or a proof assistant. Such a framework might offer some form of theory-based resolution—as mentioned above—, or might already incorporate inferential capabilities, implemented in their turn in terms of other decision methods. Hence, the new goal to be faced consists in realizing a combination of inferential capabilities into a single mechanism. The issue is not simply the one of achieving an acceptable overall complexity: unfortunately, it often happens that decidability gets disrupted when decidable theories, or fragments thereof, are put together. Suffice it to recall, as a striking illustration of this state of affairs, that Presburger arithmetic becomes undecidable when extended with a single uninterpreted monadic predicate symbol [58].

Nelson and Oppen's proposal [81, 82] for combining decidable theories, which relies of Craig's interpolation theorem, is the first—perhaps, to date, still the most significant—effective technique designed to solve this task. When the theories to be combined have a decidable satisfiability problem and they meet a number of precise requirements (in particular the signatures of their languages can only share the equality sign), the Nelson–Oppen combination technique provides a method for deciding the validity of universal formulae in the union of the underlying languages. The combination method normalizes the given universal sentence to be proved into a conjunction of formulae, each of which belongs to one of the component theories. Each component decision procedure is then exploited as a black box to extract information from the conjunct pertaining to it. The method exploits the interpolant formulae, as guaranteed by Craig's theorem, as *communication means* between pairs of theories. Various refinements of the method, partially relaxing the requirements about the initial theories, have been proposed, for instance in [6, 120, 68, 83].

An alternative paradigm for combining decision procedures was proposed by Shostak in [111]. This method is less general than Nelson–Oppen's since it presupposes stronger requirements are met by the component theories, in order to realize a tighter integration between the solvers and to achieve a better overall performance. Shostak's method constitutes the basic ingredient of different (semi-)automated systems, mainly conceived to support (semi-)automated verification (PVS, STeP, SVC, to mention a few).

A rather different approach, essentially based on refinements of tableaux-based decision procedures for non-disjoint theories is taken in [128, 77]. The proposed method gets exploited to combine decision algorithms for theories of aggregates (sets, multisets, lists, etc.) with theories about elements, integers, cardinals, etc. [125, 126, 127, 129, 130]. Despite the apparently limited expressive power of these theories, they are of practical interest in fields such as automated hardware verification, software protocol certification, model checking, etc. This is because the availability of decision methods for combination of weak theories of integers, bit vectors, arrays, enables the formal verification of hardware and software components, by directly exploiting (mixed) domain-specific knowledge on such structured entities.

For an up-to-date reading on many aspects of decision algorithms, we refer the reader to [75], inside which many useful references to the area can be found.

6   Conclusions

The bibliographic references of this paper include an impressive number of contributions of Italian researchers to the automated deduction field. Many more could have been cited, but it would have been hard to reconcile fairness to everybody with unity in the material of this paper. By focusing mainly on those contributions that have had a direct echo inside the GULP community, we could identify a few steady research threads which then have formed the backbone of this paper.

The annual GULP meetings, for many years, and then, more recently, satellite workshops like the CILC ones, have regularly hosted presentations about satisfiability decision algorithms and to the development of proof-methods specifically oriented to set theory. Papers on those topics written for GULP-related events often foreran publications on valuable scientific journals ([32], and [31, 131] and [47, 48] to mention a few) or contributed to disseminating novel ideas [83], or have explored alternative uses in Logic Programming of ideas originated elsewhere [86, 114, 92, 56].

It must be stressed here a parallel between the theory-oriented developments of resolution inside the automated deduction field *per se* and the scheme for extending Logic Programming into Constraint Logic Programming proposed in [71]. In particular, T-resolution has migrated from the area of theorem-proving into Logic Programming, cf. [49, 50].

Conversely, ideas originated inside Logic Programming promise to play a role also in the design of inference rules for a proof-checker. E.g., set-unification algorithms [52, 4, 57] could be exploited to boost a behind-the-scenes 'proof-by-computation' paradigm advocated for in [93, p. 229], aimed at enhancing theorem-proving by means of the ability to perform symbolic computations efficiently in specialized contexts of algebra and analysis.

Since the year 2000, various satisfiability decision algorithms and various proof-methods for set theory have begun to be put together in a large-scale proof-verifier. This more technological aim poses new challenges and is creating new trends, whose flavor this paper has tried to convey. These pages intend to be a homage of ours to the Italian authors sparsely cited in this paper, and an encouragement to them and to others in casting the new goals in the terse formal setting germane to logic: abstract ideas usually offer, in fact, the right framework for practical long-term undertakings; and valuable algorithms can often be distilled from disappointingly intractable search methods directly stemming from theoretical investigations.

Acknowledgements

We are indebted to prof. Jacob T. Schwartz for his encouragement in pursuing research on proof-verification. The conception of an extensible set-based proof verifier owes very much to him, as also does the implementation of ÆtnaNova/Referee, cf. http://setl.dyndns.org/EtnaNova/login/. Thanks are also due to our colleagues Domenico Cantone and Marianna Nicolosi Asmundo of the University of Catania, with whom we are pleasantly carrying out our project. We are grateful to the anonymous referees for helpful advice and, to the editors for their precious work.

This research has been partially funded by PRIN 2006 project *'Large-scale development of certified mathematical proofs'*, by PRIN 2008 project *'Innovative and multi-disciplinary approaches for constraint and preference reasoning'*, and by GNCS-INdAM project *'Tecniche innovative per la programmazione con vincoli in applicazioni strategiche'*. Contacts among participants to the

ÆtnaNova/Referee project have been fostered by the Gruppo Nazionale per il Calcolo Scientifico of the Istituto Nazionale di Alta Matematica 'Francesco Severi'.

References

[1] Bundy, A. (ed.): CADE 1994. LNCS (LNAI), vol. 814, pp. 238–251. Springer, Heidelberg (1994); The QED Manifesto

[2] Andrews, P.B.: Resolution in type theory. The J. of Symbolic Logic 36, 414–432 (1971)

[3] Andrews, P.B., Longini Cohen, E.: Theorem proving in type theory. In: Proc. of IJCAI 1977, pp. 566–566 (1977)

[4] Arenas-Sánchez, P., Dovier, A.: Minimal set unification. In: Alpuente, M., Sessa, M.I. (eds.) GULP-PRODE 1995, Marina di Vietri, Italy, September 11-14, pp. 447–458 (1995)

[5] Asperti, A., Geuvers, H., Natarajan, R.: Social processes, program verification, and all that. Math. Struct. in Comp. Science 19(5), 877–896 (2009)

[6] Baader, F., Tinelli, C.: Combining equational theories sharing non-collapse-free constructors. In: Kirchner, H., Ringeissen, C. (eds.) FroCos 2000. LNCS (LNAI), vol. 1794, pp. 260–274. Springer, Heidelberg (2000)

[7] Basu, S., Pollack, R., Roy, M.-F.: Algorithms in real algebraic geometry. Algorithms and computation in mathematics, vol. 10. Springer, Heidelberg (2006)

[8] Behmann, H.: Beiträge zur Algebra der Logik, insbesondere zum Entscheidungsproblem. Math. Annalen 86, 163–220 (1922)

[9] Belinfante, J.G.F.: Reasoning about iteration in Gödel's class theory. In: Baader, F. (ed.) CADE 2003. LNCS (LNAI), vol. 2741, pp. 228–242. Springer, Heidelberg (2003)

[10] Bellè, D., Parlamento, F.: Decidability and completeness for open formulas of membership theories. Notre Dame J. of Formal Logic 36 (1995)

[11] Bellè, D., Parlamento, F.: The decidability of the $\forall^*\exists$ class and the axiom of foundation. Notre Dame J. of Formal Logic 42 (2001)

[12] Bellè, D., Parlamento, F.: Truth in V for $\exists^*\forall\forall$-sentences is decidable. J. of Symbolic Logic 71 (2006)

[13] Bledsoe, W.W.: Non-resolution theorem proving. Artificial Intelligence 9, 1–35 (1977)

[14] Breban, M., Ferro, A., Omodeo, E.G., Schwartz, J.T.: Decision Procedures for Elementary Sublanguages of Set Theory II. Formulas involving Restricted Quantifiers, together with Ordinal, Integer, Map, and Domain Notions. Comm. Pure Appl. Math. 34, 177–195 (1981)

[15] Buchberger, B., Winkler, F.: Gröebner bases and Applications. London Mathematical Society Lecture Note Series, vol. 251. Cambridge University Press, Cambridge (1998)

[16] Burstall, R., Goguen, J.: Putting theories together to make specifications. In: Reddy, R. (ed.) Proc. 5th International Joint Conference on Artificial Intelligence, Cambridge, MA, pp. 1045–1058 (1977)

[17] Cantone, D., Chiaruttini, C., Nicolosi Asmundo, M., Omodeo, E.G.: Cumulative hierarchies and computability over universes of sets. Le Matematiche 63, 31–84 (2008)

[18] Cantone, D., Cincotti, G.: Decision algorithms for some fragments of analysis and related areas. Comm. Pure Appl. Math. 40, 281–300 (1987)

[19] Cantone, D., Cincotti, G., Gallo, G.: Decision algorithms for fragments of real analysis. I. Continuous functions with strict convexity and concavity predicates. J. of Symbolic Computation 41(7), 763–789 (2006)

[20] Cantone, D., Cutello, V., Ferro, A.: Decision procedures for elementary sublanguages of set theory. XIV. Three languages involving rank related constructs. In: Gianni, P. (ed.) ISSAC 1988. LNCS, vol. 358, pp. 407–422. Springer, Heidelberg (1989)

[21] Cantone, D., Cutello, V., Schwartz, J.T.: Decision problems for Tarski's and Presburger's arithmetics extended with sets. In: Schönfeld, W., Börger, E., Kleine Büning, H., Richter, M.M. (eds.) CSL 1990. LNCS, vol. 533, pp. 95–109. Springer, Heidelberg (1991)

[22] Cantone, D., Ferro, A.: Some recent decidability results in set theory. Atti degli incontri di Logica Matematica III, 383–387 (1985)

[23] Cantone, D., Ferro, A., Omodeo, E.G.: Computable set theory, Vol.1. Oxford Science Publications of International Series of Monographs on Computer Science, vol. no.6. Clarendon Press (1989)

[24] Cantone, D., Ferro, A., Omodeo, E.G., Policriti, A.: Scomposizione sillogistica disgiuntiva. In: Mello [78], pp. 199–209

[25] Cantone, D., Ferro, A., Schwartz, J.T.: Decision procedures for elementary sublanguages of set theory. V. Multilevel syllogistic extended by the general union operator. J. of Computer and System Sciences 34(1), 1–18 (1987)

[26] Cantone, D., Formisano, A., Omodeo, E.G., Schwartz, J.T.: Various commonly occurring decidable extensions of multi-level syllogistic. In: Ranise, S., Tinelli, C. (eds.) Pragmatics of Decision Procedures in Automated Reasoning, PDPAR 2003 (CADE-19), Electronic proceedings, Miami, USA (2003)

[27] Cantone, D., Nicolosi Asmundo, M.: On the satisfiability problem for a 3-level quantified syllogistic. In: Complexity, Expressibility, and Decidability in Automated Reasoning – CEDAR 2008, Sydney, Australia, pp. 31–46 (2008)

[28] Cantone, D., Omodeo, E.G.: On the decidability of formulae involving continuous and closed functions. In: Sridharan, N.S. (ed.) Proc. of the 11th International Joint Conference on Artificial Intelligence, pp. 425–430. Morgan Kaufmann, San Francisco (1989)

[29] Cantone, D., Omodeo, E.G., Policriti, A.: Set Theory for Computing. From Decision Procedures to Declarative Programming with Sets. Monographs in Computer Science. Springer, Heidelberg (2001)

[30] Cantone, D., Omodeo, E.G., Schwartz, J.T., Ursino, P.: Notes from the logbook of a proof-checker's project. In: Dershowitz (ed.) [45], pp. 182–207

[31] Cantone, D., Schwartz, J.T., Zarba, C.G.: Decision procedures for fragments of set theory with monotone and additive functions. In: Rossi, Jayaraman [106], pp. 1–8

[32] Cantone, D., Ursino, P., Omodeo, E.G.: Formative processes with applications to the decision problem in set theory: I. Powerset and singleton operators. Inf. Comput. 172(2), 165–201 (2002); Appeared as Transitive Venn diagrams with applications to the decision problem in set theory. In: [79]

[33] Cantone, D., Zarba, C.G.: A new fast tableau-based decision procedure for an unquantified fragment of set theory. In: Caferra, R., Salzer, G. (eds.) FTP 1998. LNCS (LNAI), vol. 1761, pp. 126–136. Springer, Heidelberg (2000)

[34] Cantone, D., Zarba, C.G.: A tableau-based decision procedure for a fragment of set theory involving a restricted form of quantification. In: Murray, N.V. (ed.) TABLEAUX 1999. LNCS (LNAI), vol. 1617, pp. 97–112. Springer, Heidelberg (1999)

[35] Caviness, B.F., Johnson, J.R.: Quantifier elimination and cylindrical algebraic decomposition. Texts and Monographs in Computer Science. Springer, Heidelberg (1998)

[36] Chou, S.C.: Mechanical Geometry Theorem Proving. Reidel Publ. Comp., Dordrecht (1988)

[37] Collins, G.E.: Quantifier elimination for real closed fields by cylindric algebra decomposition. In: Brakhage, H. (ed.) GI-Fachtagung 1975. LNCS, vol. 33, pp. 134–183. Springer, Heidelberg (1975)

[38] D'Agostino, G., Omodeo, E.G., Schwartz, J.T., Tomescu, A.I.: Self-applied proof verification (Extended abstract). In: Cordón-Franco, A., Fernández-Margarit, A., Lara-Martin, F.F. (eds.) JAF, 26èmes Journées sur les Arithmétiques Faibles, pp. 113–117. Fénix Editora, Sevilla, Spain (2007), http://www.cs.us.es/glm/jaf26

[39] Davis, M.: A program for Presburger's algorithm. In: Summary of talks presented at the Summer Institute for Symbolic Logic, pp. 215–233. Cornell University (1957); In: [112]

[40] Davis, M.: Eliminating the irrelevant from mechanical proofs. In: Proc. of Symposia in Applied Mathematics, vol. 15, pp. 15–30. AMS (1963); Reprinted in [112]

[41] Davis, M.: The early history of automated deduction. In: Handbook of Automated Reasoning, pp. 3–13. Elsevier, Amsterdam (2001)

[42] Davis, M., Putnam, H.: A computing procedure for quantification theory. J. of the ACM 7(3), 201–215 (1960)

[43] Davis, M., Schwartz, J.T.: Correct-program technology / Extensibility of verifiers – Two papers on Program Verification with Appendix of Edith Deak. Technical Report No. NSO-12, Courant Institute of Mathematical Sciences, New York University (1977)

[44] Davis, M., Schwartz, J.T.: Metatheoretic extensibility for theorem verifiers and proof-checkers. Computers and Mathematics with Applications 5, 217–230 (1979)

[45] Dershowitz, N. (ed.): International symposium on verification (Theory and Practice) celebrating Zohar Manna's 1000000_2^{th} birthday. LNCS, vol. 2772. Springer, Heidelberg (2003)

[46] Dershowitz, N., Jouannaud, J.-P.: Rewrite systems. In: van Leeuwen, J. (ed.) Handbook of Theoretical Computer Science. Formal Models and Semantics, vol. B, pp. 243–320. Elsevier and MIT Press (1990)

[47] Dovier, A., Formisano, A., Omodeo, E.G.: Provable $\exists^*\forall$-sentences about sets with atoms. In: Rossi, Jayaraman [106], pp. 9–17

[48] Dovier, A., Formisano, A., Omodeo, E.G.: Decidability results for sets with atoms. ACM Transactions on Computational Logic 7(2), 269–301 (2006)

[49] Dovier, A., Formisano, A., Policriti, A.: On T-logic programming. In: Falaschi, M., Navarro, M., Policriti, A. (eds.) Joint Conference on Declarative Programming, AGP 1997, Grado, Italy, June 16-19, pp. 457–466 (1997)

[50] Dovier, A., Formisano, A., Policriti, A.: On T-logic programming. In: Proc. of ILPS 1997, pp. 323–337 (1997); A preliminary version appeared in [49]

[51] Dovier, A., Omodeo, E.G., Policriti, A.: Solvable set/hyperset contexts: II. A goal-driven unification algorithm for the blended case. Appl. Algebra Eng. Commun. Comput. 9(4), 293–332 (1999)

[52] Dovier, A., Omodeo, E.G., Pontelli, E., Rossi, G.: {log}: A logic program-ming language with finite sets. In: Furukawa, K. (ed.) ICLP 1991, pp. 111–124. MIT Press, Cambridge (1991)

[53] Dovier, A., Omodeo, E.G., Pontelli, E., Rossi, G.: {log}: A logic programming language with finite sets. In: Asirelli, P. (ed.) Sesto convegno nazionale di pro-grammazione logica, GULP 1991, Pisa, pp. 241–355 (1991)

[54] Dovier, A., Omodeo, E.G., Pontelli, E., Rossi, G.: Embedding finite sets in a logic programming language. In: Lamma, E., Mello, P. (eds.) ELP 1992. LNCS (LNAI), vol. 660, pp. 150–167. Springer, Heidelberg (1993)

[55] Dovier, A., Omodeo, E.G., Pontelli, E., Rossi, G.: A language for programming in logic with finite sets. J. of Logic Programming 28(1), 1–44 (1996); See also [52, 54, 53]

[56] Dovier, A., Piazza, C., Rossi, G.: Narrowing the gap between set-constraints and CLP(SET)-constraints. In: Freire-Nistal, J.L., Falaschi, M., Ferro, M.V. (eds.) Joint Conference on Declarative Programming, AGP 1998, A Coruña, Spain, July 20-23, pp. 43–56 (1998)

[57] Dovier, A., Pontelli, E., Rossi, G.: Set unification. Theory and Practice of Logic Programming 6(6), 645–701 (2006)

[58] Downey, P.J.: Undecidability of Presburger arithmetic with a single monadic predicate letter. Technical Report 18-72, Harvard University Center for Research in Computing Technology (1972)

[59] Dreben, B., Goldfarb, W.D.: The Decision Problem. Solvable classes of quantifi-cational formulas. Addison-Wesley, Reading (1979)

[60] Ershov, Y.L., Lavrov, I.A., Taimanov, A.D., Taitslin, M.A.: Elementary theories. Russ. Math. Survey 20, 35–106 (1965)

[61] Farmer, W.M., Guttman, J.D., Thayer, F.J.: IMPS: An interactive mathematical proof system. J. Automated Reasoning 11, 213–248 (1993)

[62] Ferro, A., Omodeo, E.G., Schwartz, J.T.: Decision procedures for some fragments of set theory. In: Bibel, W., Kowalski, R. (eds.) CADE 1980. LNCS, vol. 87, pp. 88–96. Springer, Heidelberg (1980)

[63] Fisher, M.J., Rabin, M.O.: Super-exponential complexity of Presburger arith-metic. In: Complexity and computation, vol. VII, pp. 27–41. SIAM-AMS, Philadelphia (1974)

[64] Formisano, A., Omodeo, E.G.: An equational re-engineering of set theories. In: Caferra, R., Salzer, G. (eds.) FTP 1998. LNCS (LNAI), vol. 1761, pp. 175–190. Springer, Heidelberg (2000)

[65] Formisano, A., Omodeo, E.G., Temperini, M.: Instructing equational set-reasoning with Otter. In: Goré, R.P., Leitsch, A., Nipkow, T. (eds.) IJCAR 2001. LNCS (LNAI), vol. 2083, pp. 152–167. Springer, Heidelberg (2001)

[66] Formisano, A., Policriti, A.: T-resolution: Refinements and model elimination. J. Automated Reasoning 22(4), 433–483 (1999)

[67] Geuvers, H.: Proof assistants: History, ideas and future. Sādhanā 34, 3–25 (2009)

[68] Ghilardi, S., Nicolini, E., Ranise, S., Zucchelli, D.: Decision procedures for exten-sions of the theory of arrays. Ann. Math. Artif. Intell. 50(3-4), 231–254 (2007)

[69] Grigoriev, D.: Complexity of deciding Tarski algebra. J. of Symbolic Computa-tion 5(1/2), 65–108 (1988)

[70] Gurevich, Y.: Elementary properties of ordered Abelian groups. Translations of AMS 46, 165–192 (1965)

[71] Jaffar, J., Maher, M.J.: Constraint logic programming: a survey. J. of Logic Programming (19/20), 503–581 (1994)

[72] Kharlampovich, O., Myasnikov, A.: Elementary theory of free non-Abelian groups. J. of Algebra 302(2), 451–552 (2006)

[73] Knuth, D.E., Bendix, P.B.: Simple word problems in universal algebras. In: Leech, J. (ed.) Computational Problems in Abstract Algebra, pp. 263–267. Pergamon Press, Oxford (1970)

[74] Kohlenbach, U.: Applied Proof Theory: Proof Interpretations and their Use in Mathematics. Springer Monographs in Mathematics. Springer, Heidelberg (2008)

[75] Kroening, D., Strichman, O.: Decision procedures: an algorithmic point of view. Texts in Theoretical Computer Science. Springer, Heidelberg (2008)

[76] Loveland, D.W.: Automated theorem proving: A quarter century review. In: Bledsoe, W.W., Loveland, D.W. (eds.) Contemporary Mathematics: Automated Theorem Proving - After 25 Years, pp. 1–45. AMS (1984)

[77] Manna, Z., Zarba, C.G.: Combining decision procedures. In: Aichernig, B.K., Maibaum, T. (eds.) Formal Methods at the Cross Roads: From Panacea to Foundational Support. LNCS, vol. 2757, pp. 381–422. Springer, Heidelberg (2003)

[78] Mello, P. (ed.): Quarto convegno nazionale di programmazione logica. In: GULP 1989, Bologna (1989)

[79] Meo, M.C., Vilares Ferro, M. (eds.): Joint Conference on Declarative Programming, AGP 1999, L'Aquila, Italy, September 6-9. GTE (1999)

[80] Montagna, F., Mancini, A.: A minimal predicative set theory. Notre Dame J. of Formal Logic 35(2), 186–203 (1994)

[81] Nelson, G., Oppen, D.C.: Simplification by cooperating decision procedures. ACM Transaction on Programming Languages and Systems 1(2), 245–257 (1979)

[82] Nelson, G., Oppen, D.C.: Fast decision procedures based on congruence closure. J. of the ACM 27(2), 356–364 (1980)

[83] Nicolini, E., Ringeissen, C., Rusinowitch, M.: Satisfiability procedures for combination of theories sharing integer offsets. In: Kowalewski, S., Philippou, A. (eds.) TACAS-ETAPS 2009. LNCS, vol. 5505, pp. 428–442. Springer, Heidelberg (2009); Also in CILC 2009: 24-esimo Convegno Italiano di Logica Computazionale

[84] Niemelä, I.: Stable models and difference logic. Ann. Math. Artif. Intell. 53(1-4), 313–329 (2008)

[85] Omodeo, E.G.: The Linked Conjunct method for automatic deduction and related search techniques. Computers and Mathematics with Applications 8, 185–203 (1982)

[86] Omodeo, E.G., Bossi, A., Sambin, G.: Tre possibili orientamenti per una programmazione dichiarativa basata sulla teoria degli insiemi. In: Demo, B. (ed.) Secondo convegno nazionale di programmazione logica, GULP 1987, Torino, pp. 265–276 (1987)

[87] Omodeo, E.G., Cantone, D., Policriti, A., Schwartz, J.T.: A computerized Referee. In: Stock, O., Schaerf, M. (eds.) Reasoning, Action and Interaction in AI Theories and Systems. LNCS (LNAI), vol. 4155, pp. 117–139. Springer, Heidelberg (2006)

[88] Omodeo, E.G., Parlamento, F., Policriti, A.: A derived algorithm for evaluating ε-expressions over abstract sets. J. of Symbolic Computation 15(5-6), 673–704 (1993)

[89] Omodeo, E.G., Parlamento, F., Policriti, A.: Decidability of $\exists^*\forall$-sentences in membership theories. Mathematical Logic Quarterly (formerly Zeitschrift für Mathematische Logik und Grundlagen der Mathematik) 42 (1996)

[90] Omodeo, E.G., Policriti, A.: Solvable set/hyperset contexts: I. Some decision procedures for the pure, finite case. Comm. Pure Appl. Math. 48(9-10), 1123–1155 (1995); Special Issue in honor of J.T. Schwartz

[91] Omodeo, E.G., Policriti, A.: The Bernays-Schönfinkel-Ramsey class for set theory: semidecidability. J. of Symbolic Logic (2010)

[92] Omodeo, E.G., Policriti, A., Rossi, G.: Che genere di insiemi/multi-insiemi/iper-insiemi incorporare nella programazione logica? In: Saccà, D. (ed.) GULP 1993, pp. 55–70 (1993)

[93] Omodeo, E.G., Schwartz, J.T.: A 'Theory' mechanism for a proof-verifier based on first-order set theory. In: Kakas, A.C., Sadri, F. (eds.) Computational Logic: Logic Programming and Beyond. LNCS (LNAI), vol. 2408, pp. 214–230. Springer, Heidelberg (2002)

[94] Omodeo, E.G., Tomescu, A.I.: Using ÆtnaNova to formally prove that the Davis-Putnam satisfiability test is correct. Le Matematiche 63, 85–105 (2008); A preliminary version was presented at CILC 2007 (Messina)

[95] Papadimitriou, C.: On the complexity of integer programming. J. of the ACM 28 (1981)

[96] Parlamento, F., Policriti, A.: Decision procedures for elementary sublanguages of set theory. IX: Unsolvability of the decision problem for a restricted subclass of the Δ_0-formulas in set theory. Comm. Pure Appl. Math. XLI, 221–251 (1988)

[97] Parlamento, F., Policriti, A.: Decision procedures for elementary sublanguages of set theory: XIII. Model graphs, reflection and decidability. J. Automated Reasoning 7(2), 271–284 (1991)

[98] Paulson, L.C.: Set Theory for Verification. II: Induction and Recursion. J. Automated Reasoning 15(2), 167–215 (1995)

[99] Policriti, A., Schwartz, J.T.: T-theorem proving. I. J. of Symbolic Computation 20(3), 315–342 (1995)

[100] Prawitz, D., Prawitz, H., Voghera, N.: A mechanical proof procedure and its realization in an electronic computer. J. of the ACM 7, 102–128 (1960); Reprinted in [112]

[101] Presburger, M.: Über die vollständigkeit eines gewissen systems der aritmethik ganzer zahlen, in welchem die addition als einzige operation hervortritt. In: Comptes Rendus du premier Congrès des Mathématiciens des Pays slaves, Warsaw, pp. 92–101 (1929)

[102] Quaife, A.: Automated Deduction in von Neumann-Bernays-Gödel Set Theory. J. Automated Reasoning 8(1), 91–147 (1992)

[103] Renegar, J.: A faster PSPACE algorithm for deciding the existential theory of the reals. In: 29th Annual Symposium on Foundations of Computer Science (FOCS 1988), Los Angeles, Ca., USA, pp. 291–295. IEEE Computer Society Press, Los Alamitos (1988)

[104] Robinson, J.A.: A machine-oriented logic based on the resolution principle. J. of the ACM 12(1), 23–41 (1965); Reprinted in [112]

[105] Robu, J.: Geometry Theorem Proving in the Frame of the Theorema Project. Technical Report 02-23, RISC Report Series, University of Linz, Austria. PhD Thesis (2002)

[106] Rossi, G., Jayaraman, B. (eds.): Proc. of the Workshop on Declarative Programming with Sets, DPS 1999, Paris. Technical Report N. 200, Dipartimento di Matematica, Università di Parma, Italy (1999)

[107] Schwartz, J.T.: Instantiation and decision procedures for certain classes of quantified set-theoretic formulae. Technical Report 78-10, Institute for Computer Applications in Science and Engineering, NASA Langley Research Center, Hampton, Virginia (1978)

[108] Schwartz, J.T., Dewar, R.K.B., Dubinsky, E., Schonberg, E.: Programming with sets: An introduction to SETL. Texts and Monographs in Computer Science. Springer, Heidelberg (1986)

[109] Shankar, N., Rueß, H.: Combining Shostak theories. In: Tison, S. (ed.) RTA 2002. LNCS, vol. 2378, pp. 1–18. Springer, Heidelberg (2002)

[110] Shostak, R.E.: A practical decision procedure for arithmetic with function symbols. J. of the ACM 26(2), 351–360 (1979)

[111] Shostak, R.E.: Deciding combinations of theories. J. of the ACM 31, 1–12 (1984)

[112] Siekmann, J., Wrightson, G.: Automation of Reasoning I and II. Springer, Heidelberg (1983)

[113] Siekmann, J.H.: Unification theory. J. of Symbolic Computation 7(3-4), 207–274 (1989)

[114] Sigal, R.: Desiderata for logic programming with sets. In: Mello [78], pp. 127–141

[115] Stickel, M.E.: Automated deduction by theory resolution. J. Automated Reasoning 1(4), 333–355 (1985)

[116] Szmielew, W.: Elementary properties of Abelian groups. Fundamenta Mathematicae 41, 203–271 (1954)

[117] Tarski, A.: Sur les ensembles fini. Fundamenta Mathematicae VI, 45–95 (1924)

[118] Tarski, A.: A decision method for elementary algebra and geometry. Berkeley University Press (1951)

[119] Tarski, A.: What is elementary geometry? In: Hintikka, J. (ed.) The philosophy of mathematics — Oxford readings in philosophy, pp. 164–175. Oxford University Press, Oxford (1969); First published in 1959

[120] Tinelli, C., Ringeissen, C.: Unions of non-disjoint theories and combinations of satisfiability procedures. Theoretical Computer Science 290(1), 291–353 (2003)

[121] Tinelli, C., Zarba, C.G.: Combining nonstably infinite theories. J. Automated Reasoning 34(3), 209–238 (2005)

[122] Vaught, R.L.: On a theorem of Cobham concerning undecidable theories. In: Nagel, E., Suppes, P., Tarski, A. (eds.) Proc. of the 1960 International Congress on Logic, Methodology, and Philosophy of Science, pp. 14–25. Stanford University Press (1962)

[123] Wos, L.: The problem of finding an inference rule for set theory. J. Automated Reasoning 5(1), 93–95 (1989)

[124] Wu, W.-T.: On the decision problem and the mechanization of theorem-proving in elementary geometry. Scientia Sinica 21(2), 159–172 (1978); Also in Selected works of Wen-Tsün Wu. World Scientific Publishing, Singapore (2008)

[125] Zarba, C.G.: Combining lists with integers. In: Goré, R., Leitsch, A., Nipkov, T. (eds.) International Joint Conference on Automated Reasoning, IJCAR 2001 (Short Papers), Technical Report DII 11/01, pp. 180–189. University of Siena, Italy (2001)

[126] Zarba, C.G.: Combining multisets with integers. In: Voronkov, A. (ed.) CADE 2002. LNCS (LNAI), vol. 2392, p. 363. Springer, Heidelberg (2002)

[127] Zarba, C.G.: Combining sets with integers. In: Armando, A. (ed.) FroCos 2002. LNCS (LNAI), vol. 2309, pp. 103–116. Springer, Heidelberg (2002)

[128] Zarba, C.G.: A tableau calculus for combining non-disjoint theories. In: Egly, U., Fermüller, C. (eds.) TABLEAUX 2002. LNCS (LNAI), vol. 2381, pp. 315–329. Springer, Heidelberg (2002)

[129] Zarba, C.G.: Combining sets with elements. In: Dershowitz [45], pp. 762–782

[130] Zarba, C.G.: Combining sets with cardinals. J. Automated Reasoning 34(1), 1–29 (2005)

[131] Zarba, C.G., Cantone, D., Schwartz, J.T.: A decision procedure for a sublanguage of set theory involving monotone, additive, and multiplicative functions, I: The two-level case. J. Automated Reasoning 33(3-4), 251–269 (2004)

Constraint Logic Programming

Marco Gavanelli[1] and Francesca Rossi[2]

[1] Dipartimento di Ingegneria-Università di Ferrara
[2] Dipartimento di Matematica Pura e Applicata - Università di Padova

Abstract. Constraint Logic Programming (CLP) is one of the most successful branches of Logic Programming; it attracts the interest of theoreticians and practitioners, and it is currently used in many commercial applications. Since the original proposal, it has developed enormously: many languages and systems are now available either as open source programs or as commercial systems.

Also, CLP has been one of the technologies able to recruit researchers from other communities to the declarative programming cause. Current CLP engines include technologies and results developed in other communities, which themselves discovered logic as an invaluable tool to model and solve real-life problems.

1 The CLP Paradigm

Constraint Logic Programming (CLP) [7] represents a successful attempt to merge the best features of logic programming (LP) and constraint solving.

Constraint solving [127, 6, 56, 31] includes a variety of expressive modelling frameworks and efficient solving tools for real-life problems that can be described via a set of variables and constraints over them. A constraint is just a restriction imposed over the combination of values of some variables of the problem. Solving a problem with constraints means finding a way to assign values to all its variables such that all constraints are satisfied. Constraint solving methods have been successfully applied to many application domains, such as scheduling, planning, resource allocation, vehicle routing, computer networks, and bioinformatics [137, 127, 51].

Embedding the notion of constraint into a high-level programming language allows for a more flexible and practical constraint processing environment, where constraints can be represented as formulae and can be incrementally accumulated. Moreover, the presence of constraints in a programming language usually augments its expressive power, in the sense that some complex relations can be defined easily by means of constraints, and there are also efficient techniques to prove them.

For these reasons, constraints have been embedded in many programming environments, but some are more suitable than others. For example, the fact that constraints can be seen as relations or predicates, that constraint solving can be seen as a generalized form of unification, that their conjunction can be seen as a *logical and*, and that backtracking search is the base methodology to

A. Dovier, E. Pontelli (Eds.): 25 Years of Logic Programming, LNCS 6125, pp. 64–86, 2010.

solve them, makes them very compatible with logic programming, which is based on predicates, unification, logical conjunctions, and depth-first search.

These observations led to the development of the CLP paradigm, where constraints are embedded in the logic programming paradigm. The main goal is to maintain a declarative programming paradigm while increasing expressivity and efficiency via the use of specific constraint sorts and algorithms.

The first CLP language was Prolog II [42], designed by Colmerauer in the early 80's. Prolog II could treat term equations like Prolog, but in addition could also handle term disequations. After this, Jaffar and Lassez observed that both term equations and disequations were just a special form of constraints, and developed the concept of a constraint logic programming scheme in 1987 [99].

Syntactically, constraints are added to logic programming by considering a specific constraint sort (e.g., linear equations over the reals) and then allowing constraints of this type in the body of the usual logic programming clauses. Beside the classical resolution engine of logic programming, a (complete or incomplete) constraint solving system is added, able to check the consistency of constraints of the considered sort. Moving from LP to CLP, the concept of unification is generalized to constraint solving: the relationship between a goal and a clause (to be used in a resolution step) can be described not only via term equations but via more general statements, i.e., constraints. This allows for a more general and flexible way to control the flow of the computation. Also, the presence of an underlying constraint solver, usually based on incomplete constraint propagation of some sort, allows one to alternate backtracking search (as in classical LP) with efficient constraint propagation, thus generating a more efficient solver, that is nevertheless complete, being based on systematic search.

More precisely, a CLP clause is just like an LP clause, except that its body may contain also constraints of the considered sort. For example, if we can use linear inequations over the reals, a CLP clause could be:

$$\texttt{p(X,Y) :- X < Y+1, q(X), r(X,Y,Z).}$$

Logically speaking, this clause states that $\texttt{p(X,Y)}$ is true if $\texttt{q(X)}$ and $\texttt{r(X,Y,Z)}$ are true, and if the value of x is smaller than that of $y + 1$.

From the operational point of view, in an LP resolution step, we have to check the existence of a most general unifier between the selected subgoal and the head of a clause. In CLP, instead, we also have to check the consistency of the current set of constraints (called the *constraint store*) with the constraints in the body of the clause. Thus two solvers are involved: unification, as usual in LP, and the specific constraint solver for the constraints in use. To make it more efficient, this constraint solver may be not complete, that is, it may fail to discover some inconsistencies.

While in LP a computation state consists of a goal and a substitution, in CLP we have a goal and a constraint store. While in LP we just accumulate substitutions during a computation, in CLP we also accumulate constraints. Given a state $\langle G, S \rangle$, where G is the current goal (the resolvent) and S is the current constraint store, assume G consists of an atom A (that we want to rewrite) and a rest R, i.e., $G = (A, R)$. Then, at each step:

– if A is a constraint, A is added to S and its consistency is checked through a transition that checks if $consistent(A \wedge S)$; if it is, the new state is $\langle R, prop(S \wedge A) \rangle$, where $prop(C)$ is the result of applying some constraint propagation algorithm (like arc-consistency) to the constraint store C;

– if instead A is a literal, and there is a clause $H : -B$ with the same head-predicate as A, then we add the constraint $A = H$ to the constraint store, check its consistency, and replace A with B in the resolvent: the new goal is then $\langle (B, R), prop(S \wedge \{A = H\}) \rangle$.

A CLP computation is successful if there is a way to get from the initial state $\langle G, true \rangle$ to the goal $\langle G', S \rangle$, where G' is the empty goal and S is satisfiable.

Derivation trees are defined as in LP, except that each node in the tree now represents both the current goal and the current constraint store. Also, in practical CLP systems, the usual depth-first Prolog traversal mode is retained, with subgoals selected from left to right, and clauses from the first to the last one. Early detection of failing computations is achieved by checking the consistency of the current constraint store. At each node, the underlying constraint system is automatically invoked (via function *prop* above) to check consistency and the computation along this path continues only if the check is successful (although the check itself could be incomplete). Otherwise, backtracking is performed.

Although CLP significantly extends LP in expressive power and application domains, it maintains its semantic properties, such as the existence of equivalent operational, model-theoretic, and fixpoint semantics [99]. Several semantics, describing different observable properties of CLP programs, have been presented in the literature, with significant contributions from Italian researchers [84, 94, 52, 74, 43, 115, 19]. Properties of such semantics, such as fully abstraction, compositionality, and correctness, have been studied in depth. The power of CLP has also been exploited to treat negation in LP, by allowing constraints that are equalities or inequalities over the Herbrand domain [29]. Also, constraint solving in LP was compared with the equivalent notions in automated deduction [8].

Finally, abstract interpretation has been applied to CLP [11], but we will not discuss the issue because it is subject of another chapter of this book [59].

2   Constraint Sorts

CLP is not a programming language, but a programming *paradigm*, which is parametric with respect to the class (sort) of constraints used in the language. Working with a particular CLP language means choosing a specific class of constraints (for example, finite domains, linear, or arithmetic) and a suitable constraint system for that class. Notice also that unification is not *replaced*, rather it is assisted by the specific constraint solver, since every CLP language also needs to perform usual LP-style unification over its variables.

Denoting a CLP language over a constraint class X as CLP(X), we can say that logic programming is just $CLP(Trees)$, where Trees identifies the class of term equalities, with the unification algorithm to solve them. Other examples of instances of the CLP scheme are Prolog III [41], that treats constraints over terms, strings, booleans, and real linear arithmetic, and the language CLP(R) [100], that works with both terms and arithmetic constraints over the reals.

The possibility to instantiate the CLP scheme with many constraint sorts is one of the features that made CLP successful, since in this way the variety of solvers added to a LP language becomes almost unlimited (e.g., [110]).

2.1   Finite Domains

A popular class of constraints used with the CLP scheme is the class of constraints with variables ranging over finite domains. Constraint logic programming using finite domain constraints is a useful language scheme, referred to as CLP(FD). Its applicability is very large, since many real-life problems can be modelled via imposing a set of constraints over variables with finite domains (for example, the wide class of Constraint Satisfaction Problems [56]). Examples can be found in configuration, scheduling, and resource allocation [56, 127, 12, 51].

Finite domain constraints, as used within CLP languages, are usually intended to be arithmetic constraints over finite integer domain variables. Thus a CLP(FD) language needs a constraint system which is able to perform consistency checks and projection over this kind of constraints. Usually, the consistency check is based on some kind of constraint propagation, such as arc-consistency [105], some weaker version, like bound-consistency [20], or, more rarely, path-consistency [117] (see also Section 3.1).

Many CLP(FD) languages or environments have been developed, either in academic or commercial environments. Constraint logic programming over finite domains was first implemented in the late 80's by Pascal Van Hentenryck [135] within the language CHIP. Since then, more sophisticated constraint propagation algorithms have been developed and added to more recent CLP(FD) languages, like ECL^iPS^e [37], GNU Prolog [60], CIAO [32], B-Prolog [141], SWI-Prolog [139] and SICStus Prolog [35].

One of the main features of CLP(FD) languages is that they have a specific mechanism for defining the initial finite domains of the variables: usually as an interval over the integers. For example, a typical CLP(FD) syntax to say that the domain of variable x contains all integers between 1 and 10 is `X in [1..10]`, or `X::[1,10]`, or `fd_domain(X,1,10)`.

Another feature of all CLP(FD) languages is the use of a built-in predicate called `labeling` defined over a list of variables, and which finds values for them such that all constraints in the current store are satisfied. The `labeling` predicate provides a mechanism to generate solutions, that is, variable assignments that satisfy all accumulated constraints. More precisely, this predicate triggers backtracking search over a set of variables. For example, the following clause defines a problem with three finite domain variables (x, y, and z), each with domain containing the integers from 1 to 10, and sets a constraint over them ($x+y = 9-z$). After this, it triggers backtracking search via predicate `labeling`:

```
p(X,Y,Z) :- [X,Y,Z]::[1,10], X + Y = 9 - Z, labeling([X,Y,Z]).
```

The result of executing the goal :- p(X,Y,Z). is any instantiation of the three variables over their domains which satisfies the constraint $x + y = 9 - x$. Notice that without labeling, this same goal would return just the new domains obtained after applying constraint propagation (together with the constraint store). E.g., running this goal in the CLP(FD) language GNU Prolog [40] returns the answer [X,Y,Z]:[1,7], meaning that the domains have been reduced from [1..10] to [1..7] via constraint propagation. The clause above presents the typical shape of a CLP(FD) program: first the variable domains are specified, then the constraints are imposed, and finally the backtracking search is invoked via a labeling predicate. A CLP(FD) program can consist of many clauses, but the overall structure of the program always reflects this order, which refers to a methodology called *constrain and generate*, where first variables are constrained and only later (when the domains are smaller) backtracking search is invoked. This corresponds to applying constraint propagation prior to search and therefore avoiding early some dead-ends.

In many CLP(FD) systems, arc-consistency is considered too expensive: for each binary constraint one should (in general) check if for each domain element there exists a support in the other domain. So, for each constraint involving two variables with d elements in the domains, one has to do $O(d^2)$ constraint checks. Since constraints are many, and arc-consistency propagation can wake up many times the same constraint, a quicker algorithm is often adopted, at the expenses of a lower pruning. Bound consistency considers only the bounds (minimum and maximum values) of the domains, so the number of checks is drastically reduced. This means that, e.g., the propagation of the $X = 2Y$ constraint will not remove all the odd values from the domain of X, but will have to perform only 4 checks.

A powerful feature of CLP(FD) is that for each constraint one can have a different propagation algorithm: if we know an efficient algorithm to perform arc-consistency for a specific constraint, we can use it, even if for other constraints the solver performs only bound consistency. For example, consider the goal $A ::$ $[-1, 0, 1]$, $B :: [-1, 1]$, $C :: [0, 1]$, $A = B$, $A^2 \leq C$. If all the constraints have bound-consistency propagation, no pruning occurs, in fact all the extreme values in each domain are consistent with some value in each other domain. On the other hand, arc-consistency propagation for the equality constraint is very simple: one has to compute the intersection of the two domains, which has linear complexity, instead of the expensive $O(d^2)$ of the general case. By applying arc-consistency to the $A = B$ constraint we can remove value 0 from the domain of A. Now, the bound-consistency propagation of $A^2 \leq C$ detects that the value 0 in the domain of C is no longer supported and removes it, implicitly assigning 1 to C. So, by strengthening the propagation of a single constraint (in the example, the equality constraint), we can propagate removals also by constraints with a weak bound-consistency propagation.

Global constraints are non-binary constraints that appear often in applications and for which specialized constraint propagation methods are developed. Sometimes those constraints are logically equivalent to the conjunction of a set of binary constraints, but global constraints typically perform stronger propagation than applying standard arc-consistency to many binary constraints. A typical example is the `alldifferent` constraint [136], which requires that n variables have mutually different values. Although this constraint can be defined with a binary not-equal constraint for each pair of variables, such a representation does not allow for much domain pruning by arc-consistency. Since such a constraint appears very often, it is worthwhile to strengthen its propagation method by employing an ad hoc filtering algorithm. The concept of arc-consistency was suitably extended for non-binary constraints and named *Generalized Arc-Consistency (GAC)*. Most current CLP languages are equipped with a rich taxonomy of global constraints. During a computation, the current constraint store in a CLP computation may contain both binary and global constraints such as `alldifferent`. At each step, when constraint propagation is performed, each constraint propagates with its own algorithm, and achieves arc or bound-consistency. Not all non-binary constraints have a specialized constraint propagation algorithm, just those that occur more frequently in applications.

Other logic languages, such as Answer Set Programming (ASP) [27], address similar types of problems addressed by CLP(FD); there are works comparing the two approaches [63, 109], and also integrating the two [15]. We will not give more details on ASP, since it is the subject of another chapter of this book [27].

2.2   Sets

Various Italian researchers studied the integration of *sets* into logic programming. Sets are widely used in mathematics to define new objects, and they allow for a natural representation of concepts in AI and in software engineering. One of the languages that integrate sets into logic programming is $\{log\}$ [64], that later evolved into the language CLP($\mathcal{SET}$) [65]. In CLP($\mathcal{SET}$), unification is extended to deal with variables representing sets and set objects. Prolog users often represent collections of values as lists, but this is insufficient when one needs a set semantics. Sets intrinsically remove symmetries (see also Section 4.2), since $\{1,2\}$ and $\{2,1\}$ are the same set, while for lists $[1,2]$ and $[2,1]$ represent different terms (i.e., they do not unify). In CLP($\mathcal{SET}$), $\{1,2\} = \{2,1\}$ succeeds, as well as $\{1,2,3,2\} = \{3,2,1,1\}$; moreover, one can have variables and non-ground terms as elements of sets, so the unification $\{p(X), p(2)\} = \{Y\}$ succeeds, giving $Y = p(2)$ and $X = 2$. CLP($\mathcal{SET}$) supports sets, possibly partially specified and nested like e.g. $\{X, \{\emptyset\}\} \cup Y$. Moreover, set unification and set constraint solving has been analysed in a modular way so as to easily replace sets with multi sets (and other similar data structures)—see e.g. [67, 66].

CLP($\mathcal{SET}$) has been used for various applications, among which to represent actions [124], and to implement abductive reasoning [89] (see also Section 3.3). Other efforts tried to integrate reasoning on sets with the classical CLP(FD). In one case the starting point was a visual search application [45]. Visual search and image recognition are classical applications of CLP(FD) [45, 75]. Visual search is the task of finding an object (described in some formal way, called the *object model*) in an image. CLP(FD) provides the language for describing the object model: first one decides the visual features (the basic components of the image, such as lines, points, surface patches, etc.), then he/she defines the object model by means of constraints that relate the visual features (surface s_1 is orthogonal to surface s_2, etc.). Now, before CLP(FD) performs constraint propagation and subsequent search, one has to know all the visual features in the image, as they compose the domains of the variables. This task is performed by a segmentation system, that takes often most of the computing time, since it has to relate the pixels of the image with higher-level information. In order to speed up the acquisition process, one can interleave constraint propagation and value acquisition; in this way only those features actually required for solving the CSP are acquired from the segmentation system. The classical CSP model is then extended to an Interactive CSP [46], with corresponding solving algorithms. A corresponding CLP language [88] uses sets to represent the domains of FD variables. Later on, a general integration of the two sorts was proposed [50, 18], which integrates sets and finite domain variables to speedup the CLP($\mathcal{SET}$) computation.

3   Related Frameworks

3.1   Constraint Handling Rules

In classical CLP languages, solvers are embedded in the language in a *hard-wired* way: each language comes with one or more solvers for some constraint sorts. However, defining a new constraint, or even a new solver, is often tricky: one has to know (part of) the implementation of the solver itself, study the interface for defining new constraints, and implement the propagation algorithm. While usually very efficient, this approach is rather operational and not always flexible. Constraint Handling Rules (CHR) [82] represents a successful example of a high-level, logic language for designing constraint solvers. Also, usually solvers adopt arc or bound consistency, that look at one constraint at a time. For example, the constraints $[A, B] : [1..10]$, $A \leq B$, $B \leq A$, $A \neq B$ do not perform any pruning, even if we can easily see that there is no solution. If we looked at pairs of constraints, we could infer from $(A \leq B \wedge B \leq A)$ that $A = B$, and from $(A = B \wedge A \neq B)$ that there is no solution. Intuitively, looking at pairs of constraints allows one to achieve higher levels of consistency, such as path-consistency [117].

    CHR is a powerful language for modelling solvers, based on the rewriting of constraints into simpler ones until they are solved. CHR can be seen as a CLP language where clauses are multi-headed guarded rules for constraint rewriting.

    CHR rules are of two kinds, based on the notions of *simplification* and *propagation* over user-defined constraints. Simplification rules replace constraints by simpler constraints while preserving logical equivalence. Propagation rules add new, logically redundant constraints, which may cause further simplifications. More precisely, a CHR program is a finite set of CHR rules. A *simplification* CHR rule is of the form $H \Leftrightarrow G|B$ and a *propagation* CHR rule is of the form $H \Rightarrow G|B$. The multi-head H is a conjunction of CHR constraints. The optional guard G is a conjunction of built-in constraints. The body B is a conjunction of built-in and CHR constraints. An example of a simplification rule is $X \leq Y \wedge Y \leq X \Leftrightarrow X = Y$, while a possible propagation rules is $X \leq Y \wedge Y \leq Z \Rightarrow X \leq Z$.

A state of a computation is a conjunction of built-in and CHR constraints, and states evolve via derivation steps. An *initial state (or query)* is an arbitrary state. In a *final state (or answer)*, either the built-in constraints are inconsistent or no derivation step is possible anymore. A rule with head H and guard G is *applicable* to CHR constraints H' in the context of constraints D, if the underlying constraint theory entails D and $\exists \theta (H\theta = H' \wedge G\theta)$. Notice that the symbol $=$ is to be understood as built-in constraint for syntactic equality and is usually implemented by a (one-way) unification. If H' matches H, we equate H' and H. This corresponds to parameter passing in conventional programming languages, since only variables from the rule head H can be further constrained, and all those variables are new. Finally, using the variable equalities from D and $H' = H$, we check the guard G.

Any of the applicable rules can be applied, but the choice of the rule is a committed choice, thus it cannot be undone.

If an applicable simplification rule $(H \Leftrightarrow G \mid B)$ is applied to the CHR constraints H', H' is removed from the state, and the body B, the equation $H = H'$, and the guard G are added to the state. If a propagation rule $(H \Rightarrow G \mid B)$ is applied to H', we add B, $H = H'$ and G, but do not remove H'.

CHR is now implemented in most major CLP languages (e.g., SICStus, SWI or ECL^{i}PS^e), and the number of applications developed in CHR is impressive (see, e.g., the web page[1] *"The first fifty applications using CHR"*, amongst which we find many works of Italian researchers [4, 126, 22, 61].)

Beside the operational semantics briefly outlined above, several declarative semantics have been defined for CHR programs, and soundness and completeness results have been obtained. The issue of confluence has also been studied in depth, since applicable CHR rules may be applied in any order giving rise to resulting states with the same meaning but not necessarily the same syntax. This may be a problem in terms of constraint solvers, since the ability to detect the inconsistency of the current set of constraints depends also on the syntax. Another important property is compositionality [58]. This property allows to compute the semantics of a conjunctive query from the semantics of its components, and is obviously very desirable since it allows to define incremental and modular analysis and verification tools.

Various extensions of the basic CHR language have been proposed in the literature. For example, CHR has been extended with a probabilistic weighting of the rules, by specifying the probability of their application [83]. In this way, it is possible to formalise various randomised algorithms, such as simulated annealing.

3.2   Concurrent Constraint Programming

In CLP, each computation step adds new constraints to the constraint store, and checks if the resulting store is consistent. However, the constraint store could also be used to check whether it contains enough information to entail certain

---

[1]   http://www.cs.kuleuven.be/~dtai/projects/CHR/chr-appls.html

constraints. This is what is done in the concurrent constraint (cc) programming paradigm [130], where several agents work concurrently with a unique constraint store. Each agent can perform two kinds of actions: either to add (called *tell*) a new constraint to the store, and proceed if this produces a consistent new store, or to wait (called *ask*) until the current store entails a certain constraint, and proceed only after this holds. In this paradigm, the concurrent agents communicate via the shared constraint store. CLP can be seen, very abstractly, as a restriction of the cc paradigm where only tell operations are performed.

Many significant results from Italian researchers have been obtained in defining and proving properties of several different semantics for the cc paradigm [53, 73,71]. Also, the cc paradigm has been extended to work with soft constraints [26], with probabilistic actions [123], and with timed operators [54,23].

We avoid entering into the details of the various research lines related to cc, since it is the subject of another chapter of this volume [85].

3.3   Abductive Constraint Logic Programming

Logic programming is based on deductive reasoning, i.e., if we have a rule with conditions and a conclusion, and we know that the preconditions of the rule are true, we infer that also the conclusion is true. On the other hand, the human mind uses also other types of inference: for example, in medical diagnosis a physician is given a set of symptoms, that are the effects of some illness, and has to infer the illness that possibly caused such effects. The inference rule that allows one to reason from the conclusions to possible causes, or conditions, was called *abduction* by the philosopher Peirce.

Abductive Logic Programming [102,101] is an extension of LP that deals with incomplete information by performing abduction. In ALP, there are some syntactically distinguished predicates that have no definition, and cannot be proven: an abductive proof-procedure will *assume* their possible truth, and provide the abduced literal in the answer. E.g., an abductive program could be:

```
headache :- flu.
```

where *flu* is declared as an abducible predicate. Given the query `:- headache`, an abductive proof-procedure will provide as answer

```
yes, flu.
```

However, abductive reasoning has a very wide search space, and researchers soon found out that it could be reduced by means of constraints [103]. Obviously the integration also provides more expressivity to the abductive language, as the user can now write constraints in his/her programs. This opened the path to the development of a series of proof-procedures that integrate abductive reasoning with constraint propagation [104,69,3]. Abductive constraint programming languages have been used for a variety of applications, including agents, planning, web service composition [2,1], web sites verification [107] and two-player games [87].

More on Abductive Logic Programming can be found in the chapter [95].

3.4   Soft Constraints and Preferences

Classical constraints are statements that have to be satisfied in order to obtain a feasible solution. Thus the role of a constraint solver is to find a variable assignment that satisfies all constraints. In several real-life scenarios, this approach is too rigid, since there may be no variable assignment that satisfies all constraints. These scenarios often occur when constraints are used to formalize desired properties rather than requirements that cannot be violated. Such desired properties are not faithfully represented by constraints, but should rather be considered as *preferences*, whose violation should be avoided as far as possible. *Soft constraints* [24] provide one way to model such preferences, by extending the classical constraint notion into a more general and flexible one.

A soft constraint is just like a constraint, but instead of being only satisfied or violated, it may have several levels of satisfiability. Historically, first a variety of specific extensions of the basic constraint formalism have been introduced, such as fuzzy constraints [129]. Later, these extensions have been generalized using more abstract frameworks, which have been crucial in proving general properties and in identifying the relationship among the specific frameworks [24, 133]. Moreover, for each of the specific classes, algorithms for solving problems specified in the corresponding formalisms have been defined. In fact, many techniques and approaches to solve classical constraints, included constraint propagation, have been generalized to work also with soft constraints.

In the semiring-based formalism [24], a soft constraint is a cost function, where each assignment of the variables of the constraint is associated to an element coming from an ordered set, whose properties are similar to those of a semiring. This set contains all possible levels of preference (or costs, or quality, etc.), of a variable assignment in the considered constraint class. For example, for fuzzy constraints, the preference levels are values between 0 and 1, and higher values are more preferred. Classical constraints can also be cast in this general framework: in this case the preference set contains just two elements (*true* and *false*, or *satisfied* and *violated*). The preference set also comes with an operation to combine preference levels. This is useful to compute the satisfiability level of a complete variable assignment from those given by the constraints to the portion of the assignment relevant to them. For example, in fuzzy constraints the combination takes the minimum preference level, while in classical constraints it is just a *logical and*, since *all* constraints need to be satisfied. A survey of the various approaches to deal with soft constraints can be found in [113].

The notion of global constraints has been exploited also in the context of soft constraints. For example, in [97] a general method to soften global constraints is presented, which is based on the notion of a flow in a graph, and several global constraints are defined in their soft version. Also, in [140] efficient algorithms are proposed to achieve generalized arc consistency for the soft global cardinality constraint.

Classical CLP handles only standard constraint solving. Thus it is natural to try to extend the CLP formalism in order to handle also soft constraints. A first attempt was the *hierarchical CLP* (HCLP) system [28], a CLP language where each constraint has a level of importance and a solution of a constraint problem is found by respecting the hierarchy of constraints. The finite domain CLP language clp(fd) [40] has been extended to handle semiring-based constraints, obtaining a language paradigm called clp(fd,S) [93] where S is any semiring, chosen by the user. By choosing one particular semiring, the user uses a specific class of soft constraints: fuzzy, optimized, probabilistic, or even classical hard constraints.

The language SCLP [25] treats in a uniform way, and with the same underlying machinery, all constraints that can be seen as instances of the semiring-based approach: from optimization to satisfaction problems, from fuzzy to probabilistic, prioritized, or uncertain constraints, and also multi-criteria problems, while still being able to handle classical constraints. Syntactically, SCLP extends CLP by allowing the presence of preference levels as the body of a clause. E.g., the clause p(X,Y,N) :- (X+Y)/N. states that $(X + Y)/N$ is the preference level to be given to the assignment (X, Y, N) for constraint p. The usual three equivalent semantics (model-theoretic, fix-point, and operational) can be defined also for the SCLP paradigm, although suitably generalized to handle soft constraints.

4   Improvements, Solution Techniques

4.1   Integration with Operations Research

CLP(FD) is an effective language to model and solve combinatorial problems. However, there are other frameworks that address the same problems, such as meta-heuristics, integer linear programming, population-based methods, etc. CLP(FD) has unique advantages: there are many types of available constraints, compared to integer linear programming that accepts only linear inequalities. It supports complete solving algorithms, while local search or genetic algorithms are usually incomplete (i.e., they might fail to produce a solution even if it exists). On the other hand, there are some types of problems in which other techniques are more efficient. For this reason, various efforts tried to merge algorithms and solvers, in order to improve on both of them. The fact that CLP(FD) is very general makes it the ideal playground to test the integration of different techniques.

One type of integration, already mentioned, is global constraints. In general, the (generalized) arc-consistency propagation of an n-ary constraint is very expensive (see, e.g., [116]): since an n-ary constraint can encode a whole CSP, removing all values that do not belong to a solution is in general NP-hard. However, despite this worst-case complexity, there exist significant constraints of practical use that have polynomial-time, specific propagation algorithms. For example, the `alldifferent` constraint uses results from graph theory, the global cardinality constraint `gcc` computes the maximum flow of a graph, all techniques borrowed from Operations Research (OR). In OR there are very efficient algorithms to solve very specific tasks, however a slight change in the problem formulation (e.g., a new constraint added by the user) can make a very good algorithm inapplicable. CLP(FD), instead, is very general-purpose. In OR, combining a graph algorithm with a maximum flow is a rather complex task, while in CLP(FD) it is trivial: just a matter of adding two constraints (alldifferent and gcc) to the program, and they will automatically communicate through the constraint store and the domains of the variables. The user does not even need to know the details of the propagation algorithm.

Another key observation is that CLP(FD), being based on the concept of consistency, is very oriented to solve satisfiability problems, and optimization problems are often converted into (sequences of) satisfiability ones. OR, instead, has a wide literature focussed on optimization problems, using bounds, relaxations, and cuts, to remove sub-optimal parts of the search space. Moreover, arc-consistency reasons about one constraint at a time, meaning that if no constraint is able to perform pruning alone, no propagation occurs. This can be partially solved using higher levels of consistency, also supported by languages like CHR (Section 3.1), but this is not always a solution, since higher levels of consistency require more computation time. Linear programming algorithms, instead, navigate a polytope focussing only on the vertices carrying the best values of objective function, so they have a more global view.

So, an interesting way to integrate CLP and OR is by trying to exploit both the satisfaction-based techniques of CLP and the optimization-based tools of OR. A simple idea is to use both a linear model and a CLP(FD) model at the same time: if either of the two detects inconsistency, we can fail and backtrack. An important information a linear solver provides is a *bound*: by giving up the integrality constraint, the linear solver is able to compute an over-optimal solution. So, if the linear relaxation of the current node gives a worse bound than the best solution found so far, the current node can be pruned [34]. Moreover, the linear solver is able to provide another piece of information, namely *reduced costs*. For each variable x_i in the linear model, the reduced cost r_i is the derivative of the objective function with respect to x_i. Suppose we have a minimization problem $min(f)$, and that the linear relaxation provides a value LB (Lower Bound). Suppose that we already know a solution with cost UB (Upper Bound). Of course, if $LB \geq UB$, we can fail and backtrack. Otherwise, suppose that there is some variable x_i that in the optimal solution of the linear relaxation takes value 0, and suppose the reduced cost is 10. This means that, if we change the value of x_i to 1, the value of the objective function will increase of at least 10. If $LB + 10 \geq UB$, then I cannot add 1 to x_i, because that would mean going to a worse solution than the current best, so we can remove the value 1 from the domain of x_i. This is called *cost-based filtering* [80, 90].

Other techniques from (integer) linear programming have been adapted to include constraint programming. Column generation is a technique used in linear programming to solve very large problems. The basic idea is that the simplex algorithm uses a tableaux to represent the linear program, and uses reduced costs to drive the search. Since reduced costs are the derivatives of the objective function with respect to the variables in the current solution, if all reduced costs are positive, then there is no way to reduce the value of the objective function, i.e., we are in the optimal solution (global minimum). Otherwise, if there is at least a negative reduced cost, increasing the value of the corresponding variable will reduce the objective function, and the search continues. However, if the tableaux contains a huge number of columns, finding a negative cost may become a constraint satisfaction problem itself that can be solved with various techniques, including constraint programming [96].

Bender's decomposition is another technique used to solve very large problems. The whole problem is decomposed into a master problem and a subproblem, that will then communicate. One of the two could be more easily solvable by an FD solver, while the other by a linear solver; this gives an interesting pattern to have the two solvers communicate [70, 17, 98].

Finally, various methods exist to integrate local search with CLP [38, 112, 78, 39].

4.2 Symmetry Breaking

In CLP and constraint reasoning in general, there are several techniques that try to change the problem formulation to improve the efficiency of the solution process. For example, some approaches include rewriting (through folding and unfolding steps) a constraint logic program [77], to make it more efficient for a specific instance or a query. We will not go into further details, as the interested reader will find an exhaustive exposition in another chapter of this book [122].

Another interesting and useful idea is to try to remove some symmetrical parts of the search space, by rewriting the constraint program or by adding (by hand or automatically [108], in the CLP program) so-called *symmetry breaking constraints*. In fact, the presence of symmetries can expand exponentially the size of the search space. Consider, for example, a graph coloring problem: each node of a graph should be assigned a color from a finite palette (the same one for all nodes), with the constraint that two nodes connected with an arc should have different colors. Backtracking search will try to assign a value to a first node, for example color red to node N_1. Suppose that, after constraint propagation and a long search, we find out that there is no solution with $N_1 = red$: backtrack search will now choose the second value in the domain of N_1, say *blue*. However, since the colors are symmetric, there is no solution with blue as well. This observation can be used to reduce significantly the search space. Other problems have many more symmetries than the graph coloring. The classical benchmark problem in this research area is the social golfer, which is an abstraction of many real-life scenarios: N golf players want to play golf every week, in groups of M golfers; we have to find a schedule for W weeks such that no two players play in the same group more than once.

A first way to tackle this problem is by changing the constraint model, by switching to a representation with no symmetries, or with a reduced number of symmetries. The first solution to the social golfer problem was implemented by Stefano Novello in CLP [120]. The idea was to use a set representation (see also Section 2.2): the position of elements in a set is immaterial, so the intrinsic symmetry related to the order of the elements no longer exists.

Other solutions include finding the equivalence classes for the symmetries, and adding constraints that are satisfied only by one representative of each equivalence class. In the graph coloring example, one can leave only one element in the domain of a given node. Of course, this simple constraint will not always remove all the symmetries, but it usually greatly reduces the search space. When the constraint problem is represented by a sequence of symmetric variables (i.e., every permutation of a solution is still a solution), one can impose that the variables are ordered. If the problem contains a matrix of variables, and exchanging two lines or two columns of a solution yields another solution, a lexicographic ordering between the rows/columns can be imposed [81].

In some cases, one has a very powerful heuristics for solving a CSP, and the heuristic can become less effective if we change the constraint model; in particular the heuristic could be deceived by the addition of symmetry breaking constraints. In those cases, one can revert to algorithms that break the symmetries during search: i.e., after exploring (unsuccessfully) some part of the search space, they prune the symmetrical parts of the already explored zones [114, 92, 79, 72].

All these methods assume that the symmetries are already known; however, there are also approaches trying to identify the symmetries from the specifications [108]. In some cases, one tries to detect the symmetries from the general model [33], without looking at the specific instance. E.g., the graph coloring problem has symmetries in general, irrespectively of the particular graph we are considering. In other cases, one tries to detect symmetries that hold only in the given instance we are about to solve [86].

5   Applications

CLP has shown to be successfully used in many application domains. For space reasons, we will just mention few of them, not intending to give a complete survey. The reader can refer to existing surveys on CLP applications [137], as well as on the chapter on applications of LP in this book [51].

In recent years, biology has been the source of interesting application problems for the whole of computer science, due to the large volume of data and the combinatorial nature of many scenarios. CLP, and constraint programming in general, has been recently applied to some of these problems [10]. In particular, CLP has been used to tackle the protein structure prediction problem, which is one of the most challenging problems in biological sciences, and which can be seen as an optimization problem [48, 55]. The complexity of constraint propagation was also studied [49]. The results obtained on small proteins show that CLP can be employed for studying protein simplified models. The advantage of CLP over other approaches lies in the rapid software prototyping, in the easy way of encoding heuristics, and in the several efficient constraint-based techniques, such as constraint propagation, to prune huge search spaces.

Constraint logic programming was also used to reason about spatial and temporal data, and a CLP solver was integrated with a geographical information system. One practical applications was the study of the mating habit of the crested porcupine [125], in which information is gathered through radio-collars and processed by a CLP program.

Planning and scheduling have always been two of the main application areas for constraint-based approaches [12]. Scheduling is the problem of assigning a timing to the various tasks composing a complex activity, and often, other resources. As such, it has various specializations: in sport scheduling [131] one wants to fix the matches of a tournament; in school timetabling [132, 86] the aim is deciding when and where lessons take place, in crew rostering we have to find a sequencing of a given set of duties into rosters satisfying operational constraints [34], etc. [30, 36]. CLP(FD) has proved to be very successful in this area mainly because of an important global constraint, called `cumulative`. This constraint relates the start times, the durations, and the resource consumptions of a set of tasks, and it ensures that in any instant of time, the total resource consumption of the tasks being executed does not exceed a given limit. So, for a school timetabling, one can state that the rooms are resources: if in a school there are R available rooms, there cannot be more that R lessons at the same time. Teachers can also be considered as resources: two contemporary lessons cannot involve the same teacher, and so on. There are various implementations of the `cumulative` constraint, that give different balances of computational complexity (usually from $O(n^2)$ to $O(n^3)$) and achieved pruning.

Planning, instead, is the problem of finding a sequence of actions that, taken in the correct order, achieve a given goal. Each action has pre-conditions and post-conditions, and the automatic planner must ensure that the post-conditions of some action do not invalidate the pre-conditions of the subsequent actions. CLP(FD) is useful to detect such possible situations, called threats [14, 13], and to implement the temporal reasoning [121]. Also, some notable works propose to implement action description languages in CLP(FD) [62].

A remarkable amount of work in CLP is connected with database theories and applications. Considering the theory, the semantics of the U-Datalog language is cast through a CLP semantics, and, in particular, updates in rule bodies are specified through constraints [118]. Constraints are also used to schedule the transactions in a distributed database [111]. Constraints are also useful to represent incomplete information, e.g., in temporal-probabilistic databases [106].

CLP(FD) was used to find an optimal placement of sirens to alert the population in Venice of the high tide [9]. The map of the city is divided by a grid into cells, and for each cell a number of features is recorded, such as the average and maximum height of the buildings, their density, etc. The authors use a simulator to compute the sound propagation, and they relate the sound propagation with the position of the sirens through constraints. The objective is to find the best placement (that minimizes the number of sirens) such that in each cell the signal strength is greater than or equal to a given threshold.

Other authors [76] tackle the problem of detecting excess of pollution in the Venice lagoon. Every day, information is acquired through sensors, and fed to a decision support system. The system is implemented in CLP, and uses constraints to model the propagation of pollutants in the lagoon; it is able to provide suggestions to the Venice Water Magistracy on which implants to close, which to relocate, etc, to keep the level of pollution within acceptable levels.

The system LODE [91] applies CLP to reason about temporal information in an e-learning software devoted to deaf children. Deaf people can have difficulties in understanding temporal relations in textual information, and such software helps them by proposing stories and exercises.

Verification is a very important application of CLP, that has been deeply studied by many authors in Italy and abroad. It is also a vast discipline, that includes important applications of theoretical and practical importance, such as security verification [57, 44, 16]. We will not delve into this fascinating discipline, because it is the subject of another chapter of this book [59].

6   Conclusions

Constraint Logic Programming is a computation paradigm that joins the theoretical features of Logic Programming (declarative semantics, soundness, completeness) with an important range of practical applications. However, additional efforts are needed to make it more widely applicable. Features such as uncertainty, multi-agent reasoning, lack of data, and vast amounts of information, just to cite few examples, should be fully integrated and satisfactorily handled in CLP-style languages if we want CLP to be successfully used also in more modern applications. We also see two other threats to the spreading of the CLP technology into the industrial world. One is the lack of a common syntax: as already hinted in Section 2.1, every CLP(FD) solver has itw own syntax for defining domains, and same constraints can have different names. There are standardisation efforts, and new modelling languages such as (Mini)Zinc [119] become more and more supported by CLP systems, but still the goal of a commonly agreed language seems far away. A second threat comes from imperative and object-oriented languages: many solvers are now available also with C++ (ILOG[2], Gecode [134]) or Java syntax (Choco, Jacop[3], JsetL [128]), giving up the gains coming from logic programming, but with the advantage of an easier integration into already developed applications. To keep up with those solvers, CLP languages should either provide new features, unapplicable to imperative/OOP languages, or have better integration with real world applications, with the ability to develop attractive user interfaces, access to web services, and so on. Finally, CLP languages are usually tailored for the experienced user: one can develop extremely efficient search strategies, and heuristics for solving a specific problem, even with integration of different solvers, but these technologies are often out of reach for the naive user. CLP has taken the opposite viewpoint with respect to, e.g., SAT, MIP or ASP solvers: in those languages the user has only to state the problem, and the solver will choose a good strategy to solve it. Research trying to bridge the gap between the unexperienced user and the state-of-the-art technology could really boost the widespreading of the CLP word.

Acknowledgements.   This research has been partially funded by PRIN 2008 project *'Innovative and multi-disciplinary approaches for constraint and preference reasoning'*.

---

[2] ILOG: www.ilog.com/products/cp/

[3] Choco: http://choco.emn.fr/, Jacop: http://jacop.osolpro.com/

References

1. Alberti, M., Cattafi, M., Gavanelli, M., Lamma, E., Chesani, F., Montali, M., Mello, P., Torroni, P.: Integrating abductive logic programming and description logics in a dynamic contracting architecture. In: IEEE Int. Conf. on Web Services (2009)
2. Alberti, M., Chesani, F., Gavanelli, M., Lamma, E., Mello, P., Montali, M.: An abductive framework for a-priori verification of web services. In: PPDP (2006)
3. Alberti, M., Chesani, F., Gavanelli, M., Lamma, E., Mello, P., Torroni, P.: Verifiable agent interaction in abductive logic programming: the SCIFF framework. ACM Transactions on Computational Logics 9(4) (2008)
4. Alberti, M., Lamma, E.: Synthesis of object models from partial models: A CSP perspective. In: van Harmelen, F. (ed.) ECAI, pp. 116–120. IOS Press, Amsterdam (2002)
5. Alpuente, M., Sessa, M. (eds.): GULP-PRODE 1995 (1995)
6. Apt, K.R.: Principles of Constraint Programming. Cambridge Univ. Press, Cambridge (2003)
7. Apt, K.R., Wallace, M.G.: Constraint Logic Programming Using ECL^iPS^e. Cambridge University Press, Cambridge (2006)
8. Armando, A., Melis, E., Ranise, S.: Constraint solving in logic programming and in automated deduction: A comparison. In: Giunchiglia, F. (ed.) AIMSA 1998. LNCS (LNAI), vol. 1480, pp. 28–38. Springer, Heidelberg (1998)
9. Avanzini, F., Rocchesso, D., Belussi, A., Dal Palù, A., Dovier, A.: Designing an urban-scale auditory alert system. IEEE Computer 37(9), 55–61 (2004)
10. Backofen, R., Gilbert, D.: Bioinformatics and constraints. In: Rossi, et al [127]
11. Bagnara, R., Gori, R., Hill, P.M., Zaffanella, E.: Finite-tree analysis for constraint logic-based languages. In: Cousot, P. (ed.) SAS 2001. LNCS, vol. 2126, pp. 165–184. Springer, Heidelberg (2001)
12. Baptiste, P., Laborie, P., Le Pape, C., Nuijten, W.: Constraint-based scheduling and planning. In: Rossi, et al [127]
13. Barruffi, R., Milano, M., Montanari, R.: Planning for security management. IEEE Intelligent Systems 16(1), 74–80 (2001)
14. Barruffi, R., Milano, M., Torroni, P.: Planning while executing: A constraint-based approach. In: Ohsuga, S., Raś, Z.W. (eds.) ISMIS 2000. LNCS (LNAI), vol. 1932, pp. 228–236. Springer, Heidelberg (2000)
15. Baselice, S., Bonatti, P., Gelfond, M.: Towards an integration of answer set and constraint solving. In: Gabbrielli, M., Gupta, G. (eds.) ICLP 2005. LNCS, vol. 3668, pp. 52–66. Springer, Heidelberg (2005)
16. Bella, G., Bistarelli, S.: Soft constraint programming to analysing security protocols. TPLP 4(5-6), 545–572 (2004)
17. Benini, L., Lombardi, M., Mantovani, M., Milano, M., Ruggiero, M.: Multi-stage Benders decomposition for optimizing multicore architectures. In: Perron, L., Trick, M.A. (eds.) CPAIOR 2008. LNCS, vol. 5015, pp. 36–50. Springer, Heidelberg (2008)
18. Bergenti, F., Dal Palù, A., Rossi, G.: Generalizing finite domain constraint solving. In: Formisano, A. (ed.) CILC 2008 (2008)
19. Bertolino, B., Bonatti, P.A., Montesi, D., Pelagatti, S.: Correctness and completeness of logic programs under the CLP schema. In: Asirelli, P. (ed.) Proc. Sixth Italian Conference on Logic Programming, Pisa, Italy, pp. 391–405 (1991)
20. Bessiere, C.: Constraint propagation. In: Rossi, et al. [127]

21. Bessière, C. (ed.): CP 2007. LNCS, vol. 4741. Springer, Heidelberg (2007)
22. Bistarelli, S., Frühwirth, T.W., Marte, M.: Soft constraint propagation and solving in chrs. In: SAC, pp. 1–5. ACM, New York (2002)
23. Bistarelli, S., Gabbrielli, M., Meo, M., Santini, F.: Timed soft concurrent constraint programs. In: Lea, D., Zavattaro, G. (eds.) COORDINATION 2008. LNCS, vol. 5052, pp. 50–66. Springer, Heidelberg (2008)
24. Bistarelli, S., Montanari, U., Rossi, F.: Semiring based constraint solving and optimization. Journal of the ACM 44(2), 201–236 (1997)
25. Bistarelli, S., Montanari, U., Rossi, F.: Semiring-based constraint logic programming. In: IJCAI 2001, pp. 352–357 (2001)
26. Bistarelli, S., Montanari, U., Rossi, F.: Soft concurrent constraint programming. In: Le Métayer, D. (ed.) ESOP 2002. LNCS, vol. 2305, pp. 53–67. Springer, Heidelberg (2002)
27. Bonatti, P., Calimeri, F., Leone, N., Ricca, F.: Answer Set Programming. In: Dovier, A., Pontelli, E. (eds.) 25 Years of Logic Programming, ch.8. LNCS, vol. 6125, pp. 159–182. Springer, Heidelberg (2010)
28. Borning, A., Maher, M., Martindale, A., Wilson, M.: Constraint hierarchies and logic programming. In: Levi, G., Martelli, M. (eds.) ICLP (1989)
29. Bruscoli, P., Levi, F., Levi, G., Meo, M.: Compilative constructive negation in constraint logic programs. In: Tison, S. (ed.) CAAP 1994. LNCS, vol. 787, pp. 52–67. Springer, Heidelberg (1994)
30. Brusoni, V., Console, L., Lamma, E., Mello, P., Milano, M., Terenziani, P.: Resource-based vs. task-based approaches for scheduling problems. In: Michalewicz, M., Raś, Z.W. (eds.) ISMIS 1996. LNCS, vol. 1079. Springer, Heidelberg (1996)
31. Buscemi, M.G., Montanari, U.: A survey of constraint-based programming paradigms. Computer Science Review 2(3), 137–141 (2008)
32. Cabeza, D., Hermenegildo, M.: Implementing distributed concurrent constraint execution in the CIAO system. In: Lucio, P., Martelli, M., Navarro, M. (eds.) APPIA-GULP-PRODE (1996)
33. Cadoli, M., Mancini, T.: Using a theorem prover for reasoning on constraint problems. In: Bandini, S., Manzoni, S. (eds.) AI*IA. Springer, Heidelberg (2005)
34. Caprara, A., Focacci, F., Lamma, E., Mello, P., Milano, M., Toth, P., Vigo, D.: Integrating constraint logic programming and operations research techniques for the crew rostering problem. Softw. Pract. Exper. 28(1), 49–76 (1998)
35. Carlsson, M., Widen, J.: SICStus Prolog User's Manual. Technical report, Swedish Institute of Computer Science (SICS) (1999)
36. Carraresi, P., Gallo, G., Rago, G.: A hypergraph model for constraint logic programming and applications to bus drivers' scheduling. AMAI 8(3-4) (1993)
37. Cheadle, A., Harvey, W., Sadler, A., Schimpf, J., Shen, K., Wallace, M.: ECLiPSe: a tutorial introduction (2003), http://eclipse-clp.org/doc/tutorial
38. Cipriano, R., Di Gaspero, L., Dovier, A.: Hybrid approaches for rostering: A case study in the integration of constraint programming and local search. In: Almeida, F., Blesa Aguilera, M.J., Blum, C., Moreno Vega, J.M., Pérez Pérez, M., Roli, A., Sampels, M. (eds.) HM 2006. LNCS, vol. 4030, pp. 110–123. Springer, Heidelberg (2006)
39. Cipriano, R., Di Gaspero, L., Dovier, A.: A hybrid solver for large neighborhood search: Mixing Gecode and EasyLocal^{++}. In: Sampels, M. (ed.) HM 2009. LNCS, vol. 5818, pp. 141–155. Springer, Heidelberg (2009)
40. Codognet, P., Diaz, D.: Compiling constraints in clp(fd). J. Log. Prog. (1996)

41. Colmerauer, A.: An introduction to Prolog-III. Communication of the ACM (1990)
42. Colmerauer, A.: Prolog II reference manual and theoretical model. Technical report, Groupe Intelligence Artificielle, Universitè Aix-Mareseille II (October 1982)
43. Colussi, L., Marchiori, E., Marchiori, M.: A dataflow semantics for constraint logic programs. In: Alpuente, Sessa [5], pp. 557–568
44. Corin, R., Etalle, S.: An improved constraint-based system for the verification of security protocols. In: Hermenegildo, M.V., Puebla, G. (eds.) SAS 2002. LNCS, vol. 2477, pp. 326–341. Springer, Heidelberg (2002)
45. Cucchiara, R., Gavanelli, M., Lamma, E., Mello, P., Milano, M., Piccardi, M.: Extending CLP(FD) with interactive data acquisition for 3D visual object recognition. In: Proc. PACLP 1999, pp. 137–155 (1999)
46. Cucchiara, R., Gavanelli, M., Lamma, E., Mello, P., Milano, M., Piccardi, M.: From eager to lazy constrained data acquisition: A general framework. New Generation Computing 19(4), 339–367 (2001)
47. Dahl, V., Niemelä, I. (eds.): ICLP 2007. LNCS, vol. 4670. Springer, Heidelberg (2007)
48. Dal Palù, A., Dovier, A., Fogolari, F.: Constraint logic programming approach to protein structure prediction. BMC Bioinformatics 5 (2004)
49. Dal Palù, A., Dovier, A., Pontelli, E.: Computing approximate solutions of the protein structure determination problem using global constraints on discrete crystal lattices. Int'l Journal of Data Mining and Bioinformatics 4(1) (January 2010)
50. Dal Palù, A., Dovier, A., Pontelli, E., Rossi, G.: Integrating finite domain constraints and CLP with sets. In: PPDP 2003, pp. 219–229. ACM, New York (2003)
51. Dal Palù, A., Torroni, P.: 25 Years of Applications of Logic Programming. In: Dovier, Pontelli [68], vol. 6125, ch.14, pp. 298–325 (2010)
52. de Boer, F.S., Di Pierro, A., Palamidessi, C.: An algebraic perspective of constraint logic programming. Journal of Logic and Computation 7(1), 1–38 (1997)
53. de Boer, F.S., Gabbrielli, M.: Infinite computations in concurrent constraint programming. Electr. Notes Theor. Comput. Sci. 6 (1997)
54. de Boer, F.S., Gabbrielli, M., Meo, M.C.: A timed concurrent constraint language. Inf. Comput. 161(1), 45–83 (2000)
55. De Maria, E., Dovier, A., Montanari, A., Piazza, C.: Exploiting model checking in constraint-based approaches to the protein folding. In: WCB 2006 (2006)
56. Dechter, R.: Constraint Processing. Morgan Kaufmann, San Francisco (2003)
57. Delzanno, G., Etalle, S.: Proof theory, transformations, and logic programming for debugging security protocols. In: Pettorossi, A. (ed.) LOPSTR 2001. LNCS, vol. 2372, p. 76. Springer, Heidelberg (2002)
58. Delzanno, G., Gabbrielli, M., Meo, M.: A compositional semantics for CHR. In: PPDP 2005, pp. 209–217. ACM, New York (2005)
59. Delzanno, G., Giacobazzi, R., Ranzato, F.: Analysis, Abstract Interpretation, and Verification in (Constraint Logic) Programming. In: Dovier, Pontelli [68], vol. 6125, ch. 7, pp. 136–158 (2010)
60. Díaz, D., Codognet, P.: GNU Prolog: Beyond compiling Prolog to C. In: Pontelli, E., Santos Costa, V. (eds.) PADL 2000. LNCS, vol. 1753, p. 81. Springer, Heidelberg (2000)
61. Dondossola, G., Ratto, E.: GRF temporal reasoning language. Technical report, CISE, Milano (1993)
62. Dovier, A., Formisano, A., Pontelli, E.: Multivalued action languages with constraints in CLP(FD). In: Dahl, Niemelä [47], pp. 255–270

63. Dovier, A., Formisano, A., Pontelli, E.: An empirical study of constraint logic programming and answer set programming solutions of combinatorial problems. J. Exp. Theor. Artif. Intell. 21(2) (2009)
64. Dovier, A., Omodeo, E., Pontelli, E., Rossi, G.: {log}: A logic programming language with finite sets. In: ICLP, pp. 111–124 (1991)
65. Dovier, A., Piazza, C., Pontelli, E., Rossi, G.: Sets and constraint logic programming. ACM Trans. Program. Lang. Syst. 22(5), 861–931 (2000)
66. Dovier, A., Piazza, C., Rossi, G.: A uniform approach to constraint-solving for lists, multisets, compact lists, and sets. ACM Trans. Comput. Log. 9(3) (2008)
67. Dovier, A., Policriti, A., Rossi, G.: A uniform axiomatic view of lists, multisets, and sets, and the relevant unification algorithms. Fundam. Inform. 36(2-3) (1998)
68. Dovier, A., Pontelli, E. (eds.): 25 Years of Logic Programming. LNCS, vol. 6125. Springer, Heidelberg (2010)
69. Endriss, U., Mancarella, P., Sadri, F., Terreni, G., Toni, F.: The CIFF proof procedure for abductive logic programming with constraints. In: Alferes, J.J., Leite, J. (eds.) JELIA 2004. LNCS (LNAI), vol. 3229, pp. 31–43. Springer, Heidelberg (2004)
70. Eremin, A., Wallace, M.: Hybrid Benders decomposition algorithms in constraint logic programming. In: Walsh [138], pp. 1–15
71. Etalle, S., Gabbrielli, M., Meo, M.: Transformations of CCP programs. ACM Trans. Program. Lang. Syst. 23(3), 304–395 (2001)
72. Fahle, T., Schamberger, S., Sellman, M.: Symmetry breaking. In: Walsh [138]
73. Falaschi, M., Gabbrielli, M., Marriott, K., Palamidessi, C.: Confluence in concurrent constraint programming. Theor. Comput. Sci. 183(2), 281–315 (1997)
74. Falaschi, M., Gabbrielli, M., Marriott, K., Palamidessi, C.: Constraint logic programming with dynamic scheduling: A semantics based on closure operators. Information and Computation 137(1), 41–67 (1997)
75. Farenzena, M., Fusiello, A., Dovier, A.: Reconstruction with interval constraints propagation. In: CVPR, pp. 1185–1190. IEEE Computer Society, Los Alamitos (2006)
76. Festa, G., Sardu, G., Felici, R.: A decision support system for the Venice lagoon. In: Herold, A. (ed.) Handbook of parallel constraint logic programming applications (1995)
77. Fioravanti, F., Pettorossi, A., Proietti, M.: Transformation rules for locally stratified constraint logic programs. In: Bruynooghe, M., Lau, K.-K. (eds.) Program Development in Computational Logic. LNCS, vol. 3049, pp. 291–339. Springer, Heidelberg (2004)
78. Focacci, F., Laburthe, F., Lodi, A.: Local search and constraint programming: LS and CP illustrated on a transportation problem. In: Milano, M. (ed.) Constraint and Integer Programming. Towards a Unified Methodology, pp. 137–167. Kluwer Academic Publishers, Dordrecht (2003)
79. Focacci, F., Milano, M.: Global cut framework for removing symmetries. In: Walsh [138], pp. 77–92
80. Focacci, F., Milano, M., Lodi, A.: Soving TSP with time windows with constraints. In: International Conference on Logic Programming, pp. 515–529 (1999)
81. Frisch, A., Hnich, B., Kızıltan, Z., Miguel, I., Walsh, T.: Propagation algorithms for lexicographic ordering constraints. Artif. Int. 170(10), 803–834 (2006)
82. Frühwirth, T.: Theory and practice of constraint handling rules. Journal of Logic Programming 37, 95–138 (1998)
83. Frühwirth, T., Di Pierro, A., Wiklicky, H.: An implementation of probabilistic constraint handling rules. In: Comini, M., Falaschi, M. (eds.) WFLP (2002)

84. Gabbrielli, M., Dore, G.M., Levi, G.: Observable semantics for constraint logic programs. J. Log. Comput. 5(2), 133–171 (1995)
85. Gabbrielli, M., Palamidessi, C., Valencia, F.D.: Concurrent and Reactive Constraint Programming. In: Dovier, Pontelli [68], vol. 6125, ch. 11, pp. 225–248 (2010)
86. Gavanelli, M.: University timetabling in ECLiPSe. ALP Newsletter 19(3) (2006)
87. Gavanelli, M., Alberti, M., Lamma, E.: Integration of abductive reasoning and constraint optimization in SCIFF. In: Hill, P.M., Warren, D.S. (eds.) ICLP 2009. LNCS, vol. 5649, pp. 387–401. Springer, Heidelberg (2009)
88. Gavanelli, M., Lamma, E., Mello, P., Milano, M.: Dealing with incomplete knowledge on CLP(FD) variable domains. ACM TOPLAS 27(2) (2005)
89. Gavanelli, M., Lamma, E., Mello, P., Torroni, P.: An abductive framework for information exchange in multi-agent systems. In: Dix, J., Leite, J. (eds.) CLIMA 2004. LNCS (LNAI), vol. 3259, pp. 34–52. Springer, Heidelberg (2004)
90. Gavanelli, M., Milano, M.: Cost-based filtering for determining the Pareto frontier. In: Junker, U., Kießling, W. (eds.) Multidisciplinary Workshop on Advances in Preference Handling, in conjunction with ECAI 2006 (2006)
91. Gennari, R., Mich, O.: Constraint-based temporal reasoning for e-learning with LODE. In: Bessiere [21]
92. Gent, I.P., Smith, B.M.: Symmetry breaking in constraint programming. In: Horn, W. (ed.) ECAI, pp. 599–603. IOS Press, Amsterdam (2000)
93. Georget, Y., Codognet, P.: Compiling semiring-based constraints with clp(fd,s). In: Maher, M.J., Puget, J.-F. (eds.) CP 1998. LNCS, vol. 1520, p. 205. Springer, Heidelberg (1998)
94. Giacobazzi, R., Debray, S., Levi, G.: Generalized semantics and abstract interpretation for constraint logic programs. J. Log. Program. 25(3) (1995)
95. Giordano, L., Toni, F.: Knowledge representation and non-monotonic reasoning. In: Dovier, Pontelli [68], vol. 6125, ch. 5, pp. 86–110 (2010)
96. Gualandi, S., Malucelli, F.: Constraint programming-based column generation. 4OR: A Quarterly Journal of Operations Research 7(2), 113–137 (2009)
97. Van Hoeve, W.J., Pesant, G., Rousseau, L.-M.: On global warming: Flow-based soft global constraints. Journal of Heuristics 12(4-5), 347–373 (2006)
98. Hooker, J.: Logic-Based Methods for Optimization: Combining Optimization and Constraint Satisfaction. John Wiley & Sons, Chichester (2000)
99. Jaffar, J., Lassez, J.-L.: Constraint logic programming. In: Proc. 14th symp. on Principles of programming languages. ACM, New York (1987)
100. Jaffar, J., Michaylov, S., Stuckey, P., Yap, R.: The CLP(R) Language and System. ACM Transactions on Programming Languages and Systems (1992)
101. Kakas, A.C., Kowalski, R.A., Toni, F.: Abductive Logic Programming. Journal of Logic and Computation 2(6), 719–770 (1993)
102. Kakas, A.C., Mancarella, P.: On the relation between Truth Maintenance and Abduction. In: Fukumura, T. (ed.) PRICAI (1990)
103. Kakas, A.C., Michael, A., Mourlas, C.: ACLP: Abductive Constraint Logic Programming. Journal of Logic Programming 44(1-3), 129–177 (2000)
104. Kakas, A.C., van Nuffelen, B., Denecker, M.: A-System: Problem solving through abduction. In: Nebel, B. (ed.) Proc. of IJCAI 2001, pp. 591–596 (2001)
105. Mackworth, A.: Consistency in networks of relations. Artif. Intell. 8(1) (1977)
106. Majkic, Z.: Constraint logic programming and logic modality for event's valid-time approximation. In: 2nd Indian Int. Conf. on Artificial Intelligence (2005)
107. Mancarella, P., Terreni, G., Toni, F.: Web sites verification: An abductive logic programming tool. In: Dahl, Niemelä [47]

108. Mancini, T., Cadoli, M.: Detecting and breaking symmetries by reasoning on problem specifications. In: Zucker, J.-D., Saitta, L. (eds.) SARA 2005. LNCS (LNAI), vol. 3607, pp. 165–181. Springer, Heidelberg (2005)

109. Mancini, T., Micaletto, D., Patrizi, F., Cadoli, M.: Evaluating ASP and commercial solvers on the CSPLib. Constraints 13(4), 407–436 (2008)

110. Manco, G., Turini, F.: A structural (meta-logical) semantics for linear objects. In: Alpuente, Sessa [5], pp. 421–434

111. Mascardi, V., Merelli, E.: Agent-oriented and constraint technologies for distributed transaction management. In: Parenti, R., Masulli, F. (eds.) Proc. Int. ICSC Symposia IIA 1999 and SOCO 1999 (1999)

112. Merelli, E., De Leone, R., Martelli, M., Panti, M.: Embedding constraint logic programming formula in a local search algorithm for job shop scheduling. In: EURO XVI, Bruxelles (July 1998)

113. Meseguer, P., Rossi, F., Schiex, T.: Soft constraints. In: Rossi, et al [127]

114. Meseguer, P., Torras, C.: Exploiting symmetries within constraint satisfaction search. Artificial Intelligence 129(1-2), 133–163 (2001)

115. Mesnard, F., Ruggieri, S.: On proving left termination of constraint logic programs. ACM Trans. Comput. Log. 4(2) (2003)

116. Mohr, R., Masini, G.: Good old discrete relaxation. In: ECAI (1988)

117. Montanari, U.: Networks of constraints: Fundamental properties and applications to picture processing. Information Science 7, 95–132 (1974)

118. Montesi, D., Bertino, E., Martelli, M.: Transactions and updates in deductive databases. IEEE Trans. Knowledge and Data Engineering 9(5), 784–797 (1997)

119. Nethercote, N., Stuckey, P., Becket, R., Brand, S., Duck, G., Tack, G.: MiniZinc: Towards a standard CP modelling language. In: Bessiere [21], pp. 529–543

120. Novello, S.: ECL^{i}PS^e examples (1998),
http://eclipse-clp.org/examples/golf.ecl.txt

121. Orlandini, A.: Model-based rescue robot control with ECLiPSe framework. In: Oddi, A., Cesta, A., Fages, F., Policella, N., Rossi, F. (eds.) CSCLP (2008)

122. Pettorossi, A., Proietti, M., Senni, V.: The Transformational Approach to Program Development. In: Dovier, Pontelli [68], vol. 6125, ch. 6, pp. 111–135 (2010)

123. Pierro, A.D., Wiklicky, H.: An operational semantics for probabilistic concurrent constraint programming. In: ICCL, pp. 174–183 (1998)

124. Provetti, A., Rossi, G.: Action specifications in {log}. In: Falaschi, M., Navarro, M., Policriti, A. (eds.) APPIA-GULP-PRODE (1997)

125. Raffaetà, A., Ceccarelli, T., Centeno, D., Giannotti, F., Massolo, A., Parent, C., Renso, C., Spaccapietra, S., Turini, F.: An application of advanced spatio-temporal formalisms to behavioural ecology. Geoinformatica 12(1), 37–72 (2008)

126. Raffaetà, A., Frühwirth, T.W.: Spatio-temporal annotated constraint logic programming. In: Ramakrishnan, I.V. (ed.) PADL 2001. LNCS, vol. 1990, pp. 259–273. Springer, Heidelberg (2001)

127. Rossi, F., van Beek, P., Walsh, T. (eds.): Handbook of Constraint Programming. Elsevier, Amsterdam (2006)

128. Rossi, G., Panegai, E., Poleo, E.: JSetL: a Java library for supporting declarative programming in Java. Softw. Pract. Exper. 37(2), 115–149 (2007)

129. Ruttkay, Z.: Fuzzy constraint satisfaction. In: FUZZ-IEEE 1994, Orlando, FL (1994)

130. Saraswat, V.A.: Concurrent Constraint Programming. MIT Press, Cambridge (2003)

131. Schaerf, A.: Scheduling sport tournaments using constraint logic programming. Constraints 4(1), 43–65 (1999)

132. Schaerf, A.: A survey of automated timetabling. Artif. Intell. Review 13(2) (1999)
133. Schiex, T., Fargier, H., Verfaillie, G.: Valued constraint satisfaction problems: hard and easy problems. In: IJCAI 1995, pp. 631–637 (1995)
134. Schulte, C., Stuckey, P.: Efficient constraint propagation engines. In: ToPLaS 2008 (2008)
135. Van Hentenryck, P.: Constraint Satisfaction in Logic Programming. MIT, Cambridge (1989)
136. van Hoeve, W.-J.: The all different constraint: a survey. In: Sixth Annual Workshop of the ERCIM Working Group on Constraints (2001)
137. Wallace, M.: Practical applications of constraint programming. Constraints (1996)
138. Walsh, T. (ed.): CP 2001. LNCS, vol. 2239. Springer, Heidelberg (2001)
139. Wielemaker, J., Huang, Z., Van der Meij, L.: SWI-Prolog and the web. Theory and Practice of Logic Programming 8(3), 363–392 (2008)
140. Zanarini, A., Milano, M., Pesant, G.: Improved algorithm for the soft global cardinality constraint. In: Beck, J.C., Smith, B.M. (eds.) CPAIOR 2006. LNCS, vol. 3990, pp. 288–299. Springer, Heidelberg (2006)
141. Zhou, N.-F.: Programming finite-domain constraint propagators in action rules. Theory and Practice of Logic Programming 6(5), 483–507 (2006)

Knowledge Representation and Non-monotonic Reasoning

Laura Giordano[1] and Francesca Toni[2]

[1] Università del Piemonte Orientale, Italy
laura@mfn.unipmn.it
[2] Imperial College London, UK
ft@imperial.ac.uk

Abstract. Logic programming has been deployed to support non-monotonic reasoning since the late '80s. In this paper, we review semantics, formalisms and computational mechanisms for logic programming for non-monotonic reasoning. We also discuss some formalisms that have emerged from the cross fertilization between the two areas and some applications in as diverse areas as reasoning about dynamic domains, security, diagnosis and legal reasoning.

1 Introduction

Since the beginning of the '80s, when non-monotonic logics first came into existence [114,93,94,96], members of the AI community have started their investigation in the field of commonsense reasoning. Modelling commonsense reasoning requires the ability to jump to conclusions in the presence of incomplete knowledge as well as the ability to revise knowledge to deal with new, possibly conflicting evidence. Non-monotonic logics exhibit these abilities and allow to model non-monotonic reasoning (NMR). The topic soon aroused the interest of the logic programming (LP) community. Indeed, since its onset in the '70s [79], LP was meant as a paradigm which combines the use of logic for knowledge representation with efficient goal directed proof procedures, and the construct of negation as failure (NAF) [27], which was originally introduced as a procedural feature of Prolog, is inherently non-monotonic. This has given rise, from the end of the '80s, to a challenging research activity on the semantics for NAF and, more generally, on NMR in LP.

The development of new logical semantics for capturing the non-monotonic features of logic programs, and, in particular, to provide a logical characterization of NAF, has resulted in the cross fertilization between the areas of NMR and LP. On the one hand, the features of non-monotonic logics have been adapted and tuned to the need of LP giving rise to new logical semantics and extensions of LP. On the other hand, the non-monotonic solutions proposed in the context of LP have led to the definition of new logics for NMR (notably argumentation theory [19,44,18] and inductive definitions [37]).

The aim of this paper is to provide a short survey of the work in the field of NMR and LP, with a special attention to the use of NMR for knowledge

A. Dovier, E. Pontelli (Eds.): 25 Years of Logic Programming, LNCS 6125, pp. 87–111, 2010.

representation and with focus on putting the contribution by the Italian LP community and by Italian researchers abroad within the context of this field.

The paper is organized as follows. In Section 2 we give an overview of semantics for LP seen as a mechanism for NMR. In Section 3 we discuss some extensions of LP which have stemmed from the understanding of LP as a mechanism for NMR. In Section 4 we discuss applications and in Section 5 we conclude.

2 LP and Non-monotonicity: Semantics for NAF

NAF has been introduced in LP as an extra-logical feature which allows to capture non-derivability. It has soon become the main tool for performing NMR in LP and a lot of efforts has been devoted to the specification of a declarative logical semantics for it. This section is devoted to providing an outline of the main approaches to the definition of a semantics of logic programs with NAF that have been presented in the literature.

General logic programs extend the formalism of *definite clauses* by allowing the new connective *not*, NAF, to occur in the body of clauses. A clause of a general logic program has the form:

$$A_0 \leftarrow A_1 \wedge \ldots \wedge A_n \wedge not\, B_{n+1} \wedge \ldots \wedge not\, B_{n+m} \qquad (n \geq 0, m \geq 0)$$

where the A_j's and B_i's are atomic formulas, and all variables in them are (implicitly) universally quantified with scope the whole clause. If $n = 0 = m$, the body of the clause corresponds to the special atom *true*. A_0 is referred to as the head of the clause.

To deal with NAF, SLD resolution [80] has been extended to SLDNF resolution [89]. From the operational point of view (assuming B_i is ground), a goal *not* B_i succeeds when the goal B_i fails finitely. Conversely, the goal *not* B_i fails finitely when the goal B_i succeeds. To provide a formulation of SLDNF in the propositional case, let us introduce the following abstract syntax for goals G and for clauses C:

$$G ::= true \mid A \mid not\, A \mid G_1 \wedge G_2$$
$$C ::= A \leftarrow G$$

where the body of a clause is defined to be a goal, that is, a conjunction of literals (atoms or negated atoms). We introduce two relations, $\vdash_t$ to denote success and $\vdash_f$ to denote finite failure. The operational semantics is defined by the following rules, which determine when a goal G *succeeds* ($\vdash_t G$) or *finitely fails* ($\vdash_f G$) from a (general logic) program P:

$\vdash_t true$
$\vdash_t A$ if there is a clause $A \leftarrow G \in P$ such that $\vdash_t G$
$\vdash_t not\, A$ if $\vdash_f A$
$\vdash_t G_1 \wedge G_2$ if $\vdash_t G_1$ and $\vdash_t G_2$

$\vdash_f A$ if, for all clauses $A \leftarrow G \in P, \vdash_f G$
$\vdash_f not\, A$ if $\vdash_t A$
$\vdash_f G_1 \wedge G_2$ if $\vdash_f G_1$ or $\vdash_f G_2$

When the above operational semantics is extended to the first order case, suitable conditions have to be introduced to guarantee that when a negative goal is selected all its free variables are groundly instantiated. This is needed to guarantee the soundness and completeness of the proof procedure. Let us consider P_1 with the following clauses:

$flies(X) \leftarrow bird(X) \wedge not\ abnormal\_bird(X)$
$abnormal\_bird(Y) \leftarrow penguin(Y)$
$abnormal\_bird(Y) \leftarrow ostrich(Y)$
$bird(Z) \leftarrow penguin(Z)$
$bird(tweety)$
$penguin(opus)$

The goal $flies(tweety)$ succeeds from P_1 ($\vdash_t flies(tweety)$) while the goal $flies(opus)$ fails finitely ($\vdash_f flies(opus)$). Observe, however, that the proof of the goal $not\ abnormal\_bird(X)$ (where X is implicitly existentially quantified) would *flounder*, as X is not ground. To avoid *floundering* situations, a notion of *allowedness* has been defined for programs and goals [83]. In particular, the allowedness condition on clauses requires that each variable occurring anywhere in a clause occurs in at least one positive literal in its body. This condition is satisfied by program P_1 above.

Starting from this operational behavior and in order to fulfil the requirements of knowledge representation, the LP community has developed different semantics for capturing NAF, but also alternative proof procedure for LP with NAF, and, more generally, alternative ways for modelling non-monotonicity in LP. The first way of providing a logical understanding for NAF is due to Clark [27], who introduced a construction, known as *Clark's completion*, which is based on the idea of interpreting the program as a set of sufficient and necessary conditions rather than simply as a set of *sufficient* conditions. More precisely, each predicate in a program is taken to be completely defined by the set of all clauses where it occurs in the head, and to be false in all other cases. For instance, the Clark's completion of the earlier program P_1 is:

$(\forall X) flies(X) \leftrightarrow bird(X) \wedge not\ abnormal\_bird(X)$
$(\forall Y) abnormal\_bird(Y) \leftrightarrow penguin(Y) \vee ostrich(Y)$
$(\forall Z) bird(Z) \leftrightarrow Z = tweety \vee penguin(Z)$
$(\forall W) penguin(W) \leftrightarrow W = opus$
$(\forall S) ostrich(S) \leftrightarrow false$

In classical logic, from the above completion, we can conclude:

$penguin(opus),\ bird(opus),\ abnormal\_bird(opus),$
$bird(tweety),\ not\ penguin(tweety),\ not\ abnormal\_bird(tweety),$
$not\ flies(opus),\ flies(tweety).$

Clark's completion is strongly related to McCarthy's Circumscription [93], as proven in [115]. One problem of Clark's completion is that in general the completion of a program is not guaranteed to be consistent. As shown by Shepherdson [124,125], Clark's completion, as well as Reiter's Closed World Assumption

(CWA) [113], cannot provide an exact characterization of NAF in SDLNF. However, as we will see below, the idea of defining a completion of the program has had a big impact on the field and has been proved to be crucial for the definition of other approaches to non-monotonicity in LP.

One of the approaches to define a semantics for NAF is based on the idea of putting some syntactic restrictions on the logic program. Chandra and Harel [23] introduced a notion of *stratification*, which was further generalized by Przymusinski, by defining a notion of *local stratification* and the notion of *perfect model* [111]. The idea of stratification is that of partitioning the predicate symbols of the program into different strata in such a way that the definition of each predicate can depend positively on predicates defined in lower strata or in the same stratum, while it can depend negatively only on predicates defined in lower strata.

Another direction in the definition of a semantics for NAF is based on the idea of using a three-valued logic rather than a two-valued logic. Fitting [50] and Kunen [82] have introduced a three-valued semantics for LP with NAF based on Kleene's strong three-valued logic. They showed that the least three-valued model of the program completion can be computed as the fix-point of a certain operator. In this approach, loops causing non-terminating computations are captured by means of the truth-value *undefined*. For instance, given the program P_2 with clauses

$$a \leftarrow a$$
$$b \leftarrow not\ b$$
$$c \leftarrow true$$
$$c \leftarrow not\ d$$

this semantics assigns the truth-value *true* to c, the truth-value *false* to d and the truth-value *undefined* to both a and b, as both the proofs of goal a and of goal b produce non-terminating computations.

A different three-valued semantics for logic programs, the *well-founded semantics*, has been defined by Van Gelder, Ross and Schlifp [56,129]. Differently from the semantics of Fitting and Kunen, which captures a notion of finite failure, the well-founded semantics captures a notion of infinite failure. For instance, given the program P consisting solely of clause $a \leftarrow a$, differently from Fitting's and Kunen's semantics, the well-founded semantics assigns *false* to a, as the proof of goal a infinitely fails.

Assume that P is the Herbrand instantiation for a general logic program. The definition of the well-founded semantics is based on the notion of unfounded set. A set of atoms A is an *unfounded set* of a program P with respect to a (partial) interpretation I (i.e. a set of positive and negative literals) if, informally, for each atom $p \in A$, no clause in P can be used to support p, given what is already known (I). More precisely, for each atom $p \in A$, each clause for p either contains a literal in the body which is false in I, or [unfounded condition] it contains a positive literal in the body which is contained in A. The unfounded condition says that "of all the rules that might still be usable to derive something in the set A, each requires an atom to be true" [129]. By this condition, atoms (say a, b, c) among which there is a circular positive dependency (for instance, a depends positively on b, b depends positively on c and c depends positively on a) and that do not have any alternative independent support, are taken to belong to the unfounded set.

The well-founded model is defined by taking the least fix-point of the operator $W_P(I) = T_P(I) \cup U_P(I)$, where $T_P(I)$ is the *immediate consequence operator* and $U_P(I)$ is the greatest unfounded set of P wrt I. As the operator $W_P(I)$ is monotone with respect to subset inclusion, the least fix-point can be computed by transfinite iteration. At each iteration step, the atoms in $T_P(I)$ are taken to be true, while the atoms in $U_P(I)$ are taken to be false. The value of the remaining atoms is *unknown*. As an example, given the program P_2 above, the well-founded semantics would assign the truth-values *true* to c, *false* to d and a and *undefined* to b. Van Gelder et al [129] establish a precise relationship between the well-founded semantics and Fitting/Kunen three-valued models of the completed program. They also prove that, for locally stratified programs, the well-founded model coincides with the perfect model. An efficient implementation of the well-founded semantics is provided by the XSB system[1], which makes use of tabling. Tabling allows non-floundering datalog (function-free) programs with NAF to terminate with polynomial data complexity.

Fitting/Kunen semantics as well as the well-founded semantics are defined for any (general) logic program and have polynomial data complexity. However, they only allow skeptical forms of reasoning. In 1988, Gelfond and Lifschitz proposed a new declarative semantics for LP with NAF [57]: the stable model semantics (SMS). Although very simple, SMS is applicable also to programs which are not stratified (and not even locally stratified). According to the SMS, a program may have many different models, in the same way that a default theory may have many different extensions [114]. To define stable models, Gelfond and Lifschitz define a suitable transformation P^I of a program P, given an interpretation I (a subset of the Herbrand base of P). We assume that each clause of P containing variables is replaced by all its ground instances, so that all atoms in P are ground. The transformed program P^I is defined as the program obtained from P by deleting: (a) each clause that has a literal *not A* in its body, with $A \in I$; (b) all the literals *not A* in the bodies of the remaining clauses. The program P^I is a definite set of clauses and has a unique minimal model $\Gamma(P^I)$. I is defined to be a *stable model* of program P if $I = \Gamma(P^I)$.

Let us consider again the earlier program P_1 and let P be the ground program containing all the ground instances of the clauses in P_1. The interpretation $I = \{penguin(opus), bird(opus), abnormal\_bird(opus), bird(tweety), flies(tweety)\}$ is a stable model of P. Observe that the transformed program P^I contains the clause $flies(tweety) \leftarrow bird(tweety)$, as $abnormal\_ bird(tweety) \notin I$, while P^I does not contain the clause $flies(opus) \leftarrow bird(opus)$, as $abnormal\_bird(opus) \in I$.

A general program may have more than one stable model or no stable model at all. For instance, the program P_3 with clauses

---

[1] http://xsb.sourceforge.net/

$p \leftarrow not\ q$
$q \leftarrow not\ p$

has two extensions, $I_1 = \{p\}$ and $I_2 = \{q\}$, while P_4 consisting solely of clause $p \leftarrow not\ p$ has no stable model. However, Gelfond and Lifschitz showed [57] that any locally stratified program has a unique stable model, which is identical to its perfect model.

The stable model semantics has strong relations with default logic and autoepistemic logic. In particular, [57] relates stable models to the translation of logic programs into autoepistemic theories [96]. Concerning the relationships between the stable model semantics and the well-founded semantics, [129] shows that if a program P has a well-founded total model (i.e. a model in which no atom is undefined), then that model is the unique stable model of P. Moreover, the well-founded (partial) model of a program P is a subset of every stable model of P. In [112] an extended stable model semantics has been proposed, based on a three-valued semantics, where the well-founded model is the least extended stable model. In [31], Costantini has provided a characterization of stable models in terms of their difference with the WFM, while in [32] she has defined a syntactic characterization of the class of logic programs for which a stable model exists.

The stable model semantics has been widely studied and successfully extended to a wider class of programs, including programs with explicit negation as well as disjunction in the head of clauses, thus leading to the development of *Answer Set Programming* (ASP). A lot of work has been devoted to the development of efficient techniques for computing stable models, e.g. [24,20,98]. We refer to Section 3 for a short description of ASP and to [14] for a detailed treatment of the subject. Concerning the study of the stable models in the first-order case, we mention the work in [16], where a class of normal logic programs is studied whose consequences under the stable model semantics can be effectively computed, despite the fact that they admit function symbols (hence, infinite domains) and recursion.

Another line of research, initiated by Eshghi and Kowalski [49], is based on the recognition of the close relationship between NAF and abduction and the possibility to give an abductive semantics to LP with NAF. In the abductive approach, negative literals are regarded as abducibles or assumptions (see Section 3 for details on these notions). Eshghi and Kowalski have defined an abductive proof procedure which extends SLDNF resolution. In addition to the usual yes/no answers of SLDNF, the abductive procedure also provides an abductive explanation. In this way, alternative abductive explanations may be feasible for a given query. Dung [46] has proven the correctness of the Eshghi and Kowalski's abductive proof procedure with respect to the preferred extension semantics for LP with NAF [46]. This is equivalent [75] to the partial stable model semantics of [118]. The seminal work of Eshghi and Kowalski has given rise to the development of the area of Abductive Logic Programming (ALP), discussed in Section 3.

Due to space limitations, we are unable to make a comprehensive survey of all the relevant work on LP and NAF. We conclude this section by mentioning the autoepistemic extension of LP in [15] , the logic MBNF introduced by Lifschitz [87], the work on NAF in LP with hypothetical implications [64], the work on ordered LP in [84], and, also, all the work on disjunctive LP, for which we refer to the survey in [95].

3   Beyond NAF: Non-monotonic LP Extensions

Starting from the study of the semantics of NAF, new LP languages and frameworks have been developed, explicitly focused on the task of knowledge representation and NMR. We have already mentioned above how the work on the stable model semantics has led to the development of ASP, which constitutes one of the main current trends of the work in LP and NMR (see [14]). Basically, ASP relies upon

- the representation of knowledge in terms of logic programs with NAF (and possibly other features, e.g. explicit negation, disjunction, constraints etc);
- the interpretation of these logic programs under the stable model semantics [57] and its extensions (to deal with explicit negation, disjunction, constraints, etc);
- efficient computational mechanisms (ASP solvers) to compute stable models for propositional logic programs, typically based upon SAT solvers, and efficient "grounders" to turn non-propositional logic programs into propositional ones.

In a nutshell, in order to solve a problem, an ASP programmer designs a (possibly extended) logic program so that stable models of the program correspond to desired solutions to the problem. ASP solvers can then be used to compute these solutions. A simple example is the 3-graph-colouring problem, which can be modelled using the ASP program

$$clrd(V, 1) \leftarrow not\ clrd(V, 2) \wedge not\ clrd(V, 3) \wedge vtx(V)$$
$$clrd(V, 2) \leftarrow not\ clrd(V, 1) \wedge not\ clrd(V, 3) \wedge vtx(V)$$
$$clrd(V, 3) \leftarrow not\ clrd(V, 1) \wedge not\ clrd(V, 2) \wedge vtx(V)$$
$$\leftarrow edge(V, U) \wedge clrd(V, C) \wedge clrd(U, C)$$

Here, the last clause can be seen as a clause defining $false$, where $false$ can never hold in any stable model. It is an example of *denial integrity constraint*, further discussed below. Each answer set of this ASP program gives a valid coloring for the graph defined by the vtx and $edge$ predicates. For example, assume that the ASP program is extended by a simple graph description $vtx(a)$, $vtx(b)$, $vtx(c)$, $edge(a, b)$, $edge(a, c)$, then the colouring $\{clrd(a, 1), clrd(b, 2), clrd(c, 2)\}$ can be obtained from an answer set. In order to detect this colouring, the full answer set needs to be computed.

ALP [72,73,39] is a methodology for knowledge representation and reasoning, emerged in the late '80s, which relies upon

- the representation of knowledge in terms of abductive logic programs, which consist of logic programs (possibly extended with NAF, explicit negation, constraints etc) and *abducibles*, namely atoms representing information about which there is incomplete (possibly no) knowledge;
- the interpretation of these logic programs with abducibles under abductive extensions of some LP semantics (catering for NAF, explicit negation, constraints etc, if these are features of the abductive logic programs);
- abductive proof procedures to compute abductive answers to queries, given the abductive logic programs, typically based upon extensions of SLD resolution.

Integrity constraints are often also included in abductive logic programs, to provide partial information about the abducibles. These may be in the form of *denials* (as for the earlier ASP example) or implications, of the form $L_1 \wedge \ldots \wedge L_m \rightarrow A$, where the L_i's are literals (possibly *true*), A is an atom (possibly *false*, in which case we have a denial), and all variables are implicitly universally quantified with scope the whole implication.

Intuitively, a logic program provides definitions for certain predicates, while abducibles can be used to extend these definitions to form possible explanations for queries, which can be regarded as observations against the background of the world knowledge encoded in the given abductive logic program. Integrity constraints, on the other hand, restrict the range of possible explanations. Informally, given an abductive logic program $\langle P; A; IC \rangle$ and a *query* (i.e. implicitly existentially quantified conjunction of literals) Q, an *explanation* for Q is a set of abducible atoms Δ that, together with P, both "entails" (an appropriate instantiation of) Q, with respect to some notion of "entailment", and "satisfies" the set of integrity constraints IC (see [72,73] for possible notions of integrity constraint "satisfaction").

The notion of "entailment" depends on the semantics associated with the logic program P (as we have seen in Section 2, there are many different possible choices for such semantics). For example, Eshghi and Kowalski [49] use the preferred extension/partial stable model semantics of [46,118], Kakas and Mancarella [74] use the stable model semantics, and Console et al. [29], Fung and Kowalski [51], Denecker and De Schreye [41], Alberti et al. [1] and Mancarella et al. [90] all use the completion semantics [27] or some of its variant (as discussed in Section 2). For example, Kunen's three-valued completion semantics [82] is used in [51,90].

In a nutshell, in order to solve a problem, an ALP programmer designs an abductive logic program and a query so that the explanations for the query correspond to desired solutions to the problem. Abductive proof procedures can then be used to compute these solutions.

A simple example is the following abductive logic program modelling a primitive planning agent a (more sophisticated agent-based applications of ALP are mentioned in Section 4). Let P be

$has(a, X) \leftarrow buy(a, X)$
$has(a, X) \leftarrow borrow(a, X, Y) \wedge friend(Y)$
$friend(b)$

A consist of all instances of $buy(a, X)$ and $borrow(a, X, Y)$, and IC be

$$buy(a, X) \wedge not\ money(a) \rightarrow false$$

Then, given the query $has(a, r)$, for some resource r, there is only one possible explanation, namely $\{borrow(a, r, b)\}$. If P also includes $money(a)$, then there is an additional explanation for the query, namely $\{buy(a, r)\}$. Here explanations correspond to possible courses of actions for agent a, given that the query is the goal of a.

Note that, differently from ASP, ALP only focuses on "relevant" bits of the given knowledge base (abductive logic program). Indeed, in our simple example, the computation of explanations for the given goal $has(a, r)$ completely disregards the possibility that a may want to obtain other resources. Moreover, if P also contains clauses corresponding to other potential goals of a (e.g. for registering for a conference, arranging for travel, etc), the computation of explanations for the given goal $has(a, r)$ ignores the other parts of P. Finally, abductive proof procedures may compute non-ground, bound or unbound, explanations. This is a useful feature in open and partially specified environments, e.g. the web and multi-agent systems. For example, consider the "web repair" problem of [90], where P is

$$is\_node(N, T) \leftarrow node(N, T) \wedge node\_type(T)$$
$$is\_node(N, T) \leftarrow add\_node(N, T) \wedge node\_type(T)$$
$$is\_link(N1, N2) \leftarrow link(N1, N2) \wedge link\_check(N1, N2)$$
$$is\_link(N1, N2) \leftarrow add\_link(N1, N2) \wedge link\_check(N1, N2)$$
$$link\_check(N1, N2) \leftarrow is\_node(N1, \_) \wedge is\_node(N2, \_) \wedge N1 \neq N2$$
$$book\_links(B) \leftarrow is\_node(B, book) \wedge is\_node(R, review) \wedge is\_link(B, R) \wedge$$
$$is\_node(L, lib) \wedge is\_link(B, L)$$

together with some concrete definition of node-types, nodes, and links, e.g. $node\_type(lib)$, $node\_type(book)$, $node(n1,book)$, $link(n1,n3)$, $node\_type(review)$, $node(n3, review)$, IC is

$$add\_node(N, T1) \wedge node(N, T2) \rightarrow false$$
$$add\_link(N1, N2) \wedge link(N1, N2) \rightarrow false$$
$$is\_node(N, T1) \wedge is\_node(N, T2) \wedge T1 \neq T2 \rightarrow false$$
$$is\_node(B, book) \rightarrow book\_links(B)$$

and the abducibles are all atoms in the predicates add_node and add_link. CIFF [90] computes two answers to the empty query:

$$\{add\_link(n1, L), add\_node(L, lib), L \neq n3, L \neq n1\}$$

corresponding to the addition of some new node of type lib and a link from $n1$ to it, and

$$\{add\_link(n1, L), add\_node(L, lib), add\_link(n1, R),$$
$$add\_node(R, review), L \neq n3, L \neq n1, R \neq n3, R \neq n1, R \neq L\},$$

corresponding to the addition of a new review node R with appropriate links. Both answers are partially uninstantiated.

Several systems implementing abductive proof procedures exist, e.g., recently, the A-system [76], SCIFF [1], and CIFF [90]. An abductive variant of Prolog has also been implemented [26]. Several extensions of standard ALP exist, e.g. to incorporate arithmetical constraints solving [1,90] (see also [55]), and events and protocols [1]. All state-of-the-art systems can deal with non-propositional abductive logic programs (with function symbols) and non-ground queries and can compute partially instantiated abductive explanations as for the earlier example.

Argumentation [72,73,44,18] was developed, starting in the early '90s, as a computational framework to reconcile and understand common features and differences amongst most existing approaches to NMR, including NAF in LP, theorist [104], default logic [114], autoepistemic logic [96], non-monotonic modal logic [94] and circumscription [93]. Argumentation relies upon

- the representation of knowledge in terms of an argumentation framework, defining *arguments* and a binary *attack* relation between the arguments,
- the interpretation of these argumentation frameworks using a *dialectical semantics*, for example that of *admissibility*, whereby a set of arguments is admissible if it does not attack itself and it attacks every argument attacking it,
- a computational machinery for assessing the acceptability of a given set of arguments, according to the given dialectical semantics, or searching for acceptable sets of arguments containing the given set.

In its most abstract form [44], an argumentation framework simply consists of a pre-defined set of arguments and binary attack relation. Several more concrete argumentation frameworks have been defined, many extending LP, e.g. Assumption-based Argumentation (ABA) [18,45], DeLP [54], and the approach of [106]. For lack of space, below we focus on ABA, since this is the most general and a well-documented instance of abstract argumentation. In this approach, arguments are deductions (using inference rules in an underlying logic) supported by assumptions and an attack by one argument against another is a deduction by the first argument of the contrary of an assumption supporting the second argument. ABA is equipped with a computational machinery, in the form of *dispute derivations* [45], to determine the acceptability of (arguments supporting) claims. Dispute derivations determine the acceptability of given claims by building and exploring a dialectical structure of a proponent's argument for a claim, an opponent's counterarguments attacking the argument, the proponent's arguments attacking all the opponents' counterarguments, and so on. This computation style, which has its roots in SLDNF, has several advantages over other computational mechanisms for argumentation. These advantages are due mainly to the fine level of granularity afforded by interleaving the construction of arguments and determining their acceptability.

By instantiating the notion of arguments and the attack relations in the abstract argumentation framework of [44] or by instantiating rules, assumptions and contraries in ABA, different concrete non-monotonic frameworks can be constructed. For example, the instance of ABA where rules are LP rules, assumptions are NAF literals in these rules, and the contrary of a literal *not p* is *p*, corresponds to LP, with different LP semantics for NAF given by different dialectical semantics. As an example, the notion of admissibility given early corresponds to preferred extensions/partial stable models in LP. ALP can also be obtained as an instance of ABA, by also including abducibles amongst the assumptions and by using integrity constraints for setting the contrary of these new assumptions [128].

Several dialectical semantics for argumentation have been defined [44,18,48], all corresponding to semantics for NAF (for instances of argumentation corresponding to LP). In particular, the stability dialectical semantics [44,18] corresponds to the stable model semantics, the grounded dialectical semantics [44,18] corresponds to the well-founded semantics, the admissibility dialectical semantics [44,18] corresponds to the preferred extension/partial stable model semantics, and the ideal dialectical semantics [48] corresponds to the ideal semantics of [2].

Argumentation is a suitable knowledge representation framework when it is important to be able to "inspect" the rationale for accepting a claim/an argument, rather than solely providing a yes/no answer. For example, consider the following simple "home-buying" ABA framework, where the set of rules is

$good(H) \leftarrow in\_city(H, L) \wedge garden(H) \wedge quiet(H)$
$valid\_complaint(H) \leftarrow police\_report(H, R) \wedge relevant(H, R)$
$irrelevant(H, R) \leftarrow about(R, O) \wedge not\ owner(H, O)$

together with clauses defining a number of specific candidate homes, e.g.
$in\_city(h1, london)$, $garden(h1)$, $owner(h1, sue)$.
Let the set of assumptions consist of all instances of $quiet(H)$ and $relevant(H, R)$, as well as all ground NAF literals. Let the contraries of assumptions be given as follows[2]

$valid\_complaint(H)$ is a contrary of $quiet(H)$
$wooden\_floors(H)$ is another contrary of $quiet(H)$
$irrelevant(H, R)$ is the contrary of $relevant(H, R)$
A is the contrary of $not\ A$ for all atoms A

Then, the claim $good(h1)$ is acceptable (under any dialectical semantics) on the ground of an acceptable argument $arg1$ supported by the assumption $quiet(h1)$. If $police\_report(h1, r)$ is added to the set of rules, then $good(h1)$ is not acceptable, since $arg1$ is now defeated by an argument $arg2$ for $valid\_complaint(h1)$, supported by the assumption $relevant(h1, r)$. If however $about(r, ted)$ is also added to the set of rules, then $good(h1)$ is again acceptable, since $arg2$ is now counterattacked by an argument for $irrelevant(h1, r)$, supported by the assumption $not\ owner(h1, ted)$, which cannot be attacked. In this latter case, $arg1$ and $arg3$ form an acceptable set of arguments for the claim.

---

[2] We follow here the presentation of ABA given by [53] and allow for assumptions to have multiple contraries.

Several argumentation systems are available, e.g. CaSAPI[3] implementing dispute derivations for ABA, and DeLP[4], for the argumentation approach of [54]. Other non-monotonic extensions of LP include

- the *Inductive Definitions* of [37], providing a declarative viewpoint on LP integrated with classical logic, with semantics generalising to the well-founded semantics for LP;
- FLORA-2 [77], which is a recent knowledge representation and programming environment extending LP by incorporating object-oriented programming as in F-logic, meta-programming as in Hilog and the framework for modelling state changes and side effects given by Transaction Logic.

Also, *Defeasible Logic* [101], an important family of NMR methods, has been shown to have close links to the stable model and three-value completion semantics of LP [4].

4   Dealing with Specific Reasoning Tasks and Applications

In this section we describe how non-monotonic variants of LP formalisms have been used to cope with specific knowledge representation and reasoning tasks, and how, in some cases, new specific formalisms have been developed for this purpose.

Reasoning about dynamic domains
Reasoning about dynamic domains is one of the main tasks an intelligent agent has to perform. Much work has been done to address the problem of modelling the dynamics of systems following a logical approach.

The seminal work of McCarthy [92] on the *situation calculus* and of Kowalski and Sergot [78] on the *event calculus* have devised two main directions to define a logical approach for modelling actions. A third direction has started more recently with Gelfond and Lifschitz work on the action description language $\mathcal{A}$ [58], which has given rise to the definition of a whole family of action description languages. A further direction is given by modal and dynamic logic approaches to reasoning about actions [110,60,62].

In the context of reasoning about actions, the dynamics of a system is usually represented by providing a specification, in a domain description, of the effects of actions on the world and their executability conditions. The properties of the world which may change from a state to another are represented by propositions called *fluents*.

The situation calculus and the event calculus are based on different ontologies. While the situation calculus takes the state of the world as primary, by encoding actions as transformations on states, the event calculus takes events as primary. Gelfond and Lifschitz language $\mathcal{A}$ is essentially the propositional fragment of Pednault's ADL [102] and can be regarded as adopting the same ontology as

---

[3] http://www.doc.ic.ac.uk/ dg00/casapi.html
[4] http://lidia.cs.uns.edu.ar/delp_client/

the situation calculus. The same can be said about theories of action based on modal and dynamic action logics.

As an example, let us consider the following specification of the well known *Yale shooting problem* [70] in the situation calculus. The situation calculus is based on first-order logic: situations are sequences of actions (for instance, [load, wait, shoot]), corresponding to possible world histories, and are represented by terms such as $do(shoot, do(wait, do(load, s_0)))$, where s_0 is the initial situation. (Relational) fluents are represented by predicate symbols extended with an extra argument denoting a situation, such as, $alive(s_0)$ or $alive(do(wait, do(load, s_0)))$. The effects of actions on the state of the world are described by *effect axioms*, like:

$$loaded(do(load, s))$$
$$loaded(s) \rightarrow \neg alive(do(shoot, s)))$$

(where s is a variable denoting a situation and $\neg$ denotes classical negation) meaning that the gun is loaded after the execution of the action *load* and that the turkey is not alive after the execution of the action *shoot*, when the gun is loaded. *Preconditions* are requirements that must be satisfied in the current situation for the action to be executable. In the shooting domain, we can assume that all actions are always executable, so we have, for instance:

$$Poss(load, s) \leftrightarrow true$$

A more interesting precondition law, from [117],

$$Poss(repair(r, x), s) \leftrightarrow has\_glue(r, s) \wedge broken(x, s)$$

says that, whenever it is possible for a robot to repair an object, then the object must be broken and there must be glue available.

Let us assume that in the initial situation the turkey is alive and the gun is not loaded (i.e., $alive(s_0)$ and $\neg loaded(s_0)$). Then, will the turkey be alive after the sequence of actions [load, wait and shoot]? Namely, does $alive(do(shoot, do(wait, do(load, s_0))))$ hold?

To answer this *temporal projection problem*, we need a solution to the *frame problem*, i.e. the problem of specifying (in a parsimonious way) that all the properties of the world (fluents) that are not affected by the execution of the actions do not change. Fluent values must be assumed to persist, unless they are changed by action execution. Hence, solving the frame problem requires to cope with some form of non-monotonicity.

Monotonic and non-monotonic solutions have been proposed in the literature to deal with the frame problem in different formalisms. In particular, for the situation calculus, Reiter provides a monotonic solution to the frame problem [117], by introducing the so-called successor state axioms, which provide a suitable completion of the action laws, in a compact way. From the computational point of view, regression forms the basis for many planning procedures and for automated reasoning in the situation calculus. As a consequence of Clark's completion result, Prolog provides a natural implementation of action theories formulated in the situation calculus. In particular, Reiter proves the correctness of the Prolog implementation under a restriction to closed initial database [117].

Reiter's version of the situation calculus has then been used as the basis of the high-level language GOLOG [85], a logic programming language for reasoning about dynamic domains in which the definition of complex actions is allowed by making use of high-level algol-like constructs (including procedures). The semantics of GOLOG is defined via macro-expansion into sentences of the situation calculus and its implementation relies on the definition of a Prolog interpreter. In [34,35], GOLOG has been further extended in order to deal with concurrency (ConGolog) and with exogenous actions. In particular, the problem of executing programs including sensing action is tackled in [35] . These feature make GOLOG a LP language well suited for implementing applications in dynamic domains, like robotics, agent control, and intelligent software agents. In particular, we refer to [5] for the use of GOLOG in planning applications.

In the event calculus [78], events occur at time-points; an event can make a property true (the event *initiates* the property) or it can make a property false (the event *terminates* the property). As a difference with the situation calculus, the event calculus adopts a linear representation of time rather than a branching one (we refer to [13] for a detailed comparison of the two formalisms). Since the event calculus is defined within the LP framework, a natural solution to the frame problem is provided by the non-monotonic semantics of NAF. Sadri and Kowalski [119] compare the simpler variant of the event calculus with time-points (SEC) to the original formulation of the event calculus with time periods and proposes a new variant, which is essentially the Clark's completion of SEC augmented with integrity constraints. In [122], a formulation of the event calculus based on many-sorted first-order predicate calculus augmented with circumscription is provided. Based on this formulation, in [123] an abductive event calculus planner is developed, whose computations are closely related with those of partial-order planning algorithms. We cannot enter the details of the huge amount of work on the event calculus. Concerning the case when only partial information about the order of events is available, we mention the work by Chittaro, Montanari and Provetti [25] on the skeptical and credulous variants of the event calculus. Such variants allow the distinction between the properties which are necessarily true and those that are possibly true, when events are partially ordered. This work has lead to the development of a uniform modal framework to define a number of extensions of the event calculus of increasing expressive power [22,21]. An encoding of such calculi in the language λ-Prolog is provided in [22,21].

The action description language $\mathcal{A}$ introduced by Gelfond and Lifschitz [58] is intended to provide a simple and declarative description of actions, by means of action laws of the form A **causes** F **if** $P_1, \ldots, P_k$, saying that the execution of action A has the effect of making F true, when executed in a state in which $P_1, \ldots, P_k$ hold. Essentially, an action description in the language $\mathcal{A}$ provides an abstract description of a transition system, consisting of a set of nodes, representing the states of the domain, and a set of arcs labelled by actions connecting the nodes [59]. The semantics of the language $\mathcal{A}$ provides a mapping between an action description and the corresponding transition system.

Starting form the language $\mathcal{A}$, a whole family of action languages have been defined through the addition of new features enhancing the expressivity of the language. In particular, we mention the language $\mathcal{L}$ [11], including static constraints, sensing actions, and observable fluents, the language $\mathcal{AL}$ [10], which includes static and dynamic causal laws, executability conditions, as well as concurrent actions, and the languages $\mathcal{C}$ [66] and $\mathcal{C}^+$ [65] which provide an account of causality and deal with actions with indirect and non-deterministic effects and with concurrent actions.

From the computational point of view, [40] and [47] have proposed translation of (extensions of) the language $\mathcal{A}$ into ALP, while [86] has defined an extension of the language $\mathcal{A}$ to deal with concurrent actions, together with a sound and complete translation to ALP. Recently, ASP has been shown to be well suited for reasoning about dynamic domains, so that an action specification can be encoded into an ASP program (see for instance [10]). This technique, in particular, has been used for planning and diagnosis (see [14]).

A different approach to reasoning about actions is based on the use of modal, temporal and dynamic logics. The suitability of such non classical logics for reasoning about actions has been pointed out by several authors [110,60,62,63]. Indeed, classical dynamic logic essentially adopts the same ontology as the situation calculus, by taking the state of the world as primary, and encoding actions as transformations on states. Thus, actions can be represented in a natural way by modalities, and states as sequences of modalities. In the LP setting, [7] defines a modal LP language (Dylog) for reasoning about complex actions, which are defined using modal inclusion axioms. The language is able to handle knowledge producing actions and incomplete knowledge. An abductive goal-directed proof procedure is defined, which allows agents to reason about complex actions and to generate conditional plans.

A lot of work has been done on the *ramification problem*, i.e. the problem of dealing with the dependencies among fluents. Besides the work on the languages $\mathcal{C}$ [66] and $\mathcal{C}^+$ [65], we mention:

- [91], where a causal approach to ramification is proposed in which causal rules are represented by inference rules;
- [88], which deals with causality and indirect effects in the situation calculus;
- [62], which deals with ramification and causality in a modal action logic;
- [38], where a solution to the ramification based on inductive definitions is proposed; and, finally,
- the work by Thielscher on causality [126] leading to the definition of the Fluent Calculus [127].

Diagnosis

In [28] Console et al. define a theory of diagnosis for incomplete causal models consisting, essentially, of a set of Horn clauses and a set of hypotheses. An abductive definition of diagnosis is provided, which is equivalently formulated in terms of circumscription. The diagnostic process described in the paper was implemented in Prolog. The relationships between abduction and deduction have been explored in [29].

In [109] Preist, Eshghi and Bertolino present a definition of diagnoses which allows to use NAF in the modelling language. The definition is based on the generalized stable model semantics of abduction presented in [71]. The resulting framework naturally incorporates both abductive and consistency-based [116] diagnosis. The work is similar in spirit to the work by Console and Torasso [30] devising a spectrum of alternatives in the logical definition of diagnosis, based on the formulation of a diagnostic problem as an ALP problem. In [12] Baral et al. provide a characterization of diagnosis, diagnostic planning, and repair in an extension of the action language $\mathcal{L}$ [11].

Legal Reasoning

The important contribututution of LP with NAF for the logical analysis of legal reasoning and for legal knowledge representation is widely acknowledged [107]. Pioneering work in the '80s on the formalisation of the British Nationality Act [121] first pointed to the great potential of using LP with NAF to model exceptions in law. For example, the clause

$$brit(X) \leftarrow newborn\_found\_in\_uk(X) \wedge not \; born\_outside\_uk(X)$$

represents the information that a newborn infant found on UK ground can be deemed to be a British citizen unless it can be shown that the infant was born outside the UK.

The argumentation-based interpretation of LP (see Section 3) has then served, from the mid-'90s, as a framework for modelling legal arguments. Prakken and Sartor [105] proposed a form of extended LP (with NAF and explicit negation), augmented with preferences in turn defined by LP clauses, to formalise legal texts with contradictory rules, rules with assumptions, inapplicability statements, and defeasible priority rules. They adopt an argumentation-based semantics (see Section 3) of their augmented extended LPs to support the construction of legal arguments from these legal texts. They consider several applications, from Italian, Dutch and European law.

Kowalski and Toni [81] propose a concrete instance of ABA (see Section 3) to model legal arguments, e.g. of the form advocated in [105]. This instance is an abstraction of LP with NAF as well as default logic and autoepistemic/nonmonotonic modal logic. It incorporates, at the object-level, meta-level defeasible preferences over rules.

Nitta and Shibasaki [99] propose an extension of LP-based argumentation incorporating defeasible preferences as well as analogical reasoning, applied to Japanese law.

This line of work has continued in the last years, e.g. Prakken and Sartor [108] extend the approach of [105] to model the legal concept of "burden of proof", and show examples from the Italian and USA law.

Governatori and Rotolo propose an approach to legal reasoning based on the use of Defeasible Logic. In particular, [68] extends Defeasible Logic with the notions of agency, intention and obligation, while [67] extends RuleML with deontic and defeasible aspects for reasoning about business contracts.

Belief revision. The problems of revision and update are strongly related with the problem of reasoning about dynamic domains as they are concerned with reasoning about change. There is no room here for an extensive discussion on these topics, so we will limit our discussion to few aspects. Starting from the beginning of the '90s, LP semantics has been extended to provide a semantic characterization for reason maintenance systems (and, in particular, for Doyle's TMS [43] and De Kleer ATMS [36]) and, more generally, for capturing some form of revision or contradiction removal. TMS justifications can be seen, essentially, as clauses with NAF, which also include integrity constraints. In [61] a notion of generalized stable model has been defined to provide a semantics to the TMS accounting for the process of conflict resolution. In [130] a similar approach has been used to define a skeptical reason maintenance system based on the well-founded semantics. An extension of the well-founded semantics for logic programs with explicit negation and contradiction removal has been presented in [103]. [131] defines a tractable semantics for extended logic programs that allows for an incremental computation and forms a common core for the grounded argumentation semantics (see Section 3) and Alferes et al.'s well-founded semantics [2]. The semantics is based on the concept of iterative belief revision. [52] introduces a LP language *CondLP* which supports hypothetical and counterfactual reasoning. The language is based on a conditional logic which enables to formalize conditional updates of the knowledge base and relies on an abductive semantics. The kind of revision performed by this language is strongly related to Nebel's prioritized base revision [97]. Sadri and Toni in [120] deal with the problem of performing belief revision on-line, while reasoning is taking place, by means of an abductive proof procedure.

Security policies

Antoniou et al [3] provide an overview of rule based approaches to policy specification. They discuss the benefits of using LP with NAF to model security policies such as "if a packet of protocol X goes from hosts Y to hosts Z then [don't] let it pass". As they mention, "default decisions arise naturally in real-world security policies. For example, open policies prescribe that authorizations by default are granted, whereas closed policies prescribe that they should be denied unless stated otherwise". Bandara et al. [8] model these rules and preferences over them in the LP-based argumentation framework of [100]. We refer to [17] for an overview of the research on Semantic Web Policies.

Other application domains

There are several other areas that we have not mentioned, in which non-monotonic LP formalisms have been extensively (and successfully) used. These include, in particular, the areas of agent and multiagent systems, deductive databases and semantic web. These topics are addressed in [6] and [69].

5   Conclusions

Non-monotonic reasoning (NMR) has been an important focus for the logic programming (LP) community since Clark's work on negation as failure (NAF) [27]. Various semantics for LP with NAF have been given, and several extensions of LP with NAF have been proposed and applied in practice. The Italian LP community appears to be very active in the field and has provided relevant contributions to LP and NMR research. In part, this could be explained by the availability of Italian and European projects, which have supported this research (and, in particular, of several basic research projects developed during the '90s), but also by the good integration of the Italian research community with the international one through scientific collaborations and student exchanges.

Earlier surveys on LP and knowledge representation/NMR [95,9] had identified some open issues for this filed. We now reassess some of these issues, in the light of research in this area since the publication of those surveys. This assessment aims at identifying current challenges for the Italian LP community.

Both Minker [95] and Baral and Gelfond [9] agree on the need of *applications* and solution for *practical problems*, rather than toy problems, with features of non-monotonicity. We have discussed progress on some applications in Section 4. For lack of space, we have omitted to discuss some recent applications of non-monotonic LP frameworks, notably agent and multi-agent applications and web applications. These are discussed in[6] and [69] respectively.

Minker points to the need for understanding the relationships amongst various theories of NMR, including LP-based and non-LP-based ones, with the aim of being able to discriminate which theory to use for specific problems. These relationships are now well understood, as we have discussed, in part, in Sections 2 and 3. For example, argumentation [44,18] provides an abstract setting for understanding the relationships between different LP semantics and formalisms for NMR other than LP.

Baral and Gelfond [9] suggest the need for developing query-answering systems. These exist nowadays for several LP-based frameworks for knowledge representation, notably XSB (see Section 2), ASP (see [14]), abductive logic programming (see Section 3) and argumentation (see Section 3).

For lack of space, we have omitted to mention the substantial body of work devoted to the study of the computational complexity of non-monotonic LP formalisms. Both tractability and complexity issues of LPNMR formalisms have been studied extensively in the last two decades. This study has become of fundamental importance for comparing different formalisms. We refer to [33] for a survey of results on LP complexity and tractability.

Some challenges still exist. For example, the field would benefit from further applications, of a larger scale, possibly leading to industrial take-up. Moreover, the efficient treatment of non-ground queries and non-ground LP with infinite domains is still to a large extent open. Finally, further research is needed to characterise tractable, but useful fragment of several LP formalisms, e.g. argumentation.

Acknowledgement

The first author has been partially supported by Regione Piemonte, Bando Converging Technologies 2007, Project ICT4LAW. The authors would like to thank anonymous referees for useful comments on drafts of this work, and Marek Sergot and Alberto Martelli for helpful discussions.

References

1. Alberti, M., Chesani, F., Gavanelli, M., Lamma, E., Mello, P., Torroni, P.: Verifiable agent interaction in abductive logic programming: the SCIFF framework. ACM Transactions on Computational Logic (ToCL) 9(4) (2008)
2. Alferes, J.J., Dung, P.M., Pereira, L.M.: Scenario semantics of extended logic programs. In: LPNMR, pp. 334–348 (1993)
3. Antoniou, G., Baldoni, M., Bonatti, P.A., Nejdl, W., Olmedilla, D.: Rule-based policy specification. In: Secure Data Management in Decentralized Systems. Advances in Information Security, vol. 33, pp. 169–216. Springer, Heidelberg (2007)
4. Antoniou, G., Billington, D., Governatori, G., Maher, M.J.: Embedding defeasible logic into logic programming. TPLP 6(6), 703–735 (2006)
5. Baier, J.A., McIlraith, S.A.: On planning with programs that sense. In: KR, pp. 492–502 (2006)
6. Baldoni, M., Baroglio, C., Mascardi, V., Omicini, A., Torroni, P.: Agents, Multi-Agent Systems and Declarative Programming: What, When, Where, Why, Who, How? In: Dovier, A., Pontelli, E. (eds.) 25 Years of Logic Programming in Italy, ch. 10. LNCS, vol. 6125, pp. 204–230. Springer, Heidelberg (2010)
7. Baldoni, M., Martelli, A., Patti, V., Giordano, L.: Programming rational agents in a modal action logic. Ann. Math. Artif. Intell. 41(2-4), 207–257 (2004)
8. Bandara, A.K., Kakas, A.C., Lupu, E.C., Russo, A.: Using argumentation logic for firewall policy specification and analysis. In: State, R., van der Meer, S., O'Sullivan, D., Pfeifer, T. (eds.) DSOM 2006. LNCS, vol. 4269, pp. 185–196. Springer, Heidelberg (2006)
9. Baral, C., Gelfond, M.: Logic programming and knowledge representation. Journal of Logic Programming 19, 73–148 (1994)
10. Baral, C., Gelfond, M.: Reasoning agents in dynamic domains, pp. 257–279 (2000)
11. Baral, C., Gelfond, M., Provetti, A.: Representing actions: Laws, observations and hypotheses. J. Log. Program. 31(1-3), 201–243 (1997)
12. Baral, C., McIlraith, S., Son, T.: Formulating diagnostic problem solving using an action language with narratives and sensing. In: Proceedings of the Seventh International Conference on Principles of Knowledge Representation and Reasoning (KR 2000), Breckenridge, Colorado, USA, April 12-15, pp. 311–322 (2000)
13. Belleghem, K.V., Denecker, M., Schreye, D.D.: On the relation between situation calculus and event calculus. J. Log. Program. 31(1-3), 3–37 (1997)
14. Bonatti, P., Calimeri, F., Leone, N., Ricca, F.: Answer Set Programming. In: Dovier, A., Pontelli, E. (eds.) 25 Years of Logic Programming in Italy, ch. 8. LNCS, vol. 6125, pp. 159–182. Springer, Heidelberg (2010)
15. Bonatti, P.A.: Autoepistemic logic programming. J. Autom. Reasoning 13(1), 35–67 (1994)
16. Bonatti, P.A.: Reasoning with infinite stable models. Artif. Intell. 156(1), 75–111 (2004)

17. Bonatti, P.A., Duma, C., Fuchs, N., Nejdl, W., Olmedilla, D., Peer, J., Shahmehri, N.: Semantic web policies – A discussion of requirements and research issues. In: Sure, Y., Domingue, J. (eds.) ESWC 2006. LNCS, vol. 4011, pp. 712–724. Springer, Heidelberg (2006)
18. Bondarenko, A., Dung, P.M., Kowalski, R.A., Toni, F.: An abstract, argumentation-theoretic approach to default reasoning. Artificial Intelligence 93(1-2), 63–101 (1997)
19. Bondarenko, A., Toni, F., Kowalski, R.: An assumption-based framework for non-monotonic reasoning. In: Nerode, A., Pereira, L. (eds.) Proc. 2nd International Workshop on Logic Programming and Non-monotonic Reasoning, pp. 171–189. MIT Press, Cambridge (1993)
20. Buccafurri, F., Leone, N., Rullo, P.: Stable models and their computation for logic programming with inheritance and true negation. J. Log. Program. 27(1), 5–43 (1996)
21. Cervesato, I., Franceschet, M., Montanari, A.: A guided tour through some extensions of the Event Calculus. Computational Intelligence 16(2), 307–347 (2000)
22. Cervesato, I., Montanari, A.: A general modal framework for the Event Calculus and its skeptical and creduluos variants. J. Log. Program. 38(2), 111–164 (1999)
23. Chandra, A., Harel, D.: Horn clause queries and generalizations. Journal of Logic Programming 2(1), 1–5 (1985)
24. Chen, W., Warren, D.S.: Computation of stable models and its integration with logical query processing. IEEE Transactions on Knowledge and Data Engineering 8, 8–5 (1994)
25. Chittaro, L., Montanari, A., Provetti, A.: Skeptical and credoluos event calculi for supporting modal queries. In: ECAI 1994, pp. 361–365 (1994)
26. Christiansen, H., Dahl, V.: HYPROLOG: A new logic programming language with assumptions and abduction. In: Gabbrielli, M., Gupta, G. (eds.) ICLP 2005. LNCS, vol. 3668, pp. 159–173. Springer, Heidelberg (2005)
27. Clark, K.L.: Negation as failure. In: Logic and Data Bases. Plenum Press (1978)
28. Console, L., Dupré, D.T., Torasso, P.: A theory of diagnosis for incomplete causal models. In: IJCAI, pp. 1311–1317 (1989)
29. Console, L., Dupre, D.T., Torasso, P.: On the relationship between abduction and deduction. Journal of Logic and Computation 1(5), 661–690 (1991)
30. Console, L., Torasso, P.: A spectrum of logical definitions of model-based diagnosis. Computational Intelligence 7, 133–141 (1991)
31. Costantini, S.: Contributions to the stable model semantics of logic programs with negation. Theor. Comput. Sci. 149(2), 231–255 (1995)
32. Costantini, S.: On the existence of stable models of non-stratified logic programs. TPLP 6(1-2), 169–212 (2006)
33. Dantsin, E., Eiter, T., Gottlob, G., Voronkov, A.: Complexity and expressive power of logic programming. ACM Comput. Surv. 33(3), 374–425 (2001)
34. De Giacomo, G., Lesperance, Y., Levesque, H.J.: Reasoning about concurrent execution, prioritized interrupts, and exogenous actions in the situation calculus. In: IJCAI 1997: Proceedings of the Fifteenth international joint conference on Artifical intelligence, pp. 1221–1226. Morgan Kaufmann Publishers Inc., San Francisco (1997)
35. De Giacomo, G., Levesque, H.J.: An incremental interpreter for high-level programs with sensing. In: Logical Foundations for Cognitive Agents, pp. 86–102. Springer, Heidelberg (1998)
36. de Kleer, J.: An assumption-based tms. Artif. Intell. 28(2), 127–162 (1986)

37. Denecker, M., Bruynooghe, M., Marek, V.W.: Minimal belief and negation as failure. ACM Trans. Comput. Log. 2(4), 623–654 (2001)
38. Denecker, M., Dupré, D.T., Belleghem, K.V.: An inductive definition approach to ramifications. Electron. Trans. Artif. Intell. 2, 25–67 (1998)
39. Denecker, M., Kakas, A.C.: Abduction in Logic Programming. In: Kakas, A.C., Sadri, F. (eds.) Computational Logic: Logic Programming and Beyond, Part I. LNCS (LNAI), vol. 2407, pp. 402–436. Springer, Heidelberg (2002)
40. Denecker, M., Schreye, D.D.: Representing incomplete knowledge in abductive logic programming. In: Proc. of the International Symposium on Logic Programming, pp. 147–163. MIT Press, Cambridge (1993)
41. Denecker, M., Schreye, D.D.: SLDNFA: an abductive procedure for abductive logic programs. Journal of Logic Programming 34(2), 111–167 (1998)
42. Dovier, A., Pontelli, E. (eds.): 25 Years of Logic Programming in Italy. LNCS, vol. 6125. Springer, Heidelberg (2010)
43. Doyle, J.: A truth maintenance system. Artif. Intell. 12(3), 231–272 (1979)
44. Dung, P.: On the acceptability of arguments and its fundamental role in non-monotonic reasoning, logic programming, and n-person games. Artificial Intelligence 77(2), 321–357 (1995)
45. Dung, P., Kowalski, R., Toni, F.: Assumption-based argumentation. In: Rahwan, I., Simari, G. (eds.) Argumentation in AI: The Book. Springer, Heidelberg (2009) (to appear)
46. Dung, P.M.: Negations as hypotheses: An abductive foundation for logic programming. In: ICLP, pp. 3–17 (1991)
47. Dung, P.M.: Representing actions in logic programming and its applications in database updates. In: ICLP, pp. 222–238 (1993)
48. Dung, P.M., Mancarella, P., Toni, F.: Computing ideal sceptical argumentation. Artificial Intelligence 171(10-15), 642–674 (2007)
49. Eshghi, K., Kowalski, R.A.: Abduction compared with negation by failure. In: ICLP, pp. 234–254 (1989)
50. Fitting, M.: A Kripke/Kleene semantics for logic programs. Journal of Logic Programming 2, 295–312 (1985)
51. Fung, T.H., Kowalski, R.A.: The IFF proof procedure for abductive logic programming. Journal of Logic Programming 33(2), 151–165 (1998)
52. Gabbay, D.M., Giordano, L., Martelli, A., Olivetti, N., Sapino, M.L.: Conditional reasoning in logic programming. J. Log. Program. 44(1-3), 37–74 (2000)
53. Gaertner, D., Toni, F.: Hybrid argumentation and its properties. In: Hunter, A. (ed.) Proceedings of the Second International Conference on Computational Models of Argument (COMMA 2008), pp. 183–195. IOS Press, Amsterdam (2008)
54. García, A.J., Dix, J., Simari, G.R.: Argument-based logic programming. In: Rahwan, I., Simari, G. (eds.) Argumentation in AI: The Book. Springer, Heidelberg (2009) (to appear)
55. Gavanelli, M., Rossi, F.: Constraint Logic Programming. In: Dovier, A., Pontelli, E. (eds.) 25 Years of Logic Programming in Italy, ch. 4. LNCS, vol. 6125, pp. 64–86. Springer, Heidelberg (2010)
56. Gelder, A.V., Ross, K.A., Schlipf, J.S.: Unfounded sets and well-founded semantics for general logic programs. In: PODS, pp. 221–230 (1988)
57. Gelfond, M., Lifschitz, V.: The stable model semantics for logic programming. In: Kowalski, R., Bowen, K. (eds.) Proceedings of the Fifth International Conference and Symposium on Logic Programming, pp. 1070–1080. MIT Press, Cambridge (1988)

58. Gelfond, M., Lifschitz, V.: Representing action and change by logic programs. Journal of logic Programming 17, 301–322 (1993)
59. Gelfond, M., Lifschitz, V.: Action languages. Electronic Transactions on AI 3(16), 193–210 (1998)
60. Giacomo, G.D., Lenzerini, M.: PDL-based framework for reasoning about actions. In: AI*IA, pp. 103–114 (1995)
61. Giordano, L., Martelli, A.: Generalized stable models, truth maintenance and conflict resolution. In: ICLP, pp. 427–441 (1990)
62. Giordano, L., Martelli, A., Schwind, C.: Ramification and causality in a modal action logic. J. Log. Comput. 10(5), 625–662 (2000)
63. Giordano, L., Martelli, A., Schwind, C.: Specifying and verifying interaction protocols in a temporal action logic. J. Applied Logic 5(2), 214–234 (2007)
64. Giordano, L., Olivetti, N.: Combining negation as failure and embedded implications in logic programs. J. Log. Program. 36(2), 91–147 (1998)
65. Giunchiglia, E., Lee, J., Lifschitz, V., McCain, N., Turner, H.: Nonmonotonic causal theories. Artif. Intell. 153(1-2), 49–104 (2004)
66. Giunchiglia, E., Lifschitz, V.: An action language based on causal explanation: Preliminary report. In: AAAI/IAAI, pp. 623–630 (1998)
67. Governatori, G.: Representing business contracts in ruleml. Int. J. Cooperative Information Systems 14(2-3), 180–216 (2005)
68. Governatori, G., Rotolo, A.: Defeasible logic: Agency, intention and obligation. In: Lomuscio, A., Nute, D. (eds.) DEON 2004. LNCS (LNAI), vol. 3065, pp. 114–128. Springer, Heidelberg (2004)
69. Greco, S., Lisi, F.: Logic Programming Languages for Databases and the Web. In: Dovier, A., Pontelli, E. (eds.) 25 Years of Logic Programming in Italy, ch. 9. LNCS, vol. 6125, pp. 183–203. Springer, Heidelberg (2010)
70. Hanks, S., McDermott, D.V.: Nonmonotonic logic and temporal projection. Artif. Intell. 33(3), 379–412 (1987)
71. Kakas, A., Mancarella, P.: Generalized stable models: A semantics for abduction. In: ECAI, pp. 385–391 (1990)
72. Kakas, A.C., Kowalski, R.A., Toni, F.: Abductive logic programming. Journal of Logic and Computation 2(6), 719–770 (1993)
73. Kakas, A.C., Kowalski, R.A., Toni, F.: The role of abduction in logic programming. In: Handbook of Logic in Artificial Intelligence and Logic Programming, vol. 5, pp. 235–324. OUP (1998)
74. Kakas, A.C., Mancarella, P.: Generalized stable models: a semantics for abduction. In: Proceedings of the 9th European Conference on Artificial Intelligence, pp. 385–391 (1990)
75. Kakas, A.C., Mancarella, P.: Short note: Preferred extensions are partial stable models. J. Log. Program. 14(3&4), 341–348 (1992)
76. Kakas, A.C., Van Nuffelen, B., Denecker, M.: A-system: Problem solving through abduction. In: Proceedings of the 17th International Joint Conference on Artificial Intelligence, pp. 591–596 (2001)
77. Kifer, M.: Nonmonotonic Reasoning in FLORA-2. In: Baral, C., Greco, G., Leone, N., Terracina, G. (eds.) LPNMR 2005. LNCS (LNAI), vol. 3662, pp. 1–12. Springer, Heidelberg (2005)
78. Kowalski, R., Sergot, M.: A logic-based calculus of events. New Generation Computing 4(1), 67–95 (1986)
79. Kowalski, R.A.: Predicate logic as programming language. In: IFIP Congress, pp. 569–574 (1974)

80. Kowalski, R.A., Kuehner, D.: Linear resolution with selection function. Artif. Intell. 2(3/4), 227–260 (1971)
81. Kowalski, R.A., Toni, F.: Abstract argumentation. Artif. Intell. Law 4(3-4), 275–296 (1996)
82. Kunen, K.: Negation in logic programming. Journal of Logic Programming 4(4), 289–308 (1987)
83. Kunen, K.: Signed data dependencies in logic programs. J. Log. Program. 7(3), 231–245 (1989)
84. Leone, N., Rullo, P.: Ordered logic programming with sets. J. Log. Comput. 3(6), 621–642 (1993)
85. Levesque, H., Reiter, R., Lespérance, Y., Lin, F., Scherl, R.: Golog: A logic programming language for dynamic domains. Journal of Logic Programming 31, 59–83 (1997)
86. Li, R., Pereira, L.M.: Representing and reasoning about concurrent actions with abductive logic programs. Ann. Math. Artif. Intell. 21(2-4), 245–303 (1997)
87. Lifschitz, V.: Minimal belief and negation as failure. Artificial Intelligence 70(1-2), 53–72 (1994)
88. Lin, F.: Embracing causality in specifying the indirect effects of actions. In: IJCAI, pp. 1985–1993 (1995)
89. Lloyd, J.W.: Foundations of Logic Programming, 1st edn. Springer, Heidelberg (1984)
90. Mancarella, P., Terreni, G., Sadri, F., Toni, F., Endriss, U.: The CIFF proof procedure for abductive logic programming with constraints: Theory, implementation and experiments. Theory and Practice of Logic Programming 9(6), 691–750 (2009)
91. McCain, N., Turner, H.: A causal theory of ramifications and qualifications. In: IJCAI, pp. 1978–1984 (1995)
92. McCarthy, J.: Situations actions and causal laws. Technical Report. Stanford (1963); Reprinted in Semantic Information Processing (Minsky, M. (ed.)), pp. 410-417. MIT Press, Cambridge (1968)
93. McCarthy, J.: Circumscription—a form of non-monotonic reasoning. Artificial Intelligence 13, 27–39 (1980)
94. McDermott, D.V.: Nonmonotonic logic ii: Nonmonotonic modal theories. J. ACM 29(1), 33–57 (1982)
95. Minker, J.: An overview of nonmonotonic reasoning and logic programming. Journal of Logic Programming, Special Issue 17, 95–126 (1993)
96. Moore, R.C.: Semantical considerations on nonmonotonic logic. Artif. Intell. 25(1), 75–94 (1985)
97. Nebel, B.: Belief revision and default reasoning: Syntax-based approaches. In: KR, pp. 417–428 (1991)
98. Niemel, I., Simons, P.: Efficient implementation of the well-founded and stable model semantics. In: Proceedings of the Joint International Conference and Symposium on Logic Programming, pp. 289–303. MIT Press, Cambridge (1996)
99. Nitta, K., Shibasaki, M.: Defeasible reasoning in japanese criminal jurisprudence. Artif. Intell. Law 5(1-2), 139–159 (1997)
100. Noël, V., Kakas, A.C.: Gorgias-c: Extending argumentation with constraint solving. In: Erdem, E., Lin, F., Schaub, T. (eds.) LPNMR 2009. LNCS, vol. 5753, pp. 535–541. Springer, Heidelberg (2009)
101. Nute, D.: Defeasible logic. In: Bartenstein, O., Geske, U., Hannebauer, M., Yoshie, O. (eds.) INAP 2001. LNCS (LNAI), vol. 2543, pp. 87–114. Springer, Heidelberg (2003)

102. Pednault, E.P.D.: Adl: Exploring the middle ground between strips and the situation calculus. In: KR, pp. 324–332 (1989)
103. Pereira, L.M., Alferes, J.J., Aparício, J.N.: Contradiction removal semantics with explicit negation. In: Masuch, M., Polos, L. (eds.) Logic at Work 1992. LNCS, vol. 808. Springer, Heidelberg (1994)
104. Poole, D.: A logical framework for default reasoning. Artif. Intell. 36(1), 27–47 (1988)
105. Prakken, H., Sartor, G.: A dialectical model of assessing conflicting arguments in legal reasoning. Artif. Intell. Law 4(3-4), 331–368 (1996)
106. Prakken, H., Sartor, G.: Argument-based extended logic programming with defeasible priorities. Journal of Applied Non-classical Logics 7, 25–75 (1997)
107. Prakken, H., Sartor, G.: The Role of Logic in Computational Models of Legal Argument: A Critical Survey. In: Kakas, A.C., Sadri, F. (eds.) Computational Logic: Logic Programming and Beyond, Part II. LNCS (LNAI), vol. 2408, pp. 342–381. Springer, Heidelberg (2002)
108. Prakken, H., Sartor, G.: Formalising arguments about the burden of persuasion. In: ICAIL-The Eleventh International Conference on Artificial Intelligence and Law, Proceedings of the Conference, Stanford Law School, Stanford, California, USA, June 4-8, pp. 97–106. ACM, New York (2007)
109. Preist, C., Eshghi, K.: Consistency-based and abductive diagnoses as generalised stable models. In: FGCS, pp. 514–521 (1992)
110. Prendinger, H., Schurz, G.: Reasoning about action and change. A dynamic logic approach. Journal of Logic, Language and Information 5(2), 209–245 (1996)
111. Przymusinski, T.C.: Perfect model semantics. In: ICLP/SLP, pp. 1081–1096 (1988)
112. Przymusinski, T.C.: Extended stable semantics for normal and disjunctive programs. In: ICLP, pp. 459–477 (1990)
113. Reiter, R.: On closed world data bases. In: Logic and Data Bases, pp. 55–76 (1977)
114. Reiter, R.: A logic for default reasoning. Artificial Intelligence 13(1-2), 81–132 (1980)
115. Reiter, R.: Circumscription implies predicate completion (sometimes). In: AAAI, pp. 418–420 (1982)
116. Reiter, R.: A theory of diagnosis from first principles. Artif. Intell. 32(1), 57–95 (1987)
117. Reiter, R.: Knowledge in action. MIT Press, Cambridge (2001)
118. Saccà, D., Zaniolo, C.: Stable models and non-determinism in logic programs with negation. In: Proceedings of the Ninth ACM SIGACT-SIGMOD-SIGART Symposium on Principles of Database Systems, Nashville, Tennessee, April 2-4, pp. 205–217. ACM Press, New York (1990)
119. Sadri, F., Kowalski, R.A.: Variants of the event calculus. In: ICLP, pp. 67–81 (1995)
120. Sadri, F., Toni, F.: Interleaving belief updating and reasoning in abductive logic programming. In: Brewka, G., Coradeschi, S., Perini, A., Traverso, P. (eds.) Proceedings of the 17th European Conference on Artificial Intelligence (ECAI 2006), Riva del Garda, Italy, 28 August–1 September 2006. IOS Press, Amsterdam (2006)
121. Sergot, M.J., Sadri, F., Kowalski, R.A., Kriwaczek, F., Hammond, P., Cory, H.T.: The british nationality act as a logic program. Commun. ACM 29(5), 370–386 (1986)
122. Shanahan, M.: Solving the frame problem: A mathematical investigation of the common sense law of inertia. MIT Press, Cambridge (1997)

123. Shanahan, M.: An abductive event calculus planner. J. Log. Program. 44(1-3), 207–240 (2000)
124. Shepherdson, J.C.: Negation as failure: A comparison of clark's completed data base and reiter's closed world assumption. J. Log. Program. 1(1), 51–79 (1984)
125. Shepherdson, J.C.: Negation as failure ii. J. Log. Program. 2(3), 185–202 (1985)
126. Thielscher, M.: Ramification and causality. Artif. Intell. 89(1-2), 317–364 (1997)
127. Thielscher, M.: From situation calculus to fluent calculus: State update axioms as a solution to the inferential frame problem. Artif. Intell. 111(1-2), 277–299 (1999)
128. Toni, F.: A semantics for the Kakas-Mancarella procedure for abductive logic programming. In: Alpuente, M., Sessa, M.I. (eds.) 995 Joint Conference on Declarative Programming, GULP-PRODE 1995, Marina di Vietri, Italy, September 11-14, pp. 231–244 (1995)
129. Van Gelder, A., Ross, K., Schlifp, J.: The well-founded semantics for general logic programs. Journal of ACM 38(3), 620–650 (1991)
130. Witteveen, C., Brewka, G.: Skeptical reason maintenance and belief revision. Artif. Intell. 61(1), 1–36 (1993)
131. You, J.-H., Cartwright, R., Li, M.: Iterative belief revision in extended logic programming. Theoretical Computer Science, 170–171 (1996)

The Transformational Approach to Program Development

Alberto Pettorossi[1], Maurizio Proietti[2], and Valerio Senni[1]

[1] DISP, University of Rome Tor Vergata, Via del Politecnico 1, I-00133 Rome, Italy
{pettorossi,senni}@disp.uniroma2.it
[2] IASI-CNR, Viale Manzoni 30, I-00185 Rome, Italy
proietti@iasi.cnr.it

Abstract. We present an overview of the program transformation techniques which have been proposed over the past twenty-five years in the context of logic programming. We consider the approach based on rules and strategies. First, we present the transformation rules and we address the issue of their correctness. Then, we present the transformation strategies and, through some examples, we illustrate their use for improving program efficiency via the elimination of unnecessary variables, the reduction of nondeterminism, and the use of program specialization. We also describe the use of the transformation methodology for the synthesis of logic programs from first-order specifications. Finally, we illustrate some transformational techniques for verifying first-order properties of logic programs and their application to model checking for finite and infinite state concurrent systems.

1 Introduction

When deriving programs from specifications there are, among others, two main objectives to achieve: (i) program correctness, and (ii) program efficiency. Unfortunately, these two objectives are often in contrast with each other. Efficient programs may be rather intricate and their correctness proofs may be quite complex and long.

In order to overcome this difficulty, one can use the so called *program transformation* methodology by which starting from the given formal specifications, one derives efficient programs by applying a sequence of *transformation rules*, each of which preserves correctness. The transformation methodology is particularly appealing when programs are written in a declarative language such as a functional language or a logic language. In those cases, in fact, (i) the formal specifications are formulas which can easily be translated into an initial program which is, thus, correct by construction, and (ii) the transformation rules can be viewed as correctness preserving deduction rules in a suitable logic.

In order to get final programs which are more efficient than the initial ones, we need to apply the transformation rules according to suitable *transformation strategies*. This particular approach to program transformation, called the *rules + strategies* approach, has been first advocated in the seminal paper by Burstall and Darlington [17] in the case of functional programs. Then, as we will indicate at the beginning of the next section, it

A. Dovier, E. Pontelli (Eds.): 25 Years of Logic Programming, LNCS 6125, pp. 112–135, 2010.

has been adapted to logic programs [31,64], constraint logic programs [22,40], and the so-called functional-logic languages [1].

The program transformation methodology can also be used for performing *program synthesis* (see, for instance, [41] and also [5] for a recent survey). In that case the initial program is the declarative specification of a problem and the derived, transformed program is the encoding of an efficient algorithm for solving that problem.

In recent years program transformation has also been used as a technique for *program verification*. It has been shown that via program transformation, one can prove properties of programs [47] and also perform *model checking* for *finite* or *infinite* state systems [25].

In this paper we will focus our attention on the use of the program transformation methodology for the development of logic programs and we will mainly refer to the contributions coming from that area. In Section 2 we will present the most popular transformation rules, such as *unfolding* and *folding*, and we will mention some correctness results for those rules in various logic languages. In Section 3 we will describe some of the strategies that can be used to guide the application of the transformation rules for improving program efficiency. In Sections 4 and 5 we will present some transformational methods for program synthesis and program verification. Finally, in Section 6 we will discuss some future research directions in program transformation.

2   Transformation Rules

Various sets of program transformation rules have been proposed in the literature for several declarative programming languages. In their landmark paper [64] Tamaki and Sato considered definite logic programs and presented a set of transformation rules, including *definition*, *unfolding*, *folding*, *goal replacement*, and *clause deletion*. Under suitable restrictions, these rules are *correct* w.r.t. the least Herbrand model semantics [64]. Indeed, if from program P_0 we derive program P_n by several applications of the transformation rules, then under certain conditions the least Herbrand model is preserved, that is, $M(P_0) = M(P_n)$, where by $M(P)$ we denote the least Herbrand model of the program P. In the subsequent years, Tamaki and Sato's approach has been extended in several directions as we now indicate.

(1) Transformation rules for other logic-based programming languages, besides definite logic programs, have been considered. For instance, various rules have been presented for transforming: (i) general logic programs with *negation* [58], (ii) constraint logic programs [22,26,40], (iii) concurrent constraint logic programs [23,24], (iv) constraint handling rules [62], and functional-logic programs [1].

(2) The correctness of the transformation rules w.r.t. various semantics of logic languages has been proved. In particular, it has been shown that, under suitable conditions, the unfolding and folding transformation rules preserve: (i) the set of answer substitutions computed by SLD-resolution [6], (ii) the sequence of answer substitutions computed according to the Prolog operational semantics [49], (iii) termination properties such as finite failure [58] and left-termination [11], universal termination [7], and acyclicity [12], (iv) various semantics of general logic programs, such as the Clark completion [30], the perfect models of stratified programs [40,58], the stable models [57],

the well-founded models [59], and Kunen's and Fitting's three-valued models [10]. Systematic approaches for proving the correctness of the transformation rules based on the notions of *semantic kernel* and *argumentation semantics*, have been proposed in [4] and [65], respectively.

(3) The set of transformation rules has been extended either by adding extra rules such as *negative unfolding* and *negative folding* [26,60], and *simultaneous replacement* [10], or by relaxing the conditions under which we can apply the usual rules [48,53].

Now we present a set of transformation rules for locally stratified programs [40,45,60]. We will use these rules in the program transformations described in Sections 3, 4, and 5.

Given a locally stratified program P, throughout the paper by $M(P)$ we denote the perfect model of P [2], which is equal to the least Herbrand model in the case of definite logic programs. Given any conjunction C of one or more literals, by $vars(C)$ we denote the set of variables occurring in C. A similar notation will also be used for sets of conjunctions of literals. When applying the transformation rules we will feel free to rewrite clauses by: (i) renaming their variables, and (ii) rearranging the order and removing repeated occurrences of literals occurring in their bodies.

The transformation rules are used to construct a sequence $P_0, \ldots, P_n$ of programs, called a *transformation sequence*. The construction of that sequence is done as follows. Suppose that we have constructed the transformation sequence $P_0, \ldots, P_k$, for $0 \leq k \leq n-1$. Then the next program P_{k+1} in the transformation sequence is derived from program P_k by the application of a transformation rule among the following rules R1–R9.

Rule R1 is the *definition introduction* rule which is applied for introducing a new predicate definition by one or more clauses.

R1. Definition Introduction. Let us consider m (≥ 1) clauses of the form:

$$\delta_1 : newp(X_1, \ldots, X_h) \leftarrow B_1, \quad \ldots, \quad \delta_m : newp(X_1, \ldots, X_h) \leftarrow B_m$$

where: (i) $newp$ is a predicate symbol not occurring in $\{P_0, \ldots, P_k\}$, (ii) $X_1, \ldots, X_h$ are distinct variables occurring in $\{B_1, \ldots, B_m\}$, (iii) every predicate symbol occurring in $\{B_1, \ldots, B_m\}$ also occurs in P_0. The set $\{\delta_1, \ldots, \delta_m\}$ of clauses is called the *definition* of $newp$.

By *definition introduction* from program P_k we derive the program $P_{k+1} = P_k \cup \{\delta_1, \ldots, \delta_m\}$. For $k \geq 0$, $Defs_k$ denotes the set of clauses introduced by the definition rule during the transformation sequence $P_0, \ldots, P_k$. In particular, $Defs_0 = \{\}$.

The *unfolding* rule consists in: (i) replacing an atom A occurring in the body of a clause by a suitable instance of the disjunction of the bodies of the clauses whose heads unify with A, and (ii) applying suitable boolean laws for deriving clauses. There are two unfolding rules: (1) the *positive unfolding*, and (2) the *negative unfolding*, corresponding to the case where A occurs positively or negatively, respectively, in the body of the clause to be unfolded.

R2. Positive Unfolding. Let $\gamma : H \leftarrow G_L \wedge A \wedge G_R$ be a clause in program P_k and let P_k' be a variant of P_k without variables in common with γ. Let

$$\gamma_1 : K_1 \leftarrow B_1, \quad \ldots, \quad \gamma_m : K_m \leftarrow B_m \quad (m \geq 0)$$

be all clauses of P_k' such that, for $i = 1, \ldots, m$, A is unifiable with K_i, with most general unifier ϑ_i.

By *unfolding* γ *w.r.t.* A we derive the clauses $\eta_1, \ldots, \eta_m$, where for $i = 1, \ldots, m$, η_i is $(H \leftarrow G_L \wedge B_i \wedge G_R)\vartheta_i$. From P_k we derive the program $P_{k+1} = (P_k - \{\gamma\}) \cup \{\eta_1, \ldots, \eta_m\}$.

The *existential variables* of a clause γ are the variables occurring in the body of γ and not in its head.

R3. Negative Unfolding. Let $\gamma : \ H \leftarrow G_L \wedge \neg A \wedge G_R$ be a clause in program P_k and let P_k' be a variant of P_k without variables in common with γ. Let

$$\gamma_1 : \ K_1 \leftarrow B_1, \ \ldots, \ \gamma_m : \ K_m \leftarrow B_m \ \ (m \geq 0)$$

be all clauses of program P_k' such that A is unifiable with $K_1, \ldots, K_m$, with most general unifiers $\vartheta_1, \ldots, \vartheta_m$, respectively. Assume that:

1. $A = K_1\vartheta_1 = \cdots = K_m\vartheta_m$, that is, for $i = 1, \ldots, m$, A is an instance of K_i,
2. for $i = 1, \ldots, m$, γ_i has no existential variables, and
3. from $G_L \wedge \neg(B_1\vartheta_1 \vee \ldots \vee B_m\vartheta_m) \wedge G_R$ we get a logically equivalent disjunction $Q_1 \vee \ldots \vee Q_r$ of goals, with $r \geq 0$, by first pushing $\neg$ inside and then pushing $\vee$ outside.

By *unfolding* γ *w.r.t.* $\neg A$ we derive the clauses $\eta_1, \ldots, \eta_r$, where for $i = 1, \ldots, r$, η_i is $H \leftarrow Q_i$. From P_k we derive the new program $P_{k+1} = (P_k - \{\gamma\}) \cup \{\eta_1, \ldots, \eta_r\}$.

The *folding* rule consists in replacing instances of the bodies of the clauses which are the definition of a predicate by the corresponding head. As for unfolding, we have both the positive folding rule and the negative folding rule, depending on whether folding is applied to positive or negative occurrences of (conjunctions of) literals. Note that by the positive folding rule we may replace m (≥ 1) clauses by one clause only.

R4. Positive Folding. Let $\gamma_1, \ldots, \gamma_m$, with $m \geq 1$, be clauses in P_k and let $Defs_k'$ be a variant of $Defs_k$ without variables in common with $\gamma_1, \ldots, \gamma_m$. Let the definition of a predicate in $Defs_k'$ consist of the m clauses

$$\delta_1 : K \leftarrow B_1, \ \ldots, \ \delta_m : K \leftarrow B_m$$

where, for $i = 1, \ldots, m$, B_i is a non-empty conjunction of literals. Suppose that there exists a substitution ϑ such that, for $i = 1, \ldots, m$, clause γ_i is of the form $H \leftarrow G_L \wedge B_i\vartheta \wedge G_R$ and, for every variable $X \in vars(B_i) - vars(K)$, the following conditions hold: (i) $X\vartheta$ is a variable not occurring in $\{H, G_L, G_R\}$, and (ii) $X\vartheta$ does not occur in the term $Y\vartheta$, for any variable Y occurring in B_i and different from X.
By *folding* $\gamma_1, \ldots, \gamma_m$ *using* $\delta_1, \ldots, \delta_m$ we derive the clause $\eta: H \leftarrow G_L \wedge K\vartheta \wedge G_R$. From P_k we derive the program $P_{k+1} = (P_k - \{\gamma_1, \ldots, \gamma_m\}) \cup \{\eta\}$.

R5. Negative Folding. Let γ be a clause in P_k and let $Defs_k'$ be a variant of $Defs_k$ without variables in common with γ. Suppose that there exists a predicate in $Defs_k'$ whose definition consists of a single clause $\delta : K \leftarrow A$, where A is an atom. Suppose also that there exists a substitution ϑ such that clause γ is of the form: $H \leftarrow G_L \wedge \neg A\vartheta \wedge G_R$ and $vars(K) = vars(A)$.
By *folding* γ *using* δ we derive the clause $\eta: H \leftarrow G_L \wedge \neg K\vartheta \wedge G_R$. From P_k we derive the program $P_{k+1} = (P_k - \{\gamma\}) \cup \{\eta\}$.

The following *clause deletion* rule allows us to remove from P_k a redundant clause γ, that is, a clause γ such that $M(P_k) = M(P_k - \{\gamma\})$. Since the problem of testing whether or not $M(P_k) = M(P_k - \{\gamma\})$ is undecidable, we will consider some sufficient conditions based on decidable properties. These sufficient conditions are based on the notions of *subsumed* clause, clause *with false body*, and *useless* clause, which we now define.

A clause γ is *subsumed* by a clause of the form $H \leftarrow G_1$ if γ is of the form $(H \leftarrow G_1 \wedge G_2)\vartheta$ for some substitution ϑ and conjunction of literals G_2. A clause *has a false body* if it is of the form $H \leftarrow G_1 \wedge A \wedge \neg A \wedge G_2$.

The set of *useless predicates* in a program P is the maximal set U of predicates occurring in P such that a predicate p is in U iff every clause γ with head predicated p is of the form $p(\ldots) \leftarrow G_1 \wedge q(\ldots) \wedge G_2$ for some q in U. A clause in a program P is *useless* if the predicate of its head is useless in P. For example, in the following program:

$$p(X) \leftarrow q(X) \wedge \neg r(X)$$
$$q(X) \leftarrow p(X)$$
$$r(a) \leftarrow$$

p and q are useless predicates, while r is not useless.

R6. Clause Deletion. Let γ be a clause in P_k. By *clause deletion* we derive the program $P_{k+1} = P_k - \{\gamma\}$ if one of the following three cases occurs:

R6s. γ is subsumed by a clause in $P_k - \{\gamma\}$;

R6f. γ has a false body;

R6u. γ is useless in P_k.

The following *goal replacement* rule allows us to replace a conjunction of literals occurring in the body of a clause by an equivalent conjunction of literals.

R7. Goal Replacement. Let $\gamma: H \leftarrow G_1 \wedge Q \wedge G_2$ be a clause in P_k. Suppose that for some conjunction R of literals we have:

$$M(P_0) \models \forall X_1 \ldots \forall X_u \, (\exists Y_1 \ldots \exists Y_v \, Q \leftrightarrow \exists Z_1 \ldots \exists Z_w \, R)$$

where: (i) $\{X_1, \ldots, X_u\} = vars(\{H, G_1, G_2\})$, (ii) $\{Y_1, \ldots, Y_v\} = vars(Q) - \{X_1, \ldots, X_u\}$, and (iii) $\{Z_1, \ldots, Z_w\} = vars(R) - \{X_1, \ldots, X_u\}$.

Then by *goal replacement* from γ we derive the clause $\eta: H \leftarrow G_1 \wedge R \wedge G_2$. From P_k we derive the new program $P_{k+1} = (P_k - \{\gamma\}) \cup \{\eta\}$.

The following *equality introduction* rule R8i allows us to substitute a variable for a term occurring in a clause, by adding an equality in the body of the clause. The *equality elimination* rule R8e can be viewed as the inverse of rule R8i.

R8. Equality Introduction and Elimination. Let γ be a clause of the form $(H \leftarrow Body)\{X/t\}$, such that the variable X does not occur in t and let δ be the clause: $H \leftarrow X = t \wedge Body$.
R8i. By *equality introduction* we derive clause δ from clause γ. If γ occurs in P_k then we derive the new program $P_{k+1} = (P_k - \{\gamma\}) \cup \{\delta\}$.
R8e. By *equality elimination* we derive clause γ from clause δ. If δ occurs in P_k then we derive the new program $P_{k+1} = (P_k - \{\delta\}) \cup \{\gamma\}$.

The *clause splitting* rule allows us to reason by cases according to the truth value of a given atom.

R9. Clause Splitting. Let $\gamma : H \leftarrow G$ be a clause in P_k and A be an atom. Then from clause γ we derive the two clauses $\gamma_1 : H \leftarrow A \wedge G$ and $\gamma_2 : H \leftarrow \neg A \wedge G$. From P_k we derive the new program $P_{k+1} = (P_k - \{\gamma\}) \cup \{\gamma_1, \gamma_2\}$.

We say that a transformation sequence $P_0, \ldots, P_n$ is *correct* (w.r.t. the perfect model semantics), if $P_0 \cup Defs_n$ and P_n are locally stratified and $M(P_0 \cup Defs_n) = M(P_n)$. Note that, since we can introduce new predicate symbols by using rule R1, it may be the case that for a correct transformation sequence we have $M(P_0) \neq M(P_n)$.

Transformation sequences constructed by an unrestricted use of the transformation rules may not be correct. Consider, for instance, the program:

P_0: $p \leftarrow q$ $q \leftarrow$

The perfect model of P_0 is $M(P_0) = \{p, q\}$ and $M(P_0) \models p \leftrightarrow q$. Thus, we may apply the goal replacement rule R7 and replace q by p in $p \leftarrow q$. We derive the new program:

P_1: $p \leftarrow p$ $q \leftarrow$

The transformation sequence P_0, P_1 is *not* correct, because $M(P_1) = \{q\}$ and, thus, $M(P_0) \neq M(P_1)$. Indeed, P_0 succeeds for the goal p, while P_1 *does not terminate* for the goal p.

One can show that the correctness of a transformation sequence is guaranteed if termination is preserved, that is, if the initial program terminates then also the final program terminates. Now we will state a sufficient condition for the correctness of the transformation rules R1–R9 based on the notion of *left termination* [3]. An *LDNF derivation* is an SLDNF derivation constructed by using the *leftmost selection rule* [3].

Definition 1. *A program P is called* left terminating *if all LDNF derivations of P starting from a ground goal, are finite.*

The following Theorem 1 which follows from results presented in [3,9], states that if we consider a transformation sequence of locally stratified, non-floundering [3,39] programs, then the preservation of left termination guarantees the preservation of the perfect model.

Theorem 1 (Correctness of the Transformation Rules). *Let $P_0, \ldots, P_n$ be a transformation sequence such that, for $k = 0, \ldots, n$, program P_k is locally stratified, non-floundering, and left terminating. Then $M(P_0 \cup Defs_n) = M(P_n)$.*

In Theorem 1 we referred to the notion of left termination. However, weaker notions of termination may be considered and in [36], for instance, there is a correctness result for definite programs based on *existential termination*.

Theorem 1 is theoretically relevant because it relates the correctness of a transformation sequence and the preservation of left termination. However, this result is of limited use in practice for two reasons: (1) left termination is an undecidable property (as well as the properties of being locally stratified and non-floundering), and (2) left termination (or other notions of termination) may be too restrictive, especially in the cases where logic programs are used as specifications.

In Section 5 we will show some examples of transformation of nonterminating programs in the context of program verification and model checking. Correctness results w.r.t. the perfect model semantics which do not make explicit use of termination properties can be found in [26,40,52,58,60]. For lack of space we do not report those results here.

3   Transformation Strategies

In order to construct a transformation sequence $P_0, \ldots, P_n$ such that the final program P_n is more efficient than the initial program P_0, we need to apply suitable procedures, called *transformation strategies*.

In this section we will describe some of the strategies which have been proposed in the literature. In particular, we will present: (i) a strategy for *eliminating unnecessary variables* [50], (ii) a strategy for *reducing nondeterminism* [26], and (iii) a strategy for performing *program specialization* [46].

Several other strategies for transforming logic programs have been proposed. For instance, (i) the strategy for deriving *tail recursive* programs [20], (ii) the strategy for *compiling control* [13], and (iii) the strategy for *changing data representations* and, in particular, for replacing ordinary lists by *difference-lists* [68].

3.1   Eliminating Unnecessary Variables

Logic programs written in a declarative style often make use of *existential variables* (see Section 2) and *multiple variables*, that is, variables with multiple occurrences in the body of a clause. Existential variables and multiple variables are collectively called *unnecessary variables*. In the practice of logic programming, multiple occurrences of existential variables are often used for storing intermediate results, while multiple occurrences of non-existential variables are often used for defining predicates which perform multiple traversals of the input data structure.

The strategy presented in [50] has the objective of eliminating unnecessary variables, thereby avoiding both the construction of intermediate results and the multiple traversal of data structures. This strategy is related to the *deforestation* [67] and the *tupling* [43] strategies, which were introduced for the case of functional programs, and it is also related to *conjunctive partial deduction* [19] which is a technique for eliminating unnecessary variables that follows the *partial deduction* [37] approach, instead of the *rules + strategies* approach.

Now we show an example of application of the strategy for eliminating unnecessary variables.

Example 1 (Two Players Impartial Game). Consider two players sitting at a table. On the table there is a heap of matches. The two players play alternate moves and each move consists in taking away either one (move 1) or two matches (move 2) from the table. A player wins if after the opponent's move, he finds no matches on the table. Let us introduce the predicate $win(N, M)$ which holds iff either $N = 0$ or there are N matches on the table and the player who has to move, wins by making move M.

Given a natural number N, the following program *Game* computes a move M, if it exists, such that $win(N, M)$ holds.

1. $win(N, M) \leftarrow nat(N) \wedge move(M) \wedge w(N, M)$ 5. $nat(0) \leftarrow$
2. $w(0, M) \leftarrow$ 6. $nat(s(N)) \leftarrow nat(N)$
3. $w(s(N), 1) \leftarrow \neg w(N, 1) \wedge \neg w(N, 2)$ 7. $move(1) \leftarrow$
4. $w(s(s(N)), 2) \leftarrow \neg w(N, 1) \wedge \neg w(N, 2)$ 8. $move(2) \leftarrow$

The variable M occurs twice in the body of clause 1. Likewise, the variable N occurs twice in the body of clauses 1, 3, and 4. In particular, the multiple occurrences of N in clauses 3 and 4 leads to a computation with $O(2^n)$ time complexity for any query $win(n, M)$, where n is a natural number and M is a variable. We want to improve the efficiency of the above program *Game* by eliminating the multiple occurrences of variables. The strategy which allows us to do so consists in the iteration of the following two phases (see [50] for details).

Unfold phase: We apply the unfolding rule one or more times starting from clause 1, thereby deriving a set U of clauses;

Define-Fold phase: For each clause γ in U with multiple occurrences of variables in its body, we introduce a suitable new clause δ by rule R1, and we fold γ using δ so that the derived clause η has no multiple occurrences of variables in its body.

    For each new clause introduced during the *Define-Fold* phase, we perform one more iteration of the *Unfold* and *Define-Fold* phases. We store in a set, called *Defs*, all clauses introduced during every *Define-Fold* phase and we introduce a new clause δ only if we cannot apply the folding rule by using a clause already belonging to the set *Defs*.

    Let us see this strategy for eliminating the multiple occurrences of variables in action in our example.

First Iteration

Unfold. We apply the positive unfolding rule to clause 1 w.r.t. the leftmost atom in its body and we derive the following two clauses:

9. $win(0, M) \leftarrow move(M) \wedge w(0, M)$
10. $win(s(N), M) \leftarrow nat(N) \wedge move(M) \wedge w(s(N), M)$

By several applications of the positive unfolding rule, from clauses 9 and 10 we derive:

11. $win(0, M) \leftarrow move(M)$
12. $win(s(N), 1) \leftarrow nat(N) \wedge \neg w(N, 1) \wedge \neg w(N, 2)$
13. $win(s(N), 2) \leftarrow nat(N) \wedge w(s(N), 2)$

Define-Fold. We eliminate the multiple occurrences of the variable N from the bodies of clauses 12 and 13 by applying the definition introduction rule R1 and the positive folding rule R4 as follows. By rule R1 we introduce the following two clauses:

14. $new1(N) \leftarrow nat(N) \wedge \neg w(N, 1) \wedge \neg w(N, 2)$
15. $new2(N) \leftarrow nat(N) \wedge w(s(N), 2)$

and by folding clauses 12 and 13 using clauses 14 and 15, respectively, we derive:

16. $win(s(N), 1) \leftarrow new1(N)$
17. $win(s(N), 2) \leftarrow new2(N)$

without multiple occurrences of variables in their bodies. However, in the bodies of clauses 14 and 15 there are multiple occurrences of variables and, in order to eliminate them, we have to perform one more iteration of the *Unfold* and *Define-Fold* phases starting from those two clauses.

Second Iteration

Unfold. By unfolding clause 14 w.r.t. the leftmost atom in its body, we derive:

18. $new1(0) \leftarrow \neg w(0, 1) \wedge \neg w(0, 2)$
19. $new1(s(N)) \leftarrow nat(N) \wedge \neg w(s(N), 1) \wedge \neg w(s(N), 2)$

By negative unfolding, clause 18 is deleted because $w(0, 1)$ (and also $w(0, 2)$) holds (see clause 2). From clause 19, by negative unfolding w.r.t. $\neg w(s(N), 1)$, we derive:

20. $new1(s(N)) \leftarrow nat(N) \wedge w(N, 1) \wedge \neg w(s(N), 2)$
21. $new1(s(N)) \leftarrow nat(N) \wedge w(N, 2) \wedge \neg w(s(N), 2)$

Define-Fold. By applying rule R1, we introduce the following two clauses:

22. $new3(N) \leftarrow nat(N) \wedge w(N, 1) \wedge \neg w(s(N), 2)$
23. $new4(N) \leftarrow nat(N) \wedge w(N, 2) \wedge \neg w(s(N), 2)$

By folding clauses 20 and 21 using clauses 22 and 23, respectively, we derive:

24. $new1(s(N)) \leftarrow new3(N)$
25. $new1(s(N)) \leftarrow new4(N)$

without multiple occurrences of variables in their bodies. Since in the clauses 22 and 23 introduced by rule R1, there are multiple occurrences of variables, we continue the execution of the strategy starting from these two clauses as we have done above starting from clauses 14 and 15. After some more iterations of the *Unfold* and *Define-Fold* phases we derive the following final program $Game_F$ without multiple occurrences of variables.

11. $win(0, N) \leftarrow move(N)$	26. $new2(s(N)) \leftarrow new1(N)$
16. $win(s(N), 1) \leftarrow new1(N)$	27. $new3(0) \leftarrow$
17. $win(s(N), 2) \leftarrow new2(N)$	28. $new4(0) \leftarrow$
24. $new1(s(N)) \leftarrow new3(N)$	29. $new4(s(N)) \leftarrow new5(N)$
25. $new1(s(N)) \leftarrow new4(N)$	30. $new5(s(N)) \leftarrow new1(N)$

It can be verified that for the program derivation we have now completed, the local stratification, non-floundering, and left termination conditions of Theorem 1 are all satisfied. In particular, the final program $Game_L$ is a left terminating, *definite* program (and, hence, locally stratified and non-floundering). Thus, $M(Game) = M(Game_L)$.

Program $Game_L$ runs in nondeterministic $O(n)$ time for any query of the form $win(n, M)$. In the next section we will present the transformation from program $Game_L$ into a program running in deterministic $O(n)$ time.

3.2 Reducing Nondeterminism

In this section we will present the *Determinization strategy* [26] which can be applied for improving the efficiency of logic programs by reducing the nondeterminism of their computations. We will see this strategy in action by applying it to the program $Game_L$ we have derived at the end of the previous section.

Example 2 (Two Players Impartial Game, Continued). The program $Game_L$ is nondeterministic because, for any given query $win(n, M)$, where n is a ground term denoting a natural number, SLD-resolution may generate a call which is unifiable with the head of more than one program clause. For instance, if $n > 0$, the initial call $win(n, M)$ unifies with the heads of both clause 16 and clause 17. In other terms, these two clauses are *not mutually exclusive* with respect to calls of the form $win(n, M)$, where n is a ground term.

Non-mutually exclusive clauses can be avoided by transforming program $Game_L$ as follows. By the equality introduction rule R8i, from clauses 16 and 17 we derive:

31. $win(s(N), M) \leftarrow M = 1 \wedge new1(N)$
32. $win(s(N), M) \leftarrow M = 2 \wedge new2(N)$

By applying the definition introduction rule, we introduce the following two clauses:

33. $new6(N, M) \leftarrow M = 1 \wedge new1(N)$
34. $new6(N, M) \leftarrow M = 2 \wedge new2(N)$

By folding clauses 31 and 32 using clauses 33 and 34 we derive:

35. $win(s(N), M) \leftarrow new6(N, M)$

The predicate *win* is defined by the two clauses 11 and 35 which are mutually exclusive w.r.t. calls of the form $win(n, M)$. Indeed, for any given ground term n, there is at most one clause in $\{11, 35\}$ whose head is unifiable with $win(n, M)$.

Now we are left with the problem of transforming the two clauses 33 and 34 introduced by rule R1, into a set of mutually exclusive clauses (w.r.t. calls of the form $new6(n, M)$, where n is a ground term). The Determinization strategy proceeds similarly to the strategy for eliminating unnecessary variables presented in Section 3.1, by iterating an *Unfold* phase followed by a *Define-Fold* phase. During the *Define-Fold* phase we derive mutually exclusive clauses by introducing new predicates possibly defined by *more than one clause* (while in the strategy for eliminating unnecessary variables each new predicate is defined by precisely one clause).

Let us now see how the Determinization strategy proceeds in action in our example. For lack of space, we present the first iteration only.

First Iteration

Unfold. By positive unfolding, from clauses 33 and 34 we derive:

36. $new6(s(N), M) \leftarrow M = 1 \wedge new3(N)$
37. $new6(s(N), M) \leftarrow M = 1 \wedge new4(N)$
38. $new6(s(N), M) \leftarrow M = 2 \wedge new1(N)$

Define-Fold. Clauses 36, 37, and 38 are *not mutually exclusive*. By the definition introduction rule we introduce the following three clauses:

39. $new7(N, M) \leftarrow M = 1 \wedge new3(N)$
40. $new7(N, M) \leftarrow M = 1 \wedge new4(N)$
41. $new7(N, M) \leftarrow M = 2 \wedge new1(N)$

By folding clauses 36, 37, and 38 using clauses 39, 40, and 41 we derive:

42. $new6(s(N), M) \leftarrow new7(N, M)$

Clause 42 constitutes a set of mutually exclusive clauses for $new6$ (because it is one clause only). In order to transform the newly introduced clauses 39, 40, and 41 into mutually exclusive clauses, we continue the execution of the Determinization strategy and, after several iterations we derive the following program $Game_D$:

11. $win(0, M) \leftarrow move(M)$
35. $win(s(N), M) \leftarrow new6(N, M)$
42. $new6(s(N), M) \leftarrow new7(N, M)$ 45. $new8(0, M) \leftarrow M = 2$
43. $new7(0, M) \leftarrow M = 1$ 46. $new8(s(N), M) \leftarrow new9(N, M)$
44. $new7(s(N), M) \leftarrow new8(N, M)$ 47. $new9(s(N), M) \leftarrow new7(N, M)$

Program $Game_D$ is left terminating and all conditions of Theorem 1 are satisfied. Thus, $M(Game) = M(Game_D)$. Moreover, program $Game_D$ is a set of mutually exclusive clauses and computes the winning move, for any natural number n, in $O(n)$ deterministic time.

3.3   Program Specialization

Programs are often written in a parametric form so that they can be reused in different contexts, and when a parametric program is reused, one may want to improve its performance by taking advantage of the new context of use. This improvement can often be realized by applying a transformation methodology, called *program specialization* (see [29,32,37] for introductions).

The most used technique for program specialization is *partial evaluation*, also called *partial deduction* in the case of logic programs, where it has been first proposed by [33] (see also [14,15,28,38,55,61,63,66] for early work on this subject). Essentially, partial deduction can be performed by applying the transformation rules R1 (definition introduction), R2 (positive unfolding), R4 (positive folding), and R5 (negative folding) presented in Section 2 with the following restriction: by rule R1 we can introduce a new clause of the form $newp(X_1, \ldots, X_h) \leftarrow A$, where A is an atom and $X_1, \ldots, X_h$ are the variables occurring in A. This restriction limits also folding, as rules R4 and R5 are applied using clauses introduced by rule R1.

Program specialization techniques which make use of more powerful rules, such as unrestricted definition introduction (and, hence, unrestricted folding) and goal replacement have been first proposed in [8]. Here we will present an example of application of the specialization strategy introduced in [46], which extends partial deduction by also eliminating unnecessary variables and reducing nondeterminism. In our example we will derive a specialized pattern matcher for a given pattern, starting from a given parametric pattern matcher. In this example we will use constraint logic programs. As already mentioned, the extension of the transformation rules to the case of constraint logic programs has been studied in [22,26,40].

Example 3 (Constrained Matching). We define a matching relation between two strings of numbers called, respectively, the *pattern* P and the *string* S. We say that the pattern P *matches* the string S, and we write $m(P, S)$, iff $P = [p_1, \ldots, p_n]$ and in S there is a substring $Q = [q_1, \ldots, q_n]$ such that for $i = 1, \ldots, n$, $p_i \leq q_i$. (Much more complex matchers can be considered by allowing a matching relation which can be defined by any constraint logic program.)

The following constraint logic program *Match* can be taken as the specification of our parametric pattern matcher for the pattern P:

1. $m(P,S) \leftarrow app(B,C,S) \land app(A,Q,B) \land leq(P,Q)$
2. $app([\,],Ys,Ys) \leftarrow$
3. $app([X|Xs],Ys,[X|Zs]) \leftarrow app(Xs,Ys,Zs)$
4. $leq([\,],[\,]) \leftarrow$
5. $leq([X|Xs],[Y|Ys]) \leftarrow X \leq Y \land leq(Xs,Ys)$

Suppose that we want to specialize this pattern matcher to the specific pattern $P = [1,0,2]$. The specialization strategy we now apply has the same structure as the strategies presented in Sections 3.1 and 3.2. The improvements gained through the application of the specialization strategy are due to the fact that this strategy: (i) makes some precalculations which depend on the specific pattern $P = [1,0,2]$, (ii) eliminates unnecessary variables, and (iii) reduces nondeterminism. As already mentioned, these improvements are possible because we use more powerful transformation rules with respect to partial deduction (which would only perform the precalculations of Point (i)).

The specialization strategy starts off by introducing the following clause which defines the specialized matching relation m_{sp}:

6. $m_{sp}(S) \leftarrow m([1,0,2],S)$

Now we iterate *Unfold* and *Define-Fold* phases. The main difference with the applications of the strategies presented in Sections 3.1 and 3.2 will be that, in order to get mutually exclusive clauses, before applying the definition introduction rule and the folding rule, we will apply the clause splitting rule R9 whenever needed.

First Iteration

Unfold. We unfold clause 6 w.r.t. the atom $m([1,0,2],S)$. We derive:

7. $m_{sp}(S) \leftarrow app(B,C,S) \land app(A,Q,B) \land leq([1,0,2],Q)$

Define-Fold. In order to fold clause 7, we introduce the following definition:

8. $new1(S) \leftarrow app(B,C,S) \land app(A,Q,B) \land leq([1,0,2],Q)$

Then we fold clause 7 and we derive:

9. $m_{sp}(S) \leftarrow new1(S)$

Now the strategy continues by transforming the newly introduced clause 8.

Second Iteration

Unfold. We unfold clause 8 w.r.t. the atoms *app* and *leq* and we get:

10. $new1([X|Xs]) \leftarrow 1 \leq X \land app(Q,C,Xs) \land leq([0,2],Q)$
11. $new1([X|Xs]) \leftarrow app(B,C,Xs) \land app(A,Q,B) \land leq([1,0,2],Q)$

Clause Splitting. In order to derive mutually exclusive clauses, thereby reducing nondeterminism, we apply the clause splitting rule to clause 11, by separating the cases when $1 \leq X$ and when $1 > X$ (that is, $\neg(1 \leq X)$). We get:

12. $new1([X|Xs]) \leftarrow 1 \leq X \land app(B,C,Xs) \land app(A,Q,B) \land leq([1,0,2],Q)$
13. $new1([X|Xs]) \leftarrow 1 > X \land app(B,C,Xs) \land app(A,Q,B) \land leq([1,0,2],Q)$

Define-Fold. In order to fold clauses 10 and 12 we introduce the following two clauses defining the predicate $new2$:

14. $new2(Xs) \leftarrow app(Q, C, Xs) \wedge leq([0, 2], Q)$
15. $new2(Xs) \leftarrow app(B, C, Xs) \wedge app(A, Q, B) \wedge leq([1, 0, 2], Q)$

Then we fold clauses 10 and 12 by using the two clauses 14 and 15 and we also fold clause 13 by using clause 8. We derive the following clauses:

16. $new1([X|Xs]) \leftarrow 1 \leq X \wedge new2(Xs)$
17. $new1([X|Xs]) \leftarrow 1 > X \wedge new1(Xs)$

Note that these two clauses: (i) are specialized w.r.t. the information that the first element of the pattern is 1, (ii) have no unnecessary variables, and (iii) are mutually exclusive because of the constraints $1 \leq X$ and $1 > X$.

Now the program transformation strategy continues by transforming clauses 14 and 15, which define predicate $new2$. After a few more iterations of the *Unfold*, *Clause Splitting*, and *Define-Fold* phases, we derive the following specialized program $Match_{sp}$:

9. $m_{sp}(S) \leftarrow new1(S)$
16. $new1([X|Xs]) \leftarrow 1 \leq X \wedge new2(Xs)$
17. $new1([X|Xs]) \leftarrow 1 > X \wedge new1(Xs)$
18. $new2([X|Xs]) \leftarrow 1 \leq X \wedge new3(Xs)$
19. $new2([X|Xs]) \leftarrow 0 \leq X \wedge 1 > X \wedge new4(Xs)$
20. $new2([X|Xs]) \leftarrow 0 > X \wedge new1(Xs)$
21. $new3([X|Xs]) \leftarrow 2 \leq X \wedge new5(Xs)$
22. $new3([X|Xs]) \leftarrow 1 \leq X \wedge 2 > X \wedge new3(Xs)$
23. $new3([X|Xs]) \leftarrow 0 \leq X \wedge 1 > X \wedge new4(Xs)$
24. $new3([X|Xs]) \leftarrow 0 > X \wedge new1(Xs)$
25. $new4([X|Xs]) \leftarrow 2 \leq X \wedge new6(Xs)$
26. $new4([X|Xs]) \leftarrow 1 \leq X \wedge 2 > X \wedge new2(Xs)$
27. $new4([X|Xs]) \leftarrow 1 > X \wedge new1(Xs)$
28. $new5([X|Xs]) \leftarrow$
29. $new6([X|Xs]) \leftarrow$

This final program $Match_{sp}$ has no occurrences of unnecessary variables and is deterministic in the sense that at most one clause can be applied during the evaluation of any ground goal. The efficiency of $Match_{sp}$ is very high because it behaves like a deterministic finite automaton (see Figure 1) as the Knuth-Morris-Pratt matcher.

4   Program Synthesis

Program synthesis is a technique for the automatic derivation of programs from their formal specifications (see, for instance, [41] for the derivation of functional programs and [16,27,31] for the derivation of logic programs from first-order logic specifications).

In this section we present a transformational approach to program synthesis [26,56]. By following this approach, the synthesis of an *efficient* logic program from a first order logic specification can be performed in two steps: first (1) we translate the specification into a possibly inefficient logic program by applying the *Lloyd-Topor transformation* [39], and then (2) we derive an efficient program by applying the transformation rules and strategies described in Sections 2 and 3.

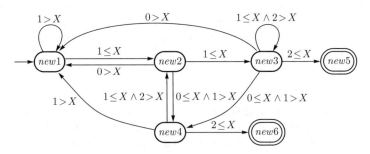

Fig. 1. The finite automaton corresponding to the program $Match_{sp}$ made out of clauses 9 and 16–29. The initial state is $new1$ and the final states are $new5$ and $new6$.

The transformational program synthesis approach will be presented through the N-queens example. This example also illustrates that powerful programming techniques such as recursion and backtracking, which are often presented in the literature for solving the N-queens problem, can indeed be automatically derived by transformation.

Example 4 (N-queens). We are required to place N (≥ 0) queens on an $N \times N$ chess board, so that no two queens attack each other, that is, they do not lie on the same row, column, or diagonal. By using the fact that no two queens should lie on the same row, we represent the positions of the N queens on the board as a permutation $L = [i_1, \dots, i_N]$ of the list $[1, \dots, N]$ which tells us that the queen on row k is placed on column i_k.

A specification of the solution L for the N-queens problem is given by the following first-order formula:

$$board(N, L) =_{def} nat(N) \land nat\_list(L) \land length(L, N) \land$$
$$\forall X \, (member(X, L) \to in(X, 1, N)) \land$$
$$\forall A \, \forall B \, \forall K \, \forall M$$
$$((1 \leq K \land K \leq M \land occurs(A, K, L) \land occurs(B, M, L))$$
$$\to (A \neq B \land A - B \neq M - K \land B - A \neq M - K))$$

where the various predicates that occur in $board(N, L)$, are defined by the following constraint logic program P:

$$nat(0) \leftarrow$$
$$nat(N) \leftarrow N = M + 1 \land M \geq 0 \land nat(M)$$
$$nat\_list([]) \leftarrow$$
$$nat\_list([H|T]) \leftarrow nat(H) \land nat\_list(T)$$
$$length([], 0) \leftarrow$$
$$length([H|T], N) \leftarrow N = M + 1 \land M \geq 0 \land length(T, M)$$
$$member(X, [H|T]) \leftarrow X = H$$
$$member(X, [H|T]) \leftarrow member(X, T)$$
$$in(X, M, N) \leftarrow X = N \land M \leq N$$
$$in(X, M, N) \leftarrow N = K + 1 \land M \leq K \land in(X, M, K)$$
$$occurs(X, I, [H|T]) \leftarrow I = 1 \land X = H$$
$$occurs(X, J, [H|T]) \leftarrow J = I + 1 \land I \geq 1 \land occurs(X, I, T)$$

In this program P we have that: (i) $in(X, M, N)$ iff $M \leq X \leq N$, and (ii) $occurs(X, I, [a_1, \ldots, a_n])$ iff $X = a_i$ and $I = i$. Now, we would like to synthesize a constraint logic program R which computes a predicate $queens(N, L)$ such that, for every N and L, the following property holds:

$$M(R) \models queens(N, L) \text{ iff } M(P) \models board(N, L) \qquad (\alpha)$$

where by $M(R)$ and $M(P)$ we denote the perfect model of the programs R and P, respectively. By applying the technique presented in [26], we start off from the formula $queens(N, L) \leftarrow board(N, L)$ (where $board(N, L)$ is the first order formula defined above) and, by applying a variant of the Lloyd-Topor transformation, we derive the following stratified program F:

$queens(N, L) \leftarrow nat(N) \wedge nat\_list(L) \wedge length(L, N) \wedge \neg aux1(L, N) \wedge \neg aux2(L)$
$aux1(L, N) \leftarrow member(X, L) \wedge \neg in(X, 1, N)$
$aux2(L) \leftarrow 1 \leq K \wedge K \leq M \wedge \neg(A \neq B \wedge A - B \neq M - K \wedge B - A \neq M - K) \wedge$
$\qquad occurs(A, K, L) \wedge occurs(B, M, L)$

It can be shown that this variant of the Lloyd-Topor transformation preserves the perfect model semantics and, thus, we have that, for every N and L:

$$M(P \cup F) \models queens(N, L) \text{ iff } M(P) \models board(N, L).$$

The derived program $P \cup F$ is not satisfactory from a computational point of view, when using LDNF resolution. Indeed, for a query of the form $queens(n, L)$, where n is a nonnegative integer and L is a variable, program $P \cup F$ works by first generating a value l for the list L and then testing whether or not $length(l, n) \wedge \neg aux1(l, n) \wedge \neg aux2(l)$ holds. This generate-and-test behavior is very inefficient and it may also lead to nontermination. Thus, the process of program synthesis proceeds by applying the definition, unfolding, folding, and goal replacement transformation rules, according to a strategy similar to the ones we have described in Section 3, with the objective of deriving a more efficient program. We derive the following definite program R:

$queens(N, L) \leftarrow new2(N, L, 0)$
$new2(N, [\,], K) \leftarrow N = K$
$new2(N, [H|T], K) \leftarrow N \geq K + 1 \wedge new2(N, T, K + 1) \wedge new3(H, T, N, 0)$
$new3(A, [\,], N, M) \leftarrow in(A, 1, N) \wedge nat(A)$
$new3(A, [B|T], N, M) \leftarrow A \neq B \wedge A - B \neq M + 1 \wedge B - A \neq M + 1 \wedge nat(B) \wedge$
$\qquad new3(A, T, N, M + 1)$

together with the clauses listed above which define the predicates in and nat.

Since the transformation rules preserve the perfect model semantics, for every N and L, we have that, $M(R) \models queens(N, L)$ iff $M(P \cup F) \models queens(N, L)$ and, thus, Property (α) holds. It can be shown that program R terminates for all queries of the form $queens(n, L)$. Program R computes a solution for the N-queens problem in a clever way: each time a new queen is placed on the board, program R tests whether or not that queen attacks any other queen already placed on the board.

5   Program Verification

Proofs of program properties are often needed during program development for checking the correctness of software components with respect to their specifications. It has been shown that the transformation rules introduced in [17,64] can be used for proving several kinds of program properties, such as equivalences of functions defined by recursive equation programs [34], equivalences of predicates defined by logic programs [44], first-order properties of predicates defined by constraint logic programs [47], and temporal properties of concurrent systems [25,54].

In this section we see the use of program transformation for proving program properties specified either by first-order logic formulas or by temporal logic formulas.

5.1 The Unfold/Fold Proof Method

Through a simple example taken from [47], now we illustrate a method, called *unfold/fold proof method*, which uses the program transformation methodology for proving first-order properties of constraint logic programs. Consider the following constraint logic program *Member* which defines the membership relation between an element and a list of elements:

$$member(X, [Y|L]) \leftarrow X = Y \qquad\qquad list([\,]) \leftarrow$$
$$member(X, [Y|L]) \leftarrow member(X, L) \qquad list([H|T]) \leftarrow list(T)$$

Suppose we want to show that every finite list of numbers has an upper bound, that is, we want to prove the following formula:

$$\forall L \, (list(L) \rightarrow \exists U \, \forall X \, (member(X, L) \rightarrow X \leq U)) \qquad\qquad (\beta)$$

The unfold/fold proof method works in two steps, which are similar to the two steps of the transformational synthesis approach presented in Section 4. In the first step, the formula β is transformed into a set of clauses by applying a variant of the Lloyd-Topor transformation, thereby deriving the following program:

$P1$: $prop \leftarrow \neg p$
$\quad\quad p \leftarrow list(L) \wedge \neg q(L)$
$\quad\quad q(L) \leftarrow list(L) \wedge \neg r(L, U)$
$\quad\quad r(L, U) \leftarrow X > U \wedge list(L) \wedge member(X, L)$

The predicate *prop* is equivalent to β in the sense that $M(Member) \models \beta$ iff $M(Member \cup P1) \models prop$. The correctness of this transformation can be checked by realizing that $M(Member) \models \beta \leftrightarrow \neg \exists L(list(L) \wedge \neg(\exists U(list(L) \wedge \neg(\exists X (X > U \wedge list(L) \wedge member(X, L))))))$.

In the second step, we eliminate the *existential variables* occurring in $P1$ (see Section 2 for a definition) by applying the transformation strategy for eliminating unnecessary variables presented in Section 3.1. We derive the following program $P2$ which defines the predicate *prop*:

$P2$: $prop \leftarrow \neg p$ $\qquad\qquad p \leftarrow p_1 \qquad\qquad p_1 \leftarrow p_1$

Now, $P2$ is a propositional program and has a *finite* perfect model, which is $\{prop\}$. Since it can be shown that all transformations we have performed preserve the perfect model, we have that $M(Member) \models \beta$ iff $M(P2) \models prop$ and, therefore, we have completed the proof of β because *prop* belongs to $M(P2)$.

The expert reader will note that the unfold/fold proof method we have now illustrated, can be viewed as an extension to constraint logic programs of the *quantifier elimination* method, which has well-known applications in the field of automated theorem proving (see [51] for a brief survey).

5.2   Infinite-State Model Checking

As indicated in [18], the behavior of a concurrent system that evolves over time according to a given protocol can be modeled as a *state transition system*, that is, (i) a set S of *states*, (ii) an *initial state* $s_0 \in S$, and (iii) a *transition relation* $t \subseteq S \times S$. We assume that the transition relation t is *total*, that is, for every state $s \in S$ there exists at least one state $s' \in S$, called a *successor state* of s, such that $t(s, s')$ holds. A *computation path* starting from a state s_1 (not necessarily the initial state) is an *infinite* sequence of states $s_1 s_2 \ldots$ such that, for every $i \geq 1$, there is a transition from s_i to s_{i+1}, that is, $t(s_i, s_{i+1})$ holds.

The properties of the evolution over time, that is, the computation paths, of a concurrent system can be specified by using a formula of a temporal logic called *Computation Tree Logic* (or CTL, for short [18]). The formulas of CTL are built from a given set of *elementary properties*, each of which may or may not hold in a particular state, by using: (i) the connectives: *not* and *and*, (ii) the quantifiers along a computation path: g ('for all states on the path' or 'globally'), f ('there exists a state on the path' or 'in the future'), x ('next time'), and u ('until'), and (iii) the quantifiers over computation paths: a ('for all paths') and e ('there exists a path'). Quantified formulas are written in a compact form and, for instance, we will write $ef(F)$ and $ag(F)$, instead of $e(f(F))$ and $a(g(F))$, respectively.

Very efficient algorithms and tools exist for verifying temporal properties of *finite state transition systems*, that is, systems where the set S of states is finite [18]. However, many concurrent systems cannot be modeled by finite state transition systems. The problem of verifying CTL properties of *infinite* state transition systems is, unfortunately, undecidable and, thus, it cannot be tackled by traditional model checking techniques. For this reason various methods based on automated theorem proving have been proposed for extending model checking so to deal with infinite state systems (see [21] for a method based on constraint logic programming). Due to the above mentioned undecidability limitation, all these methods are necessarily incomplete.

Now we present a method for verifying temporal properties of (finite or infinite) state transition systems which is based on transformation techniques for constraint logic programs [25]. As an example we consider the *Bakery* protocol [35] and we verify that it satisfies the *mutual exclusion* and *starvation freedom* properties.

Let us consider two agents A and B which want to access a shared resource in a mutually exclusive way by using the Bakery protocol. The state of the agent A is represented by a pair $\langle A1, A2 \rangle$, where $A1$, called the *control state*, is an element of the set $\{t, w, u\}$ (where t, w, and u stand for *think*, *wait*, and *use*, respectively) and $A2$, called the *counter*, is a natural number. Analogously, the state of agent B is represented by a pair $\langle B1, B2 \rangle$. The *state* of the system consisting of the two agents A and B, whose states are $\langle A1, A2 \rangle$ and $\langle B1, B2 \rangle$, respectively, is represented by the 4-tuple $\langle A1, A2, B1, B2 \rangle$. The transition relation t of the two agent system from an old state *OldS* to a new state *NewS*, is defined as follows:

$t(OldS,\ NewS) \leftarrow t_A(OldS,\ NewS)$
$t(OldS,\ NewS) \leftarrow t_B(OldS,\ NewS)$

where the transition relation t_A for the agent A is given by the following clauses whose bodies are conjunctions of constraints (see also Figure 2):

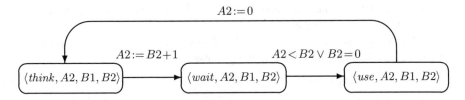

Fig. 2. The Bakery protocol: a graphical representation of the transition relation t_A for the agent A. The assignment $X := e$ on the arc from a state s_1 to a state s_2 tells us that the value of the variable X in s_2 is the value of the expression e in s_1. The boolean expression b on the arc from a state s_1 to a state s_2 tells us that the transition from s_1 to s_2 takes place iff b holds.

$$t_A(\langle t, A2, B1, B2\rangle, \langle w, A21, B1, B2\rangle) \leftarrow A21 = B2+1$$
$$t_A(\langle w, A2, B1, B2\rangle, \langle u, A2, B1, B2\rangle) \leftarrow A2 < B2$$
$$t_A(\langle w, A2, B1, B2\rangle, \langle u, A2, B1, B2\rangle) \leftarrow B2 = 0$$
$$t_A(\langle u, A2, B1, B2\rangle, \langle t, A21, B1, B2\rangle) \leftarrow A21 = 0$$

The following similar clauses define the transition relation t_B for the agent B:

$$t_B(\langle A1, A2, t, B2\rangle, \langle A1, A2, w, B21\rangle) \leftarrow B21 = A2+1$$
$$t_B(\langle A1, A2, w, B2\rangle, \langle A1, A2, u, B2\rangle) \leftarrow B2 < A2$$
$$t_B(\langle A1, A2, w, B2\rangle, \langle A1, A2, u, B2\rangle) \leftarrow A2 = 0$$
$$t_B(\langle A1, A2, u, B2\rangle, \langle A1, A2, t, B21\rangle) \leftarrow B21 = 0$$

Note that the system has an infinite number of states, because counters may increase in an unbounded way.

The temporal properties of a transition system are specified by defining a predicate $sat(S, P)$ which holds if and only if the temporal formula P is true at the state S. For instance, the following clauses define the predicate $sat(S, P)$ for the cases where P is: (i) an elementary formula F, (ii) a formula of the form $not(F)$, (iii) a formula of the form $and(F_1, F_2)$, and (iv) a formula of the form $ef(F)$:

$$sat(S, F) \leftarrow elem(S, F)$$
$$sat(S, not(F)) \leftarrow \neg sat(S, F)$$
$$sat(X, and(F_1, F_2)) \leftarrow sat(X, F_1) \wedge sat(X, F_2)$$
$$sat(S, ef(F)) \leftarrow sat(S, F)$$
$$sat(S, ef(F)) \leftarrow t(S, T) \wedge sat(T, ef(F))$$

where $elem(S, F)$ holds iff F is an elementary property which is true at state S. In particular, for the Bakery protocol we have the following clause:

$$elem(\langle u, A2, u, B2\rangle, unsafe) \leftarrow$$

that is, *unsafe* holds at a state where both agents A and B are in the control state u, that is, both agents use the shared resource at the same time. We have that $sat(S, ef(F))$ holds iff there exists a computation path π starting from state S and there exists a state S' on π such that F is true at S'.

The mutual exclusion property holds for the Bakery protocol if there is no computation path starting from the initial state such that at a state on this path the *unsafe* property holds. Thus, the mutual exclusion property holds if $sat(\langle t, 0, t, 0\rangle, not(ef(unsafe)))$ belongs to the perfect model $M(P_{mex})$, where: (i) $\langle t, 0, t, 0\rangle$ is the initial state of the system and (ii) P_{mex} is the program consisting of the clauses for the predicates t, t_A, t_B, sat, and $elem$ defined above.

In order to show that $sat(\langle t, 0, t, 0 \rangle, not(ef(unsafe))) \in M(P_{mex})$, we introduce a new predicate mex defined by the following clause:

$$mex \leftarrow sat(\langle t, 0, t, 0 \rangle, not(ef(unsafe))) \qquad (\mu)$$

and we transform the program $P_{mex} \cup \{\mu\}$ into a new program Q which contains a clause of the form $mex \leftarrow$ (see [25] for details). This transformation is performed by applying the definition, unfolding, and folding rules according to a strategy similar to the specialization strategy presented in Section 3.3, that is, a strategy that derives specialized clauses for the evaluation of the predicate mex. From the correctness of the transformation rules we have that $mex \in M(Q)$ iff $mex \in M(P_{mex} \cup \{\mu\})$ and, hence, $sat(\langle t, 0, t, 0 \rangle, not(ef(unsafe))) \in M(P_{mex})$, that is, the mutual exclusion property holds.

By applying the same methodology we can also prove the *starvation freedom* property for the Bakery protocol. This property ensures that an agent, say A, which requests the shared resource, will eventually get it. This property is expressed by the CTL formula: $ag(w_A \rightarrow af(u_A))$, which is equivalent to: $not(ef(and(w_A, not(af(u_A)))))$. The clauses defining the elementary properties w_A and u_A are:

$elem(\langle w, A2, B1, B2 \rangle, w_A) \leftarrow$
$elem(\langle u, A2, B1, B2 \rangle, u_A) \leftarrow$

The clauses defining the predicate $sat(S, P)$ for the case where P is a CTL formula of the form $af(F)$ are:

$sat(X, af(F)) \leftarrow sat(X, F)$
$sat(X, af(F)) \leftarrow ts(X, Ys) \wedge sat\_all(Ys, af(F))$
$sat\_all([], F) \leftarrow$
$sat\_all([X|Xs], F) \leftarrow sat(X, F) \wedge sat\_all(Xs, F)$

where $ts(X, Ys)$ holds iff Ys is a list of all the successor states of the state X. For instance, one of the clauses defining predicate ts in our Bakery example is:

$ts(\langle t, A2, t, B2 \rangle, [\langle w, A21, t, B2 \rangle, \langle t, A2, w, B21 \rangle]) \leftarrow A21 = B2+1 \wedge B21 = A2+1$

which says that the state $\langle t, A2, t, B2 \rangle$ has two successor states: $\langle w, A21, t, B2 \rangle$, with $A21 = B2+1$, and $\langle t, A2, w, B21 \rangle$, with $B21 = A2+1$.

Let P_{sf} denote the program obtained by adding to P_{mex} the clauses defining: (i) the elementary properties w_A and u_A, (ii) the predicate ts, (iii) the atom $sat(X, af(F))$, and (iv) the predicate sat_all. In order to verify the starvation freedom property we introduce the clause:

$$sf \leftarrow sat(\langle t, 0, t, 0 \rangle, not(ef(and(w_A, not(af(u_A)))))) \qquad (\sigma)$$

and, by applying the definition, unfolding, and folding rules according to the specialization strategy, we transform the program $P_{sf} \cup \{\sigma\}$ into a new program R which contains a clause of the form $sf \leftarrow$.

Note that the derivations needed for verifying the mutual exclusion and the starvation freedom properties can be done in a fully automatic way by using the experimental constraint logic program transformation system MAP [42].

6   Conclusions and Future Directions

We have presented the program transformation methodology and we have demonstrated that it is very effective for: (i) the derivation of correct software modules from their formal specifications, and (ii) the proof of properties of programs. Since program transformation preserves correctness and improves efficiency, it is very useful for constructing software products which are provably correct and whose time and space performance is very high.

During the past twenty-five years the research community in Italy has given a very relevant contribution to the program transformation field and, more in general, to the field of logic-based program development. The extent of this contribution is witnessed by the numerous scientific papers, a small fraction of which have been mentioned in this brief survey.

The contribution of the Italian research community has also been carried out through the participation in several national and international research projects which included as an important topic the transformation methodology of logic programs. In particular, we would like to mention the following projects: (i) ESPRIT Alpes (1984–89), (ii) Compulog I and Compulog II (1989–95), (iii) the INTAS Project 'Efficient Symbolic Computing' (1994-98), (iv) the Network of Excellence on Computational Logic, (v) the Humal Capital and Mobility Project 'Logic Program Synthesis and Transformation' (1993–96), (vi) the Italian 'Progetto Finalizzato Informatica II' (1989–93), (vii) the ANATRA Project 'Strumenti per l'analisi e la trasformazione dei programmi' (1994–95), (viii) 'Programmazione Logica: Strumenti per analisi e trasformazione di programmi, Tecniche di ingegneria del software, Estensioni con vincoli, concorrenza ed oggetti' (1995–96), (ix) Progetto Speciale 'Verifica, analisi e trasformazione di programmi logici' (1998–99), and (x) 'Tecniche formali per la specifica, l'analisi, la verifica, la sintesi e la trasformazione di sistemi software' (1998–2000). These projects were supported by the European Union, the Italian Ministry of Education, University, and Research (MIUR), and the Italian National Research Council (CNR).

All these projects gave to the research community in Italy invaluable opportunities to cooperate with other scientific groups in Europe, to strengthen their theoretical background on logic programming and to produce powerful systems and tools for logic program development, logic program analysis, knowledge representation and manipulation using logic. Research teams in Bologna, Padua, Pisa, Rome, and Venice, among others, grew considerably strong through those projects and their expertise and competence spread all over the international community and since then, their high reputation has been widely recognized.

Finally, the Italian research community has also given a very relevant contribution to the organization and the scientific success of the various meetings dedicated to the dissemination of research in logic program transformation, such as the series of Workshops and Symposia on Logic-Based Program Synthesis and Transformation (LOPSTR), held annually since 1991, and on Partial Evaluation and Semantics-Based Program Manipulation (PEPM).

Now, looking at the directions for future research, we would like to point out that, in order to make program transformation even more effective, we need to increase the level of automation of the transformation strategies for program improvement, program synthesis, and program verification. Furthermore, these strategies should be incorporated into powerful tools for program development.

Another important direction for future research is the exploration of new areas of application of the transformation methodology. In this paper we have described the use of program transformation for verifying temporal properties of infinite state concurrent systems. Similar techniques could also be devised for verifying other kinds of properties and other classes of systems, such as security properties of distributed systems, safety properties of hybrid systems, and protocol conformance of multiagent systems. A more challenging issue is the fully automatic synthesis of software systems which are guaranteed to satisfy some given properties specified by the designer.

Acknowledgements

We would like to thank the members of GULP, the Italian Association for Logic Programming, who throughout all these years have been for us of great scientific support and encouragement. Their cooperation and friendship are very much appreciated.

Many thanks also to Agostino Dovier and Enrico Pontelli, editors of this book, for their invitation to present the contributions of the program transformation methodology in the field of logic programming.

References

1. Alpuente, M., Falaschi, M., Moreno, G., Vidal, G.: A transformation system for lazy functional logic programs. In: Middeldorp, A., Sato, T. (eds.) FLOPS 1999. LNCS, vol. 1722, pp. 147–162. Springer, Heidelberg (1999)
2. Apt, K.R., Bol, R.N.: Logic programming and negation: A survey. Journal of Logic Programming 19, 20, 9–71 (1994)
3. Apt, K.R., Pedreschi, D.: Reasoning about termination of pure logic programs. Information and Computation 106, 109–157 (1993)
4. Aravindan, C., Dung, P.M.: On the correctness of unfold/fold transformation of normal and extended logic programs. Journal of Logic Programming 24(3), 201–217 (1995)
5. Basin, D., Deville, Y., Flener, P., Hamfelt, A., Fischer Nilsson, J.: Synthesis of programs in computational logic. In: Bruynooghe, M., Lau, K.-K. (eds.) Program Development in Computational Logic. LNCS, vol. 3049, pp. 30–65. Springer, Heidelberg (2004)
6. Bossi, A., Cocco, N.: Basic transformation operations which preserve computed answer substitutions of logic programs. Journal of Logic Programming 16(1&2), 47–87 (1993)
7. Bossi, A., Cocco, N.: Preserving universal termination through unfold/fold. In: Rodríguez-Artalejo, M., Levi, G. (eds.) ALP 1994. LNCS, vol. 850, pp. 269–286. Springer, Heidelberg (1994)
8. Bossi, A., Cocco, N., Dulli, S.: A method for specializing logic programs. ACM Transactions on Programming Languages and Systems 12(2), 253–302 (1990)
9. Bossi, A., Cocco, N., Etalle, S.: Transforming normal programs by replacement. In: Pettorossi, A. (ed.) META 1992. LNCS, vol. 649, pp. 265–279. Springer, Heidelberg (1992)
10. Bossi, A., Cocco, N., Etalle, S.: Simultaneous replacement in normal programs. Journal of Logic and Computation 6(1), 79–120 (1996)
11. Bossi, A., Cocco, N., Etalle, S.: Transforming left-terminating programs: The reordering problem. In: Proietti, M. (ed.) LOPSTR 1995. LNCS, vol. 1048, pp. 33–45. Springer, Heidelberg (1996)

12. Bossi, A., Etalle, S.: Transforming acyclic programs. ACM Transactions on Programming Languages and Systems 16(4), 1081–1096 (1994)
13. Bruynooghe, M., De Schreye, D., Krekels, B.: Compiling control. Journal of Logic Programming 6, 135–162 (1989)
14. Bugliesi, M., Lamma, E., Mello, P.: Partial evaluation for hierarchies of logic theories. In: Debray, S., Hermenegildo, M. (eds.) Logic Programming: Proceedings of the 1990 North American Conference, Austin, Texas, October 1990, pp. 359–376. MIT Press, Cambridge (1990)
15. Bugliesi, M., Rossi, F.: Partial evaluation in Prolog: Some Improvements about Cut. In: Lusk, E.L., Overbeek, R.A. (eds.) Logic Programming: Proceedings of the North American Conference 1989, Cleveland, Ohio, October 1989, pp. 645–660. MIT Press, Cambridge (1989)
16. Bundy, A., Smaill, A., Wiggins, G.: The synthesis of logic programs from inductive proofs. In: Lloyd, J.W. (ed.) Computational Logic, Symposium Proceedings, Brussels, November 1990, pp. 135–149. Springer, Berlin (1990)
17. Burstall, R.M., Darlington, J.: A transformation system for developing recursive programs. Journal of the ACM 24(1), 44–67 (1977)
18. Clarke, E.M., Grumberg, O., Peled, D.: Model Checking. MIT Press, Cambridge (1999)
19. De Schreye, D., Glück, R., Jørgensen, J., Leuschel, M., Martens, B., Sørensen, M.H.: Conjunctive partial deduction: Foundations, control, algorithms, and experiments. Journal of Logic Programming 41(2–3), 231–277 (1999)
20. Debray, S.K.: Optimizing almost-tail-recursive Prolog programs. In: Jouannaud, J.-P. (ed.) FPCA 1985. LNCS, vol. 201, pp. 204–219. Springer, Heidelberg (1985)
21. Delzanno, G., Podelski, A.: Constraint-based deductive model checking. International Journal on Software Tools for Technology Transfer 3(3), 250–270 (2001)
22. Etalle, S., Gabbrielli, M.: Transformations of CLP modules. Theoretical Computer Science 166, 101–146 (1996)
23. Etalle, S., Gabbrielli, M., Marchiori, E.: A transformation system for CLP with dynamic scheduling and CCP. In: PEPM 1997, pp. 137–150. ACM Press, New York (1997)
24. Etalle, S., Gabbrielli, M., Meo, M.C.: Transformations of ccp programs. ACM Transactions on Programming Languages and Systems 23(3), 304–395 (2001)
25. Fioravanti, F., Pettorossi, A., Proietti, M.: Verifying CTL properties of infinite state systems by specializing constraint logic programs. In: Proceedings of the ACM Sigplan Workshop on Verification and Computational Logic VCL 2001, Florence (Italy), Technical Report DSSE-TR-2001-3, pp. 85–96. University of Southampton, UK (2001)
26. Fioravanti, F., Pettorossi, A., Proietti, M.: Transformation rules for locally stratified constraint logic programs. In: Bruynooghe, M., Lau, K.-K. (eds.) Program Development in Computational Logic. LNCS, vol. 3049, pp. 292–340. Springer, Heidelberg (2004)
27. Flener, P., Lau, K.-K., Ornaghi, M., Richardson, J.: An abstract formalization of correct schemas for program synthesis. Journal of Symbolic Computation 30(1), 93–127 (2000)
28. Gallagher, J.P.: Transforming programs by specialising interpreters. In: Proceedings Seventh European Conference on Artificial Intelligence, ECAI 1986, pp. 109–122 (1986)
29. Gallagher, J.P.: Tutorial on specialisation of logic programs. In: Proceedings of the 1993 ACM SIGPLAN Symposium on Partial Evaluation and Semantics Based Program Manipulation, PEPM 1993, Copenhagen, Denmark, pp. 88–98. ACM Press, New York (1993)
30. Gardner, P.A., Shepherdson, J.C.: Unfold/fold transformations of logic programs. In: Lassez, J.-L., Plotkin, G. (eds.) Computational Logic, Essays in Honor of Alan Robinson, pp. 565–583. MIT, Cambridge (1991)
31. Hogger, C.J.: Derivation of logic programs. Journal of the ACM 28(2), 372–392 (1981)
32. Jones, N.D., Gomard, C.K., Sestoft, P.: Partial Evaluation and Automatic Program Generation. Prentice-Hall, Englewood Cliffs (1993)

33. Komorowski, H.J.: Partial evaluation as a means for inferencing data structures in an applicative language: A theory and implementation in the case of Prolog. In: Ninth ACM Symposium on Principles of Programming Languages, Albuquerque, New Mexico, USA, pp. 255–267 (1982)

34. Kott, L.: The McCarthy's induction principle: 'oldy' but 'goody'. Calcolo 19(1), 59–69 (1982)

35. Lamport, L.: A new solution of Dijkstra's concurrent programming problem. Communications of the ACM 17(8), 453–455 (1974)

36. Lau, K.-K., Ornaghi, M., Pettorossi, A., Proietti, M.: Correctness of logic program transformation based on existential termination. In: Lloyd, J.W. (ed.) Proceedings of the 1995 International Logic Programming Symposium (ILPS 1995), pp. 480–494. MIT Press, Cambridge (1995)

37. Leuschel, M., Bruynooghe, M.: Logic program specialisation through partial deduction: Control issues. Theory and Practice of Logic Programming 2(4&5), 461–515 (2002)

38. Levi, G., Sardu, G.: Partial evaluation of meta programs in a multiple worlds logic language. New Generation Computing 6(2&3), 227–248 (1988)

39. Lloyd, J.W.: Foundations of Logic Programming, 2nd edn. Springer, Berlin (1987)

40. Maher, M.J.: A transformation system for deductive database modules with perfect model semantics. Theoretical Computer Science 110, 377–403 (1993)

41. Manna, Z., Waldinger, R.: A deductive approach to program synthesis. ACM Toplas 2, 90–121 (1980)

42. The MAP transformation system (1995–2010),
    http://www.iasi.cnr.it/~proietti/system.html

43. Pettorossi, A.: A powerful strategy for deriving efficient programs by transformation. In: ACM Symposium on Lisp and Functional Programming, pp. 273–281. ACM Press, New York (1984)

44. Pettorossi, A., Proietti, M.: Synthesis and transformation of logic programs using unfold/fold proofs. Journal of Logic Programming 41(2&3), 197–230 (1999)

45. Pettorossi, A., Proietti, M.: Perfect model checking via unfold/fold transformations. In: Palamidessi, C., Moniz Pereira, L., Lloyd, J.W., Dahl, V., Furbach, U., Kerber, M., Lau, K.-K., Sagiv, Y., Stuckey, P.J. (eds.) CL 2000. LNCS (LNAI), vol. 1861, pp. 613–628. Springer, Heidelberg (2000)

46. Pettorossi, A., Proietti, M., Renault, S.: Derivation of efficient logic programs by specialization and reduction of nondeterminism. Higher-Order and Symbolic Computation 18(1-2), 121–210 (2005)

47. Pettorossi, A., Proietti, M., Senni, V.: Proving properties of constraint logic programs by eliminating existential variables. In: Etalle, S., Truszczyński, M. (eds.) ICLP 2006. LNCS, vol. 4079, pp. 179–195. Springer, Heidelberg (2006)

48. Pettorossi, A., Proietti, M., Senni, V.: Automatic correctness proofs for logic program transformations. In: Dahl, V., Niemelä, I. (eds.) ICLP 2007. LNCS, vol. 4670, pp. 364–379. Springer, Heidelberg (2007)

49. Proietti, M., Pettorossi, A.: Semantics preserving transformation rules for Prolog. In: 1991 ACM SIGPLAN Symposium on Partial Evaluation and Semantics Based Program Manipulation, PEPM 1991, Yale University, New Haven, Connecticut, USA, pp. 274–284. ACM Press, New York (1991)

50. Proietti, M., Pettorossi, A.: Unfolding-definition-folding, in this order, for avoiding unnecessary variables in logic programs. Theoretical Computer Science 142(1), 89–124 (1995)

51. Rabin, M.O.: Decidable theories. In: Barwise, J. (ed.) Handbook of Mathematical Logic, pp. 595–629. North-Holland, Amsterdam (1977)

52. Roychoudhury, A., Narayan Kumar, K., Ramakrishnan, C.R., Ramakrishnan, I.V.: Beyond Tamaki-Sato style unfold/fold transformations for normal logic programs. International Journal on Foundations of Computer Science 13(3), 387–403 (2002)

53. Roychoudhury, A., Narayan Kumar, K., Ramakrishnan, C.R., Ramakrishnan, I.V.: An unfold/fold transformation framework for definite logic programs. ACM Transactions on Programming Languages and Systems 26, 264–509 (2004)

54. Roychoudhury, A., Narayan Kumar, K., Ramakrishnan, C.R., Ramakrishnan, I.V., Smolka, S.A.: Verification of parameterized systems using logic program transformations. In: Schwartzbach, M.I., Graf, S. (eds.) TACAS 2000. LNCS, vol. 1785, pp. 172–187. Springer, Heidelberg (2000)

55. Safra, S., Shapiro, E.: Meta interpreters for real. In: Kugler, H.J. (ed.) Proceedings Information Processing 1986, pp. 271–278. North-Holland, Amsterdam (1986)

56. Sato, T., Tamaki, H.: Transformational logic program synthesis. In: Proceedings of the International Conference on Fifth Generation Computer Systems, pp. 195–201. ICOT (1984)

57. Seki, H.: A comparative study of the well-founded and the stable model semantics: Transformation's viewpoint. In: Proceedings of the Workshop on Logic Programming and Nonmonotonic Logic, pp. 115–123. Cornell University (1990)

58. Seki, H.: Unfold/fold transformation of stratified programs. Theoretical Computer Science 86, 107–139 (1991)

59. Seki, H.: Unfold/fold transformation of general logic programs for well-founded semantics. Journal of Logic Programming 16(1&2), 5–23 (1993)

60. Seki, H.: On inductive and coinductive proofs via unfold/fold transformations. In: De Schreye, D. (ed.) LOPSTR 2009. LNCS, vol. 6037, pp. 82–96. Springer, Heidelberg (2009)

61. Sterling, L., Beer, R.D.: Incremental flavour-mixing of meta-interpreters for expert system construction. In: Proceedings of 3rd International Symposium on Logic Programming, Salt Lake City, Utah, USA, pp. 20–27. IEEE Press, Los Alamitos (1986)

62. Tacchella, P., Gabbrielli, M., Meo, M.C.: Unfolding in CHR. In: Proceedings of the 9th International ACM SIGPLAN Conference on Principles and Practice of Declarative Programming (PPDP 2007), pp. 179–186 (2007)

63. Takeuchi, A., Furukawa, K.: Partial evaluation of Prolog programs and its application to meta-programming. In: Kugler, H.J. (ed.) Proceedings of Information Processing 1986, pp. 415–420. North-Holland, Amsterdam (1986)

64. Tamaki, H., Sato, T.: Unfold/fold transformation of logic programs. In: Tärnlund, S.-Å. (ed.) Proceedings of the Second International Conference on Logic Programming (ICLP 1984), pp. 127–138. Uppsala University, Uppsala (1984)

65. Toni, F., Kowalski, R.: An argumentation-theoretic approach to logic program transformation. In: Proietti, M. (ed.) LOPSTR 1995. LNCS, vol. 1048, pp. 61–75. Springer, Heidelberg (1996)

66. Venken, R.: A Prolog meta-interpretation for partial evaluation and its application to source-to-source transformation and query optimization. In: O'Shea, T. (ed.) Proceedings of ECAI 1984, pp. 91–100. North-Holland, Amsterdam (1984)

67. Wadler, P.L.: Deforestation: Transforming programs to eliminate trees. Theoretical Computer Science 73, 231–248 (1990)

68. Zhang, J., Grant, P.W.: An automatic difference-list transformation algorithm for Prolog. In: Proceedings 1988 European Conference on Artificial Intelligence, ECAI 1988, pp. 320–325. Pitman (1988)

Static Analysis, Abstract Interpretation and Verification in (Constraint Logic) Programming

Giorgio Delzanno[1], Roberto Giacobazzi[2], and Francesco Ranzato[3]

[1] Università di Genova, Italy
giorgio@disi.unige.it
[2] Università di Verona, Italy
roberto.giacobazzi@univr.it
[3] Università di Padova, Italy
francesco.ranzato@unipd.it

Abstract. We survey some general principles and methodologies for program analysis and verification. In particular, we focus on abstract interpretation and model checking techniques, and on their applications to constraint logic programs.

Introduction

Logic programming has served as a unique training ground for static analysis, abstract interpretation and verification. Operational and denotational semantics of logic programs feature simple and clean inductive definitions that made it possible to apply a variety of known analysis and verification techniques and tools and to define new ones tailored to solve specific problems arisen in logic programming (e.g. variable aliasing and unification). We survey here some general notions and methods — in particular abstract interpretation and model checking — for analysing and verifying programs and systems, especially focused to (constraint) logic programs.

In Section 1 we first review the principles of the abstract interpretation approach, in particular methodologies for designing abstract domains through systematic techniques such as abstract domain refinement and simplification. We then show how these methods have been applied in the systematic design of analyses and semantics in the context of logic programming.

In Section 2 we recall the main concepts underlying model checking. In model checking, the behavior of a program is described by a finite graph (a Kripke model) that describes the set of all reachable states. In this setting, temporal formulae can be used to naturally specify functional properties of the system (e.g. safety and absence of starvation). The model checking problem consists in checking the temporal specification against the model of the system. For specifications given in Computation Tree Logic (CTL), the algorithm for deciding the model checking problem is based on a fixpoint semantics of the temporal connectives. We exploit here this connection to establish a link between CTL model checking and the fixpoint semantics of logic programs. We then discuss implications of this link with a particular focus on the utilization of evaluation strategies used for logic programming as a tool for model checking of infinite-state concurrent systems.

A. Dovier, E. Pontelli (Eds.): 25 Years of Logic Programming, LNCS 6125, pp. 136–158, 2010.

In Section 3 we focus on abstract interpretation-based model checking. In abstract model checking, the verification of a temporal specification is performed in an abstract model that can be designed as an abstract interpretation of the concrete system. In particular, we concentrate on strong preservation properties of abstract models, namely on the equivalence of verifying temporal specifications in abstract and concrete models. Strong preservation is highly desirable since it allows us to draw consequences on the concrete model from negative answers on the abstract model. We survey how abstract interpretation allows to cast strong preservation as a completeness property of abstract models and consequently how this provides systematic methods to design strongly preserving abstract models through abstract domain refinements.

Finally, in Section 4 we discuss how methods used for evaluation and analysis of logic programs can be used to extend verification methods based on abstract model checking, e.g., to the case of infinite-state systems.

1 Semantics, Static Analysis and Abstract Interpretation

1.1 Abstract Interpretation Basics

One fundamental feature of abstract interpretation is that most properties in approximating semantics, like precision, completeness, and compositionality, which may involve complex operators, fixpoints etc., all depend upon the notion of *abstraction*, which is precisely and uniquely determined by the chosen domain of properties [16]. Central in the design of abstract interpretations is therefore the notion of *domain*. This is the case for instance in program analysis, in type inference and in comparative semantics, where the various abstract (approximate) semantics all correspond to suitable abstractions, namely domains.

In the following, $\langle C, \leq, \vee, \wedge, \top, \bot \rangle$ denotes a generic complete lattice C, with ordering $\leq$, lub $\vee$, glb $\wedge$, greatest element (top) $\top$, and least element (bottom) $\bot$. The downward closure of a subset $S \subseteq C$ is defined as $\downarrow S \triangleq \{x \in C \mid \exists y \in S.\ x \leq y\}$, where $\downarrow x$ is a shorthand for $\downarrow \{x\}$. The upward closure $\uparrow$ is dually defined. The notation $C \cong D$ denotes that C and D are isomorphic, possibly ordered, structures. Recall that a function $f : C \to D$ is (Scott-)continuous if f preserves lub's of (nonempty) chains iff f preserves lub's of directed subsets. In what follows, we consider abstract interpretation based on Galois connections or, equivalently, closure operators [15,16]. A pair of functions $f : A \to B$ and $g : B \to A$ between posets forms an adjunction, or Galois connection (GC for short), denoted by (A, f, B, g), if

$$\forall x \in A. \forall y \in B.\ f(x) \leq_B y \Leftrightarrow x \leq_A g(y).$$

f (resp. g) is called the left- (right-) adjoint to g (f) and it is an additive (co-additive) function, i.e., f preserves lub's (glb's) of all subsets of A (empty set included). Additive and co-additive functions f admit, respectively, right f^+ and left f^- adjoint as follows: $f^+ \triangleq \lambda x. \vee \{y \mid f(y) \leq x\}$ and $f^- \triangleq \lambda x. \wedge \{y \mid x \leq f(y)\}$. Let us also recall that $(f^+)^- = (f^-)^+ = f$. In GC-based abstract interpretation the concrete C and abstract A domains are often assumed to be complete lattices and are related by abstraction $\alpha : C \to A$ and concretization $\gamma : A \to C$ maps forming a GC (C, α, A, γ). If in addition $\forall a \in A.\ \alpha(\gamma(a)) = a$, then (C, α, A, γ) is called a Galois insertion (GI). When (C, α, A, γ) is a GI each value of the abstract domain A is useful in representing C, namely all the elements of A represent distinct members of C, being γ 1-1. Any GC may be lifted to a GI by identifying in an equivalence class those values of the abstract domain with the same concretization. This process is known as reduction of the abstract domain. An (upper) closure operator on a poset C is a map $\rho : C \to C$ which is monotone, idempotent, and extensive ($\forall x \in C.\ x \leq \rho(x)$). The set of all closure operators on C is denoted by $uco(C)$. Each closure operator ρ is uniquely determined by its image $\rho(C)$ as follows: $\rho(x) \triangleq \wedge\{y \in \rho(C) \mid x \leq y\}$. A fundamental property of closure operators is that if C is a complete lattice then both $\langle uco(C), \sqsubseteq \rangle$, where $\sqcup$ is the pointwise ordering, and $\langle \rho(C), \leq_C \rangle$ are complete lattices. It is well known since [16] that abstract domains can be equivalently specified either as Galois insertions or as closure operators on the concrete domain. In particular, a subset $X \subseteq C$ is the image of a closure ρ on C iff X is a Moore-family of C, i.e., $X = \mathcal{M}(X) \triangleq \{\wedge S \in C \mid S \subseteq X\}$ (where $\wedge\varnothing = \top \in \mathcal{M}(X)$) iff X is isomorphic to an abstract domain A in a GI (C, α, A, γ). For any subset $X \subseteq C$, $\mathcal{M}(X)$ is called the Moore-closure of X in C, i.e., $\mathcal{M}(X)$ is the least (w.r.t. set-inclusion) subset of C which contains X and it is a Moore-family of C. $\langle uco(C), \sqsubseteq \rangle$ is isomorphic to the so-called lattice $\langle \mathrm{Abs}(C), \sqsubseteq \rangle$ of abstract interpretations of C [16]. Hence, given any two abstractions $A, B \in \mathrm{Abs}(C)$, A is more precise (or conrete) than B, denoted by $A \sqsubseteq B$, when $B \subseteq A$ as Moore families of C. In the following, it is particularly convenient to identify an abstract domain $A \in \mathrm{Abs}(C)$ as (image of) a closure operator on C, which, as a function, is denoted by ρ_A.

1.2   Backward and Forward Completeness

Soundness of an abstraction can be specified in two equivalent ways [15]. Let C be a concrete domain, (C, α, A, γ) a Galois insertion, $f : C \to C$ a concrete semantic operation and $f^\sharp : A \to A$ a corresponding abstract operation. Then, (C, α, A, γ) and $f^\sharp$ give rise to a sound abstraction when $\alpha \circ f \sqsubseteq f^\sharp \circ \alpha$, or equivalently (by adjunction) when $f \circ \gamma \sqsubseteq \gamma \circ f^\sharp$. While the above two definitions of soundness are equivalent, it turns out that they are not equivalent when equality is required and they encode two different forms of completeness: in the first case, $\alpha \circ f = f^\sharp \circ \alpha$ is called backward ($\mathcal{B}$-) completeness while $f \circ \gamma = \gamma \circ f^\sharp$ is called forward ($\mathcal{F}$-) completeness — the reason for these names will be clear later in the paper. $\mathcal{B}$-completeness (see [44]) corresponds to ask that the abstract function $f^\sharp$ perfectly mimics the concrete function f when the latter is approximated in A, viz. both functions are compared in the abstract domain A. On the other hand, $\mathcal{F}$-completeness (see [37]) corresponds to ask that $f^\sharp$ perfectly mimics the function f when applied to the same abstract value, viz. they are both compared in the concrete domain C.

Recall that the best correct approximation of f on the abstract domain A is defined to be the abstract function $\alpha \circ f \circ \gamma$. It turns out (this is a simple extension of a characterization in [44]) that, given an abstract domain A, there exists an either $\mathcal{B}$- or $\mathcal{F}$-complete abstract function $f^\sharp$ defined on A iff the best correct approximation of f on A is, respectively, either $\mathcal{B}$- or $\mathcal{F}$-complete. This means that both $\mathcal{B}$- and $\mathcal{F}$-completeness are properties of abstract domains, namely a property of the GI (C, α, A, γ). Therefore, one

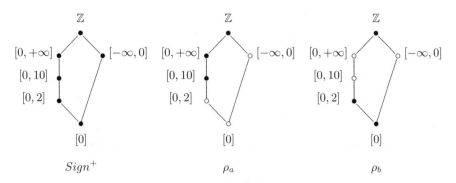

Fig. 1. The abstract domain $Sign^+$ and two abstractions

may define $\mathcal{B}$- and $\mathcal{F}$-completeness as follows: an abstract domain $A \in \mathrm{Abs}(C)$ is $\mathcal{B}$-($\mathcal{F}$-) complete for a semantic function f if $\rho_A \circ f = \rho_A \circ f \circ \rho_A$ ($f \circ \rho_A = \rho_A \circ f \circ \rho_A$).

While $\mathcal{B}$-completeness is well known in abstract interpretation and corresponds to the standard notion of completeness [44,60], the notion of forward completeness is less known. $\mathcal{B}$-completeness for a domain A means that the expessive power of A is such that no loss of precision is accumulated in A by abstracting in A itself the arguments of a semantic function f. Conversely, $\mathcal{F}$-completeness means that no loss of precision is accumulated by approximating in A the result of the function f when computed on abstract values in A. This justifies the choice of the backward and forward terminology above. We denote by, respectively, $\mathcal{F}(C, f)$ and $\mathcal{B}(C, f)$ the set of $\mathcal{F}$- and $\mathcal{B}$- complete abstractions of C for f. It is worth noting that in general $\mathcal{F}(C, f) \not\subseteq \mathcal{B}(C, f)$ and $\mathcal{F}(C, f) \not\supseteq \mathcal{B}(C, f)$, namely $\mathcal{B}$- and $\mathcal{F}$-completeness are incomparable notions.

Example 1. Let $Sign^+$ be the simple abstraction of $\langle \wp(\mathbb{Z}), \subseteq \rangle$ for analysing integer variables depicted in Fig. 1. Consider the pointwise square operation $sq : \wp(\mathbb{Z}) \to \wp(\mathbb{Z})$ defined as follows: $sq(X) \triangleq \{ x^2 \mid x \in X \}$. Let $\rho \in uco(\wp(\mathbb{Z}))$ be the closure operator associated with $Sign^+$, i.e. $\rho = \gamma_{Sign^+} \circ \alpha_{Sign^+}$, where the abstraction and concretization maps are the obvious ones. The best correct approximation of sq in $Sign^+$ is $sq^\sharp : Sign^+ \to Sign^+$ defined as $sq^\sharp(X) \triangleq \rho(sq(X))$, with $X \in Sign^+$. It is easy to note that the closure operators $\rho_a \triangleq \{ \mathbb{Z}, [0, +\infty], [0, 10] \}$ and $\rho_b \triangleq \{ \mathbb{Z}, [0, 2], [0] \}$, defined by their images — the images of ρ_a and ρ_b are depicted as bullets in Fig. 1 — are such that:

- $\rho_a \in \mathcal{F}(Sign^+, sq^\sharp)$ but $\rho_a \notin \mathcal{B}(Sign^+, sq^\sharp)$: for example, $\rho_a(sq^\sharp(\rho_a([0]))) = [0, +\infty]$ while $\rho_a(sq^\sharp([0])) = [0, 10]$;
- $\rho_b \in \mathcal{B}(Sign^+, sq^\sharp)$ but $\rho_b \notin \mathcal{F}(Sign^+, sq^\sharp)$: for example, $\rho_b(sq^\sharp(\rho_b([0, 2]))) = \mathbb{Z}$ while $sq^\sharp(\rho_b([0, 2])) = [0, 10]$. □

One key result in [44] provides a constructive characterization of the structure of abstract domains that are $\mathcal{B}$-complete for continuous functions. Given a function $f : C \to C$ and $S \subseteq C$, $f^{-1}(S)$ denotes the inverse image of f in S, i.e., $\{ x \in C \mid f(x) \in S \}$. Then, [44] shows that

$$\rho \in uco(C) \text{ is } \mathcal{B}\text{-complete for } f \Leftrightarrow \bigcup_{y \in \rho(C)} \max(f^{-1}(\downarrow y)) \subseteq \rho(C) \qquad (*)$$

Let us consider Example 1. It is easy to see that ρ_a is not $\mathcal{B}$-complete because ρ_a does not include the maximal inverse image of $sq^\sharp$ of the subset $\downarrow [0, 10]$, namely $\max(sq^{\sharp^{-1}}(\downarrow [0, 10])) = \{[0, 2]\}$.

An analogous (and trivial to prove) result can be stated for $\mathcal{F}$-completeness. In this case, $\mathcal{F}$-complete domains can be characterized for merely monotone operations as follows:

$$\rho \in uco(C) \text{ is } \mathcal{F}\text{-complete for } f \Leftrightarrow f(\rho(C)) \subseteq \rho(C) \qquad (**)$$

Thus, while $\mathcal{B}$-complete domains ρ are closed under (maximal) inverse images of the function f on $\rho(C)$, $\mathcal{F}$-complete domains ρ are closed under direct images of f on $\rho(C)$. It is easy to see in Example 1 that ρ_b is not $\mathcal{F}$-complete because ρ_b does not include the direct image of $sq^\sharp$, for instance the value $[0, 10] = sq^\sharp([0, 2])$. Characterizations $(*)$ and $(**)$ together establish a tight relationship between $\mathcal{B}$- and $\mathcal{F}$-completeness, which can be specified as an adjunction when the concrete function admits a right adjoint. In fact, it turns out that if $f : C \to C$ is an additive function (and therefore admits right adjoint f^+) then

$$\mathcal{B}(\wp(S), f) = \mathcal{F}(\wp(S), f^+). \qquad (\ddagger)$$

Moreover, it is always possible, by relying on $(*)$ and $(**)$, to associate with each continuous semantic function $f : C \to C$ a corresponding domain refinement that transforms any abstract domain A into the closest (most abstract) $\mathcal{B}$-/$\mathcal{F}$-complete domain for f which includes (i.e., is more precise than) A. This provides the notions of $\mathcal{B}$- and $\mathcal{F}$-complete shell [44]. The domain transformers $\mathcal{R}_f^{\mathcal{B}} : uco(C) \to uco(C)$ and $\mathcal{R}_f^{\mathcal{F}} : uco(C) \to uco(C)$ are defined as follows:

- $\mathcal{R}_f^{\mathcal{B}} \triangleq \lambda X \in uco(C). \ \mathcal{M}(\bigcup_{y \in X} \max(f^{-1}(\downarrow y)))$;
- $\mathcal{R}_f^{\mathcal{F}} \triangleq \lambda X \in uco(C). \ \mathcal{M}(f(X))$.

It is immediate to note that both $\mathcal{R}_f^{\mathcal{B}}$ and $\mathcal{R}_f^{\mathcal{F}}$ are monotone operators on $uco(C)$. The following equivalence, which follows from $(*)$ and $(**)$, characterizes in a unique domain-equational form the $\mathcal{B}$-/$\mathcal{F}$-complete shell of abstract domains for a continuous function $f : C \to C$. Let $A \in uco(C)$ and $\ell \in \{\mathcal{B}, \mathcal{F}\}$:

$$X \sqsubseteq A \text{ and } X \text{ is } \ell\text{-complete for } f \Leftrightarrow X = A \sqcap \mathcal{R}_f^\ell(X).$$

Therefore, the most abstract domain that includes A and is ℓ-complete for f is

$$\ell\text{-Shell}_f(A) \triangleq \text{gfp}(\lambda X. A \sqcap \mathcal{R}_f^\ell(X)).$$

This domain is called the ℓ-complete shell of A with respect to f.

1.3   Abstract Domain Refinement and Simplification

In recent years, systematic design methods of program analysis frameworks attracted a growing interest. This is mainly justified by the fact that the most successful static analyzers are parametric with respect to the property of interest [20] and therefore allow to easily handle a variety of possible analyses. Moreover, automatic methods for tuning static analyses in accuracy and cost are needed in order either to avoid reimplementation when these analyses are modified or to minimize false alarms. Similar constructions are also used in designing semantics by abstract interpretation (e.g., Hoare logic as tensor product [14] and compositional semantics as reduced power [33,42]) and in type inference (e.g., polymorphism as disjunctive completion [13,52]). Formal methods that compare/transform abstract interpretations are therefore inherently based on corresponding methods to compare/transform abstract domains. A domain, at any level of abstraction, is a set of mathematical objects which represent the properties of interest about a computational system and that are partially ordered with respect to their relative degree of precision. In program analysis, for instance, the design of a static analyzer basically corresponds to study a particular abstract domain, while modifying domains corresponds to modify analyses. As shown for instance in [71] for a reconstruction of groundness analysis in logic programming, the design of a complex abstract domain is generally the result of a number of steps which can be in some cases made systematic by applying suitable domain transformers to simpler domains for the property of interest.

The main idea behind domain transformers in abstract interpretation consists in designing abstract domains systematically from the specification of some simpler domains of basic properties of interest and then solving a recursive domain equation in order to achieve completeness with respect to some target precision level. This game can be played for most of the existing abstract domain transformers, by viewing them as instances of completeness refinements: (1) in program analysis, where a given simple (and imprecise) analysis is refined until completeness is reached by avoiding specific families of false alarms, and (2) in program semantics where a given observation is refined towards completeness in order to attain compositionality, condensation properties, etc.

The foundations of a theory of abstract domain transformers were layed by Cousot and Cousot [16] in 1979. In that seminal work the authors introduced the main structure of abstract domains enjoying Galois connections and some fundamental operators for systematically compose domains in order to achieve attribute independent and relational analyses (respectively, the reduced product and reduced power operations). Since then, a number of papers put forward novel domain transformers and studied the impact of these operations in designing abstract interpreters for specific program analysis and languages. These include Cousot and Cousot's reduced product, disjunctive completion and reduced cardinal power [16,17,18]; Nielson's tensor product [61]; Giacobazzi et al.'s dependencies, dual-Moore-set completion, complete kernels and shells, Heyting completion, and least disjunctive basis [40,44,46]; Cortesi et al.'s open product, pattern completion, and complementation [11]. The notions of domain refinement and domain simplification, introduced in [27,39], provided the very first generalization of these ideas. Intuitively, a refinement is any domain operator that performs an action of refinement with respect to the standard precision ordering $\sqsubseteq$, i.e., that adds information to domains; on the other hand, simplificators and compressors perform the dual action of "taking out" information from domains. Still these operators represent a basis for any design of abstractions.

Many domain refinements can be specified as $\mathcal{F}$-complete refinements with respect to a given semantic operation [35]. Intuitively, a domain refinement can be viewed as adding the functionalities of a given semantic operation of interest, that is, the direct image of a semantic function. As a result of the above properties of complete abstractions, this corresponds to say that a domain refinement can be specified as (greatest) solution of a $\mathcal{F}$-completeness equation. As recalled above in ($\ddagger$), whenever the semantic operation is additive, such a characterization can be put in an equivalent formulation in terms of $\mathcal{B}$-completeness.

Clearly, the construction of domains by iterative refinement (e.g., by solving a recursive domain equation) may lead to excessively complex domains for practical applications, as well as it may be interesting to isolate inner structures inside complex domains that model precisely some basic properties around which complex abstract domains are built. As observed in [39], it is possible to define a dual theory of domain simplificators and compressors, which shares with the above theory of domain refinements precisely the same, but dual, ideas and constructions. The common aspects of simplificators and compressors is that they both reduce precision in domains. A typical pattern for domain simplificators is the operation that transforms a given domain A into the most concrete (when it exists) among the abstractions of A which is complete for a given function. Like refinements, also simplificators and compressors have a constructive definition as (greatest) solutions of (systems of) recursive domain equations [44]. The main difference between simplificators and compressors can be grasped by viewing how they react when composed with the corresponding refinements, when they exist. Assume that an idempotent refinement $\mathcal{R}$ is given. $\mathcal{R}$ admits a simplificator $\mathcal{S}$ when, for any abstraction X, $\mathcal{R}(\mathcal{S}(X)) = \mathcal{S}(X)$ and $\mathcal{S}(\mathcal{R}(X)) = \mathcal{R}(X)$. This holds when both $\mathcal{R}$ and $\mathcal{S}$ transform domains to meet a given common property, like, for instance in the above case, completeness. A relevant example of domain refinement which has a corresponding simplificator is in fact the complete shell refinement in [44]. The complete shell refinement, given a domain A, returns the most abstract domain which includes A and is complete for some given semantic operation f; the corresponding simplificator, called complete core, returns the most concrete domain which is contained in A and is complete for f. Compressors, instead, act like "zip" runs on files. If $\mathcal{R}$ is a given domain refinement, $\mathcal{C}$ is a compressor for $\mathcal{R}$ if, for any abstraction X, $\mathcal{R}(X) = \mathcal{R}(\mathcal{C}(X))$ and $\mathcal{C}(\mathcal{R}(X)) = \mathcal{C}(X)$, namely when $\mathcal{C}(X)$ is the most abstract domain B such that $\mathcal{R}(B) = \mathcal{R}(X)$, and this basically holds when the whole refined domain $\mathcal{R}(X)$ can be fully reconstructed by refinement from its so-called basis $B = \mathcal{C}(X)$. A domain theoretic definition of abstract domain compressors has been introduced in [41]. Examples of domain compressors include complementation [11,28], which is the compressor associated with reduced product, and least disjunctive basis [40], which is associated with the disjunctive completion refinement. Clearly, not all refinements admit a corresponding simplificator or compressor. Moreover, as suggested by the above definitions, it is possible to relate refinements and simplificators/compressors by means adjunctions [35,39].

1.4   How to Cook an Abstract Domain or Semantics

The above methods can be used as a recipe for "cooking" an abstract domain/semantics for specific applications.

1. Specify a concrete semantics for the considered programming language, with a (possibly many sorted algebra as) concrete domain $\mathcal{C} = \langle C, op_1, \ldots, op_n \rangle$;
2. Identify, as a subset of the lattice of abstract interpretations, some basic semantics properties $\pi \subseteq \mathrm{Abs}(C)$ that are to be preserved by the abstraction process;
3. Design a suitable refinement $\mathcal{R}_\pi$ which adds to domains some functionalities of the concrete algebra $\mathcal{C}$, in such a way that $\mathcal{R}_\pi(X) = X \Rightarrow X \in \pi$;
4. Define an adequate abstract domain A that encodes the basic properties of interest (e.g. the basic properties to analyze) concerning concrete data objects;
5. Solve the (system) of recursive domain equations $X = A \sqcap \mathcal{R}_\pi(X)$.

Step (1) is common to any abstract interpretation, and corresponds to the design of a suitable base (typically collecting) semantics. Step (2) is instead a meta-level operation: The designer has to identify the common structure of any domain which shares a given semantic property that has to be preserved in the abstraction process. This may include completeness, compositionality, and any combination of semantic properties of interest for the specific application. A taxonomy of basic observable properties of semantics is essential in order to solve this problem, see e.g. [21] for a recent account on the logic programming case. Step (3) is strongly related to step (2) and is based on the theory of domain refinements described above [39]. Step (4) strongly needs a creative contribution of the designer, which has to guess a minimal domain of basic properties of interest for concrete data objects. Compressors may provide here a tool for simplifying and adapting the solutions envisaged at design time. Steps (5) is standard. Most of these steps, in particular (3) and (5), are systematic and, in most cases, constructive.

1.5   Applications in Logic Programming

Logic programming has been an ideal programming setting where the above ideas have found straight application. This because of the clean nature of the declarative semantics of a (constraint) logic program, which consists of a simple fixpoint solution of a recursive equation on predicates, where ground predicates provide the so called model-based semantics and possibly nonground predicates provide the so called computed-answer substitution semantics, also called s-semantics [26]. This motivates the use of logic programming as a natural and intelligible environment where abstract domain transformers can be tested and applied for a very first practical use, and characterized the research in abstract interpretation in the years across Y2000 mainly in Padova, Parma, Pisa and Verona. Of course, all the above definitions and notions hold on generic complete lattices and semantic structures, fulfilling the language independence feature of abstract interpretation. Here, we list some results in semantics and static program analysis obtained by applying the above mentioned domain transformers. These results are characterized by a scattered coverage of known and new properties of semantics of logic programming, all having a distinctive nature of being systematically derived by means of abstract domain transformations. The result was a puzzle of methods and techniques for handling semantics and analyses with the ambition of fully developing Strachey's programme of *"understanding of the mathematical ideas of programming languages and combine them with other principles of common sense as correctives of exaggeration, allowing the individual reader to draw as moderate conclusions as she/he will"* [74].

Analysis. The abstract domain for relational groundness analysis *Pos* has been reconstructed as solution of a completeness problem, i.e., as greatest (w.r.t. $\sqsubseteq$) solution of the simple recursive abstract domain equation $X = \mathcal{G} \sqcap (X \to X)$ over the concrete domain of downward-closed sets of idempotent substitutions with respect to variable instantiation, where $\to$ is the Heyting completion of an abstract domain [46] and $\mathcal{G}$ is the basic domain for groundness analysis, specifying whether a variable is ground or not [71]. Disjunctive completion and bases for groundness analysis have been studied in [40].

The phenomenon of so-called condensation in logic program analysis has been fully modeled as a completeness property of the underlying abstract domains in [45]. A static analysis is condensing if (bottom-up) goal-independent and (top-down) goal-dependent analyses agree, i.e., whenever it is possible to reconstruct the analysis of a given goal from the result of a goal-independent analysis without loss of precision. In this case, a condensing domain can always be systematically derived from a possibly noncondensing one A by solving the recursive domain equation $X = A \sqcap X \sqcap (X \stackrel{\wedge}{\multimap} X)$ on the concrete quantale of sets of idempotent substitutions, where the conjunction $\wedge$ in the quantale of idempotent substitutions is most general unification and where $\stackrel{\wedge}{\multimap}$ is the linear refinement with respect to $\wedge$ [45]. Condensing domains for freeness, independence, type representations, pair-independence, non-pair-sharing, and information-flow analysis have all been derived in this way in [45,56,57,58,73]. A condensing domain for sharing analysis, i.e., a solution to the equation $X = Sh \sqcap X \sqcap (X \stackrel{\wedge}{\multimap} X)$, with *Sh* being the domain for set-sharing, is still unknown. Completeness has been also used in combination with complementation to prove that set-sharing is redundant for pair-sharing [3].

Semantics. Semantics can be composed and complemented as easy as abstract domains. Applications in logic programming have shown that the semantics $\mathcal{S} \boxminus \mathcal{C}$, obtained by complementing [11] the Clark semantics of correct answer substitutions $\mathcal{C}$ with respect to the more concrete semantics of computed answer substitution $\mathcal{S}$, corresponds precisely to the fully abstract semantics for partial computed answer substitutions [38]. Similar characterizations have been obtained by domain complementation of Clark vs. Herbrand model-based semantics and call vs. success pattern semantics [38]. By considering linear refinement, the OR-compositional semantics of logic programs can be systematically derived as least solution of the recursive domain equation $X = \mathcal{S} \sqcap X \sqcap (X \stackrel{\frown}{\multimap} X)$ over the concrete quantale of SLD-traces of atoms where conjunction is trace concatenation $\frown$ [42] and $\stackrel{\frown}{\multimap}$ is the linear refinement w.r.t. $\frown$. A more general construction for arbitrary compositional semantics on traces can be found in [34].

2 Temporal Logic and Model Checking

2.1 Basics of Model Checking

Model checking (see e.g. [10]) is a technique for verifying finite state (concurrent) systems. It has been applied to verify properties of digital circuits, communication protocols, and, in the last years, to abstract models of software programs. Model checking is automatic and, if the model contains an error, it produces a counterexample that can be used to find the error in the original system. Model checking is based on the following ingredients: a specification language to describe a model of the behavior of a given system, a logic to describe the properties that the model is suppose to satisfy, and a decision procedure to test the properties against a model. The behavior of a system is described by means of a Kripke model, i.e., a finite graph in which nodes are labeled by propositions and edges represent transitions between states (the transition relation). Propositions represent local properties of a given state. Global properties are described in temporal logic, a formalism that can be used to reason on the transitive closure of the state transition relation. There exist several types of temporal logic specification languages. In this paper we focus on Computation Tree Logic (CTL).

Computation Tree Logic. CTL can be used to reason about branching time properties of a Kripke model. A CTL model is a tuple $M = \langle \text{States}, \rightarrow, \ell \rangle$ such that States is a set of states, $\rightarrow \subseteq \text{States} \times \text{States}$ is a (typically total) transition relation and $\ell :$ States $\rightarrow \wp(\text{Atoms})$ is a labeling function that defines the set of atomic predicates, taken from a finite set Atoms, that holds at each state. When a labeling function is omitted, we assume that $\ell(s) = \{s\}$ (i.e., states are used as predicates). CTL formulae extend propositional logic with temporal formulae of the form $Q_P Q_T$, where Q_P is a path quantifier and Q_T is a temporal quantifier. The path quantifier can be either A (for all paths) or E (there exists a path). The temporal quantifier can be either X (next state), F (eventually), G (always), or U (until). For instance, the formula $\text{EX}\varphi$ holds in the current state if there exists a successor in which φ holds, $\text{EF}\varphi$ holds in the current state if there exists a path in which φ eventually holds, and $\text{AG}\varphi$ holds in the current state if in all paths φ always holds. To formally define the semantics of CTL formulae, we define a path σ in M as an infinite sequence of states $s_0 s_1 \ldots s_i \ldots$ such that $s_k \rightarrow s_{k+1}$ for $k \geq 0$ and we use $\sigma[i]$ to denote the i-th state in σ. Furthermore, we use $P_M(s)$ to define the set of paths σ in M such that $\sigma[0] = s$. The satisfiability relation $M, s \models \varphi$ is defined then as follows:

- $M, s \models p$ iff $p \in \ell(s)$
- $M, s \models \neg\phi$ iff $s \not\models \phi$
- $M, s \models \varphi \vee \psi$ iff $s \models \varphi$ or $s \models \psi$
- $M, s \models \text{EX}\varphi$ iff $\exists \sigma \in P_M(s).\sigma[1] \models \varphi$
- $M, s \models \text{E}(\varphi \, \text{U} \, \psi)$ iff $\exists \sigma \in P_M(s) \, \exists j \geq 0. \, \sigma[j] \models \psi \wedge (\forall k \in [0, j). \, \sigma[k] \models \varphi)$
- $M, s \models \text{EF}\varphi$ iff $\exists \sigma \in P_M(s) \, \exists j \geq 0.\sigma[j] \models \varphi$
- $M, s \models \text{EG}\varphi$ iff $\exists \sigma \in P_M(s) \, \forall j \geq 0.\sigma[j] \models \varphi$

The semantics of the other logical/temporal operators is derived by exploiting semantic equivalences like $\neg EF\varphi \equiv AG\neg\varphi$.

Model Checking Problem. Given a CTL model M, an initial state s_0, and a CTL formula φ, the CTL model checking problem consists in checking whether $M, s_0 \models \varphi$ holds or not.

CTL formulas can be used to express functional properties of a concurrent system like mutual exclusion, termination, absence of starvation, etc. For instance, assume that proposition cs_i denotes states in which process i is in its critical section. Mutual exclusion for processes $1, \ldots, n$ is represented then by the CTL property $AG(\neg(\bigwedge_{i=1}^{n} cs_i))$, i.e., for all paths and all states, it is never the case that the formula $cs_1 \wedge cs_2 \wedge \ldots \wedge cs_n$ is satisfied. For finite-state Kripke models, the CTL model checking problem is decidable in polynomial time as discussed in the next section.

2.2   Model Checking Algorithm

The model checking decision procedure is based on a fixpoint characterization of the semantics of CTL formulae. Given a formula φ, we define its denotation as the set of states that satisfies it, namely,

$$[\![\varphi]\!] \triangleq \{s \in \text{States} \mid M, s \models \varphi\}.$$

The set of CTL formulae ordered with respect to the inclusion of their denotations forms a complete lattice. The bottom element is *false* (any unsatisfiable formula), the top element is *true* (any tautology), and $\wedge$ and $\vee$ correspond to the greatest lower bound and the least upper bound operations, respectively. Temporal connectives can be viewed as transformers of sets of states (i.e., of denotations). To clarify this point, let us recall that temporal connectives as e.g. EF satisfy expansion axioms like

$$\text{EF}\varphi \equiv \varphi \vee \text{EX}\,\text{EF}\varphi.$$

Lifting this axiom to the denotation level we obtain the fixpoint equation

$$Z = h(Z)$$

where $h : \wp(\text{States}) \to \wp(\text{States})$ is defined as

$$h \triangleq \lambda Z.[\![\varphi]\!] \cup Pre(Z)$$

where $Pre(Z)$ is the set of predecessor states of Z, i.e.,

$$Pre(Z) \triangleq \{s \in \text{States} \mid \exists s' \in Z.s \to s'\}.$$

The denotation of the formula $\text{EF}\varphi$ is the *least fixpoint* of the operator h, which is monotonic over the complete lattice $\langle \wp(\text{States}), \subseteq, \cup, \cap, \text{States}, \varnothing \rangle$. By applying Knaster-Tarski fixpoint theorem, the least fixpoint of h is the union $\bigcup_{i \geq 0} I_i$ of the sets $I_0, \ldots, I_i, \ldots$ inductively defined as $I_0 = \varnothing$ and $I_{i+1} = h(I_i)$ for $i \geq 0$. This computation corresponds to a backward visit of the graph that defines the state transition relation starting from the set of states that satisfy φ. Since the model has finitely many states this backward analysis is always guaranteed to terminate and requires a number of steps that is linear in the size of the model (in the worst case one state is added in each computation of Pre).

A similar reasoning can be applied to the other CTL connectives. The denotation of formulae that quantify over all states along a path, like AG and EG, can be computed as greatest fixpoints of their corresponding transformers, whereas the denotation of temporal formulae like AF and EF can be computed as least fixpoints. The model checking algorithm is defined then by induction on the structure of the input formula φ and computes its denotations bottom-up starting from the denotations of its subformulae. For instance, given the formula $\mathrm{AG}((\mathrm{EF}\,p) \wedge q)$ we first compute the denotation of the subformula $\mathrm{EF}p$, by means of a least fixpoint computation, and that of q. We then compute their intersection I. Finally, we compute the denotation of the transformer AG applied to I by using a greatest fixpoint computation.

The time complexity of this model checking algorithm is polynomial in the size of the input formula φ and of the model M. It is important to notice that the number of states in the transition graph is in general exponential in the description of the model which is usually given in some high level language (e.g. a collection of formulae), and this is commonly referred to as state explosion problem. Heuristics like symbolic model checking [6] attack this problem by using compact representations of sets of states, e.g., by using binary decision diagrams as a representation of sets of states.

3   Abstract Model Checking and Refinement

Approximate automated verification by abstract model checking [9] provides one important solution to the state explosion problem [8] that arises in model checking systems with parallel components. In abstract model checking, approximation is encoded by an abstract model A that hides some details of the concrete model M so that verification becomes more efficient on A rather than on M. The design of an abstract model checking framework always includes a preservation result, roughly stating that for any formula φ expressed in some language $\mathcal{L}$, if $A \models \varphi$ then $M \models \varphi$. Clearly, abstract verification of φ on A may yield false negatives due to the approximation of M to A. On the other hand, strong preservation means that a formula φ in $\mathcal{L}$ holds on A if and only if φ holds on M. Strong preservation is thus highly desirable since it allows to draw consequences from negative answers on the abstract side.

The relationship between abstract model checking and abstract interpretation has been the subject of a number of works (e.g. [9,19,22,37,43]). We recall here how the above notion of strong preservation in abstract model checking can be generalized from an abstract interpretation perspective. This abstract interpretation-based view of strong preservation allows to understand some common principles in well-known algorithms that refine abstract Kripke structures in order to make them strongly preserving for some temporal language.

3.1   Abstract Semantics of Languages

We deal with generic (temporal) languages $\mathcal{L}$ whose state formulae φ are inductively defined by:

$$\mathcal{L} \ni \varphi ::= p \mid f(\varphi_1, ..., \varphi_n)$$

where p ranges over a (typically finite) set of atomic propositions Atoms, while f ranges over a finite set Op of operators, for example standard temporal operators like existential/universal next EX/AX, until EU/AU, globally EG/AG, etc. The semantics of a language is determined by a suitable semantic structure $\mathcal{S}$, e.g. a Kripke structure, on a concrete state space States, that provides an interpretation of atoms and operators in $\mathcal{L}$ as, respectively, elements and operators on the powerset $\wp(\text{States})$. Thus, $\mathcal{S}$ determines for any formula $\varphi \in \mathcal{L}$ a concrete semantics $[\![\varphi]\!]_{\mathcal{S}} \in \wp(\text{States})$, namely the set of states making φ true w.r.t. $\mathcal{S}$. In turn, this also defines a state partition $P_{\mathcal{L}} \in \text{Part}(\text{States})$, i.e. state equivalence, induced by the language $\mathcal{L}$ as follows:

$$P_{\mathcal{L}}(s) \triangleq \{s' \in \text{States} \mid \forall \varphi \in \mathcal{L}.\ s \in [\![\varphi]\!]_{\mathcal{S}} \Leftrightarrow s' \in [\![\varphi]\!]_{\mathcal{S}}\}.$$

As shown in Section 1, abstract interpretation provides a systematic technique for approximating a concrete semantics by an abstract semantics defined on some abstract domain. We consider abstract domains of the powerset $\langle \wp(\text{States}), \subseteq \rangle$ that plays here the role of concrete semantic domain. An abstract domain $A \in \text{Abs}(\wp(\text{States}))$, defined by abstraction/concretization maps α/γ, induces an abstract semantic structure $\mathcal{S}^A$ where the interpretation of an atom $p \in \wp(\text{States})$ is abstracted to $\alpha(p)$ while a concrete semantic operator $f : \wp(\text{States})^n \to \wp(\text{States})$ is abstracted by its best correct approximation f^A on A, that is $f^A(a_1, ..., a_n) \triangleq \alpha(f(\gamma(a_1), ..., \gamma(a_n)))$. Thus, any abstract domain A systematically induces an abstract semantics $[\![\varphi]\!]^A_{\mathcal{S}} \in A$ that evaluates formulae $\varphi \in \mathcal{L}$ in the abstract domain A.

It turns out that this approach based on abstract semantics generalizes standard abstract model checking [10]. Given a Kripke structure $\mathcal{K} = (\text{States}, \to)$, a standard abstract model is specified as an abstract Kripke structure $\mathcal{A} = (\text{AStates}, \to^{\sharp})$ where the set AStates of abstract states is defined by a surjective map $h : \text{States} \to \text{AStates}$ that groups together indistinguishable concrete states whereas $\to^{\sharp}$ is the transition relation between abstract states. Thus, AStates determines a partition of States and vice versa any partition of States can be viewed as a set of abstract states.

It turns out that state partitions can be viewed as a particular class of abstract domains. On the one hand, a partition $P \in \text{Part}(\text{States})$ can be considered an abstract domain by means of the following Galois insertion $(\wp(\text{States})_{\subseteq}, \alpha_P, \wp(P)_{\subseteq}, \gamma_P)$:

$$\alpha_P(S) \stackrel{\text{def}}{=} \{B \in P \mid B \cap S \neq \varnothing\}; \quad \gamma_P(\mathcal{B}) \stackrel{\text{def}}{=} \cup_{B \in \mathcal{B}} B.$$

Hence, $\alpha_P(S)$ encodes the minimal over-approximation of S through blocks of the state partition P. On the other hand, any abstract domain $A \in \text{Abs}(\wp(\text{States}))$ induces the following partition $\text{part}(A) \in \text{Part}(\text{States})$:

$$\text{part}(A)(x) \stackrel{\text{def}}{=} \{y \in \text{States} \mid \alpha_A(\{y\}) = \alpha_A(\{x\})\}.$$

An abstract domain $A \in \text{Abs}(\wp(\text{States}))$ is called partitioning when it represents precisely a state partition, namely when $\gamma_A \circ \alpha_A = \gamma_{\text{part}(A)} \circ \alpha_{\text{part}(A)}$.

3.2   Generalized Strong Preservation

In standard abstract model checking, given a language $\mathcal{L}$ and a corresponding interpretation on a Kripke structure $\mathcal{K}$, an abstract Kripke structure $\mathcal{A}$ strongly preserves $\mathcal{L}$ when for any $\varphi \in \mathcal{L}$ and $s \in \text{States}$, we have that

$$\mathcal{A}, h(s) \models \varphi \Leftrightarrow \mathcal{K}, s \models \varphi$$

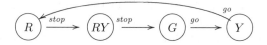

Fig. 2. A U.K. traffic light

where h : States $\rightarrow$ AStates is the abstraction map.

It turns out that strong preservation can be generalized from standard abstract Kripke structures to abstract interpretation-based models. A generalized abstract model is given as an abstract domain $A \in \text{Abs}(\wp(\text{States}))$ that systematically induces an abstract semantics $[\![\cdot]\!]_{\mathcal{S}}^{A}$. We therefore define the abstract semantics $[\![\cdot]\!]_{\mathcal{S}}^{A}$ to be strongly preserving (s.p. for short) for $\mathcal{L}$ when for any $\varphi \in \mathcal{L}$ and $S \in \wp(\text{States})$,

$$\alpha(S) \leq_A [\![\varphi]\!]_{\mathcal{S}}^{A} \Leftrightarrow S \subseteq [\![\varphi]\!]_{\mathcal{S}}.$$

Observe that strong preservation is an abstract domain property, meaning that it does not depend on the abstract interpretation of atoms and logical/temporal operators on the abstract domain A but only depends on A itself. Thus, an abstract domain $A \in \text{Abs}(\wp(\text{States}))$ is strongly preserving for $\mathcal{L}$ when $[\![\cdot]\!]_{\mathcal{S}}^{A}$ is strongly preserving for $\mathcal{L}$.

Standard strong preservation becomes a particular instance, because it turns out that an abstract Kripke structure strongly preserves $\mathcal{L}$ if and only if the corresponding partitioning abstract domain strongly preserves $\mathcal{L}$ according to the above generalized meaning. Generalized strong preservation may work where standard strong preservation may fail. In fact, it may happen that although a strongly preserving abstract semantics on a partition P always exists this abstract semantics cannot be derived from a strongly preserving abstract Kripke structure on P. The following example shows this phenomenon.

Example 2. Consider the following simple language $\mathcal{L}$:

$$\mathcal{L} \ni \varphi ::= stop \mid go \mid \text{AXX}\varphi$$

and the Kripke structure $\mathcal{K}$ depicted in Figure 2, where superscripts determine the labeling function. $\mathcal{K}$ models a four-state traffic light controller (like in the U.K.): Red $\rightarrow$ RedYellow $\rightarrow$ Green $\rightarrow$ Yellow. According to the standard semantics of AXX, we have that $\mathcal{K}, s \models \text{AXX}\varphi$ iff for any path $s_0 s_1 s_2 \ldots$ starting from $s_0 = s$, it happens that $\mathcal{K}, s_2 \models \varphi$. It turns out that $[\![\text{AXX}stop]\!]_{\mathcal{K}} = \{G, Y\}$ and $[\![\text{AXX}go]\!]_{\mathcal{K}} = \{R, RY\}$. We thus consider the state partition $P = \{\{R, RY\}, \{G, Y\}\}$. However, it turns out that there exists no abstract transition relation $\rightarrow^{\sharp}$ on the abstract state space P such that the abstract Kripke structure $\mathcal{A} = (P, \rightarrow^{\sharp})$ strongly preserves $\mathcal{L}$. Assume by contradiction that such an abstract Kripke structure $\mathcal{A}$ exists. Let $B_1 = \{R, RY\} \in P$ and $B_2 = \{G, Y\} \in P$. Since $\mathcal{K}, R \models \text{AXX}go$ and $\mathcal{K}, G \models \text{AXX}stop$, by strong preservation, it must be that $\mathcal{A}, B_1 \models \text{AXX}go$ and $\mathcal{A}, B_2 \models \text{AXX}stop$. Hence, necessarily, $B_1 \rightarrow^{\sharp} B_2$ (otherwise B_1 can never reach the state B_2 where the atom go holds) and $B_2 \rightarrow^{\sharp} B_1$ (otherwise B_2 can never reach the state B_1 where the atom $stop$ holds). This leads to the contradiction $\mathcal{A}, B_1 \not\models \text{AXX}go$. In fact, if $\rightarrow^{\sharp} = \{(B_1, B_2), (B_2, B_1)\}$ then we would have that $\mathcal{A}, B_1 \not\models \text{AXX}go$. On the other hand, if, instead, $B_1 \rightarrow^{\sharp} B_1$ (the case $B_2 \rightarrow^{\sharp} B_2$ is analogous), then we would still have that $\mathcal{A}, B_1 \not\models \text{AXX}go$. Even more, along the same lines it is not hard to check that no proper abstract Kripke structure that strongly preserves $\mathcal{L}$ can be defined, because even if either B_1 or B_2 is split (i.e., refined) we still cannot define an abstract transition relation that is strongly preserving for $\mathcal{L}$.

On the other hand, let us consider the partitioning abstract domain

$$A \triangleq \{\varnothing, \{R, RY\}, \{G, Y\}, \{R, RY, G, Y\}\}$$

that is induced by the above partition P. This abstract domain A induces a corresponding abstract semantics $[\![\cdot]\!]_{\mathcal{K}}^A : \mathcal{L} \rightarrow A$, where the best correct approximation of the operator $\mathbf{AXX} : \wp(\text{States}) \rightarrow \wp(\text{States})$ on A is as follows:

$$\alpha_A \circ \mathbf{AXX} \circ \gamma_A = \{\varnothing \mapsto \varnothing, \{R, RY\} \mapsto \{G, Y\}, \{G, Y\} \mapsto \{R, RY\},$$
$$\{R, RY, G, Y\} \mapsto \{R, RY, G, Y\}\}.$$

It is easy to check that this abstract semantics $[\![\cdot]\!]_{\mathcal{K}}^A$ is strongly preserving. As observed above, in the abstract Kripke structure $\mathcal{A}$, the formulae AXXgo and AXX$stop$ are not strongly preserved. Here, instead, we have that $\alpha_P(S) \leq_A [\![\text{AXX}go]\!]_{\mathcal{K}}^A \Leftrightarrow S \subseteq [\![\text{AXX}go]\!]_{\mathcal{K}}$ and $\alpha_P(S) \leq_A [\![\text{AXX}stop]\!]_{\mathcal{K}}^A \Leftrightarrow S \subseteq [\![\text{AXX}stop]\!]_{\mathcal{K}}$. □

3.3   Strong Preservation as Completeness

Given a language $\mathcal{L}$ and a Kripke structure $\mathcal{K} = (\text{States}, \rightarrow)$, a well-known key problem is to compute the smallest abstract state space $\text{AStates}_{\mathcal{L}}$, when this exists, such that one can define an abstract Kripke structure $\mathcal{A}_{\mathcal{L}} = (\text{AStates}_{\mathcal{L}}, \rightarrow^{\sharp})$ that strongly preserves $\mathcal{L}$. This problem admits solution for a number of well-known temporal languages like CTL (or, equivalently, the μ-calculus), ACTL and CTL-X (i.e. CTL without the next-time operator X). A number of algorithms for solving this problem exist, like those by Paige and Tarjan [62] for CTL, by Henzinger et al. [50], Tan and Cleaveland [75], Ranzato and Tapparo [66] and Gentilini et al. [32,47] for ACTL, and Groote and Vaandrager [48] for CTL-X. These are coarsest partition refinement algorithms. Given a language $\mathcal{L}$ and a state partition $P \in \text{Part}(\text{States})$ which is determined by a state labeling $\ell : \text{States} \rightarrow \wp(\text{Atoms})$ — namely, $P \triangleq \{\ell^{-1}(X) \mid X \subseteq \text{Atoms}\}$ — these algorithms can be viewed as computing the coarsest partition $P_{\mathcal{L}}$ that refines P and allows to define an abstract Kripke structure $(P, \rightarrow^{\sharp})$ that strongly preserves $\mathcal{L}$. It is worth remarking that most of these algorithms have been designed for computing well-known behavioural equivalences used in process algebra like bisimulation (for CTL), simulation (for ACTL) and divergence-blind stuttering (for CTL-X) equivalence. Our abstract interpretation-based framework allows us to provide a generalized view of these partition refinement algorithms. It turns out that the most abstract (i.e., least informative) domain, denoted by $\text{AD}_{\mathcal{L}}$, that strongly preserves a given language $\mathcal{L}$ always exists, namely the domain

$$\sqcup\{A \in \text{Abs}(\wp(\Sigma)) \mid A \text{ is s.p. for } \mathcal{L}\}$$

results to be s.p. for $\mathcal{L}$. It turns out that $\text{AD}_{\mathcal{L}}$ is a partitioning abstract domain if and only if $\mathcal{L}$ includes propositional logic, that is when $\mathcal{L}$ is closed under logical conjunction and negation. Otherwise, a proper loss of information occurs when abstracting $\text{AD}_{\mathcal{L}}$ to the corresponding partition $P_\mathcal{L}$. Moreover, for some languages $\mathcal{L}$, it may happen that one cannot define an abstract Kripke structure on the abstract state space $P_\mathcal{L}$ that strongly preserves $\mathcal{L}$ whereas the most abstract strongly preserving domain instead exists. In fact, in Example 2, the domain A actually is the most abstract s.p. domain for the language $\mathcal{L}$ whilst no s.p. abstract Kripke structure can be defined.

As discussed in Section 1, completeness in abstract interpretation encodes an ideal situation where the abstract semantics coincides with the abstraction of the concrete semantics. A precise correspondence between generalized strong preservation and completeness in abstract interpretation can be established. This is based on the notion of forward complete abstract domain. As recalled in Section 1, it turns out that forward complete abstract domains can be systematically and constructively derived from non-complete abstract domains by minimal refinements. Given any domain $A \in \mathrm{Abs}(C)$, recall that we denote by $\mathcal{F}\text{-Shell}_f(A)$ the forward complete shell of A for f. $\mathcal{F}\text{-Shell}_f(A)$ can be obtained by iteratively closing $\gamma(A)$ under direct images of f until a fixpoint is reached, i.e.,

$$\mathcal{F}\text{-}\mathrm{Shell}_f(A) \triangleq \mathrm{lfp}\left(\lambda X \subseteq C.\gamma(A) \cup X \cup f(X)\right).$$

It turns out that strong preservation is related to forward completeness as follows. As described above, the most abstract domain $\mathrm{AD}_\mathcal{L}$ that strongly preserves $\mathcal{L}$ always exists. It turns out that $\mathrm{AD}_\mathcal{L}$ coincides with the forward complete shell for the logical/temporal operators of $\mathcal{L}$ of a basic abstract domain $A_\ell \triangleq \mathcal{M}(\{\ell^{-1}(X) \mid X \subseteq \mathrm{Atoms}\})$ determined by the state labeling ℓ, i.e.,

$$\mathrm{AD}_\mathcal{L} = \mathcal{F}\text{-}\mathrm{Shell}_{\mathbf{Op}_\mathcal{L}}(A_\ell).$$

This characterization provides a generalization of partition refinement algorithms used in standard abstract model checking that can be therefore logically viewed as refinements w.r.t. forward completeness.

Example 3. Conside the above Example 2 where the labeling determines the abstract domain $A_\ell = \{\varnothing, \{R, RY\}, \{G, Y\}, \{R, RY, G, Y\}\}$. Let $\mathbf{AXX}$ be the semantic interpretation of AXX. It turns out that A_ℓ is already forward complete for $\mathbf{AXX}$ because $\mathbf{AXX}(\{R, RY\}) = \{G, Y\}$ and $\mathbf{AXX}(\{G, Y\}) = \{R, RY\}$. Thus, here

$$\mathrm{AD}_\mathcal{L} = \mathcal{F}\text{-}\mathrm{Shell}_{\mathbf{AXX}}(A_\ell) = A_\ell$$

namely A_ℓ is the most abstract strongly preserving domain for the language $\mathcal{L}$.   □

Bisimulation Equivalence. As an example, let us describe how this approach allows us to derive a novel characterization of bisimulation equivalence in terms of forward completeness of abstract domains.

Bisimulation equivalence P_{bis} on some Kripke structure $\mathcal{K}$ can be computed by the well-known Paige-Tarjan partition refinement algorithm PT. More precisely, if P_ℓ denotes the state partition determined by the labeling function ℓ then $\mathrm{PT}(P_\ell) = P_{\mathrm{bis}}$. It is well known [5] that when $\mathcal{K}$ is finitely branching, bisimulation equivalence coincides with the state equivalence induced by Hennessy-Milner logic

$$\mathrm{HML} \ni \varphi ::= p \mid \varphi_1 \wedge \varphi_2 \mid \neg\varphi \mid \mathrm{EX}\varphi$$

that is, $P_{\mathrm{HML}} = P_{\mathrm{bis}}$. As usual, the semantic interpretation of EX is the predecessor $Pre : \wp(\mathrm{States}) \to \wp(\mathrm{States})$, while conjunction and negation are, respectively, interpreted as intersection $\cap$ and complementation $\complement$ on $\wp(\mathrm{States})$.

The following characterization can then be derived in our abstract interpretation-based framework:

$$PT(P_\ell) = \mathrm{part}(\mathcal{F}\text{-}\mathrm{Shell}_{\{Pre,\complement\}}(A_\ell)).$$

Note that the forward complete shell does not need to take into account the intersection on $\wp(\mathrm{States})$ since abstract domains, being closed under intersections, are always forward complete for intersections. This characterization in turn leads to design a generalized Paige-Tarjan-like procedure for computing most abstract strongly preserving domains [67].

4 Model Checking and (Constraint) Logic Programming

In the last decade there has been a growing interest in the application of logic programming techniques to the specification, analysis, and verification of concurrent systems and software programs. For instance, in Italy the research groups in Genova and Roma have applied different types of evaluation and transformation strategies for constraint logic programming to the verification of parameterized formulations of communication protocols.

A nice example of the connections between verification and logic programming is given in [24]. In the rest of the section we briefly recall the main ideas from this paper.

4.1 Model Checking and Fixpoint Semantics in LP

As discussed in Section 2, the semantics of CTL properties is defined as a least or greatest fixpoint of a monotonic operator defined over sets of configurations, i.e., states. This property can be exploited in order to provide a link between model checking and logic programming. As an example, let us interpret an atomic formula $p(s_1, s_2, val)$ as a configuration of a system with two processes whose current states are, resp., s_1 and s_2 and with a shared variable whose current value is val. Now let P be the logic program defined as

$$p(idle, X, free) : -p(use, X, lock).$$
$$p(use, X, Y) : -p(idle, X, free).$$
$$p(X, idle, free) : -p(X, use, lock).$$
$$p(X, use, Y) : -p(X, idle, free).$$

According to the above mentioned interpretation of the predicate p, the Horn clauses in P represent one-step transitions (possible moves of one of the two processes) of a concurrent system in which the access to the critical section use is controlled via modifications to the global variable with states $lock$ and $free$.

Let us now consider the set of ground atomic predicates

$$Bad \triangleq \{p(use, use, lock), \ p(use, use, free)\}.$$

They represent violations to the mutual exclusion property for the system represented by the program P. To draw a link between the semantics of P and CTL properties like

EF, we need to resort to the fixpoint semantics of logic programs. We first recall that the immediate consequence operator of the logic program $Q \triangleq P \cup Bad$ is defined as

$$T_Q(I) \triangleq \{A\theta \mid A : -B \in Q,\ B\theta \in I,\ \theta \text{ grounding for } A, B\} \cup Bad$$

where I is a set of ground atoms with predicate p and constants taken from the set $\{idle, busy, free, lock\}$. It is immediate to see that when T_Q is applied to a set of atoms I, it computes (a representation of) the set of one-step predecessors of the configurations in I. The fixpoint semantics $\mathcal{F}_Q$ of the program Q is defined as the least fixpoint of the T_Q operator, i.e., as the set of ground atoms

$$\mathcal{F}_Q \triangleq \text{lfp}(T_Q) = \bigcup_{i \geq 0} T_Q^i(\varnothing).$$

Based on the link between T_Q and the operator Pre used in the semantics of CTL, we have that $\mathcal{F}_Q$ is a representation of the set of all predecessors of violations to mutual exclusion contained in Bad. In other words, $\mathcal{F}_Q$ is equivalent to the denotation of the CTL formula $\text{EF}(use_1 \wedge use_2)$, where use_i is the predicate that is true if and only if the process i is in the critical section. In a similar way, we can use the greatest fixpoint semantics of logic programs to characterize CTL properties like EG.

4.2 From Finite-State to Infinite-State Models

The interpretation of logic programs as a symbolic representation of transition systems paves the way to several different logic-based methods for the verification of finite-state and infinite-state systems. In [24], the s-semantics of constraint logic programs is applied to symbolically reason on infinite-state transition systems. The s-semantics of logic programs is obtained by lifting the fixpoint semantics to a domain in which interpretations are sets of nonground atoms. Going back to the previous example, we first observe that the set Bad can be represented with the single nonground atom.

$$b \triangleq p(use, use, X)$$

where X is a free variable. Furthermore, the bottom-up evaluation of the program $R \triangleq P \cup \{b\}$ can be computed symbolically by replacing the operator T_R with the corresponding nonground version S_R. The nonground immediate consequence operator S_R is obtained by replacing in the definition of T_R the grounding substitution θ with the most general unifier between B and an atom in I. More formally, given a set of nonground atoms I, the operator S_R is defined as

$$S_R(I) \triangleq \{A\theta \mid A : -B \in R,\ C \in I,\ \theta = \text{m.g.u.}(B, C)\} \cup Bad.$$

The nonground fixpoint semantics is defined as the least fixpoint of the S_R operator, i.e., as the result of a (non ground) bottom-up evaluation of the logic program R. It is important to notice that the subsumption test between nonground atoms can be used as termination test for this type of symbolic fixpoint computation. Optimizations like magic set templates can be used to specialize the bottom-up evaluation procedure with respect to a given query (e.g., a set of initial states).

As shown in [24], the s-semantics for CLP can be used to extend the link between bottom-up evaluation of logic programs and model checking to the case of infinite-state transition systems. CLP clauses can be used to symbolically represent a possibly infinite set of transition rules, and constrained atoms, i.e., atoms like $p(X,Y) : -X > Y$ can be used to symbolically represent infinite sets of configurations, i.e., all the instances of the atom $p(X,Y)$ obtained by solving the constraint $X > Y$.

4.3   Verification and Evaluation Strategies in LP

Several other types of evaluation of logic programs have been proposed for the verification of temporal properties of transition systems.

In [30,31] the transition system of counter automata (automata with guards and assignments over a finite set of counters) are symbolically represented as logic programs with linear arithmetic constraints. The bottom-up evaluation of logic programs with gap-order constraints (obtained by relaxing the linear constraints in the automata) is used to over-approximate the set of successors, i.e., the set $Post^*$, of the original automata.

In [49,76] forward and backward evaluation of CLP programs is used to verify properties of real time and hybrid systems, respectively. Constraints are used here to infer preconditions on parameters of system specifications.

Program specialization methods (e.g. partial evaluation) is another example of techniques that can be used to automatically control the abstraction required for infinite-state model checking [55,53,54]. In [29,63] program transformation techniques combined with specialized decision procedures are used to verify temporal properties of infinite-state systems.

The application of tabling to the evaluation of logic programs represents a further important research line in-between logic programming and verification. The model checker XMC based on the XSB system has been applied to several families of verification problems and concurrent models including pi-calculus and mobile process algebra [25,65,69,70,72]. For this kind of systems, tabling can be used to efficiently evaluate logic programs that encode the semantics of CTL operators. Since tabling exploits different types of subsumption mechanisms, the resulting engine can be applied both to finite-state and infinite-state systems.

Other promising approaches for logic-based verification techniques are based on logic programming frameworks based on non standard logics like linear and intuitionistic logic. For instance, in [4,23], bottom-up evaluation methods for logic programming languages like LO [1] and MSR [7] extend the use of symbolic techniques based on unification (e.g. S_P-like operators) to languages that naturally model concurrency via multiset rewriting. Other examples come from logic programming languages like Bedwyr [2] and LolliMon [59] that incorporate connectives to express least and greatest fixpoint computations. The study of evaluation strategies and abstract interpretation techniques for these powerful logic programming languages represent an interesting research direction aimed at finding new verification methods for general classes of concurrent systems.

References

1. Andreoli, J.-M., Pareschi, R.: Linear Ojects. Logical Processes with Built-in Inheritance. New Generation Comput. 9(3/4), 445–474 (1991)
2. Baelde, D., Gacek, A., Miller, D., Nadathur, G., Tiu, A.: The Bedwyr system for model checking over syntactic expressions. In: Pfenning, F. (ed.) CADE 2007. LNCS (LNAI), vol. 4603, pp. 391–397. Springer, Heidelberg (2007)
3. Bagnara, R., Hill, P., Zaffanella, E.: Set-sharing is redundant for pair-sharing. Theor. Comput. Sci. 277(1-2), 3–46 (2002)
4. Bozzano, M., Delzanno, G., Martelli, M.: Model Checking Linear Logic Specifications. TPLP 4(5-6), 573–619 (2004)
5. Browne, M.C., Clarke, E.M., Grumberg, O.: Characterizing finite Kripke structures in propositional temporal logic. Theoret. Comp. Sci. 59, 115–131 (1988)
6. Burch, J.R., Clarke, E.M., McMillan, K.L., Dill, D.L., Hwang, L.J.: Symbolic Model Checking: $10^2 0$ States and Beyond. In: Proc. IEEE LICS 1990, pp. 428–439 (1990)
7. Cervesato, I.: Typed Multiset Rewriting Specifications of Security Protocols. ENTCS 40 (2000)
8. Clarke, E.M., Grumberg, O., Jha, S., Lu, Y., Veith, H.: Progress on the state explosion problem in model checking. In: Wilhelm, R. (ed.) Informatics: 10 Years Back, 10 Years Ahead. LNCS, vol. 2000, pp. 176–194. Springer, Heidelberg (2001)
9. Clarke, E.M., Grumberg, O., Long, D.: Model checking and abstraction. ACM Trans. Program. Lang. Syst. 16(5), 1512–1542 (1994)
10. Clarke, E.M., Grumberg, O., Peled, D.A.: Model Checking. The MIT Press, Cambridge (1999)
11. Cortesi, A., Filé, G., Giacobazzi, R., Palamidessi, C., Ranzato, F.: Complementation in abstract interpretation. ACM Trans. Program. Lang. Syst. 19(1), 7–47 (1997)
12. Cortesi, A., Le Charlier, B., Van Hentenryck, P.: Combinations of abstract domains for logic programming: open product and generic pattern construction. Sci. Comput. Program. 38(1-3), 27–71 (2000)
13. Cousot, P.: Types as abstract interpretations (invited paper). In: Proc. ACM POPL 1997, pp. 316–331 (1997)
14. Cousot, P.: Constructive design of a hierarchy of semantics of a transition system by abstract interpretation. Theor. Comput. Sci. 277(1-2), 47–103 (2002)
15. Cousot, P., Cousot, R.: Abstract interpretation: A unified lattice model for static analysis of programs by construction or approximation of fixpoints. In: Proc. of Conf. Record of the 4th ACM Symp. on Principles of Programming Languages (POPL 1977), pp. 238–252. ACM Press, New York (1977)
16. Cousot, P., Cousot, R.: Systematic design of program analysis frameworks. In: Proc. of Conf. Record of the 6th ACM Symp. on Principles of Programming Languages (POPL 1979), pp. 269–282. ACM Press, New York (1979)
17. Cousot, P., Cousot, R.: Abstract interpretation and application to logic programs. J. Logic Program. 13(2-3), 103–179 (1992)
18. Cousot, P., Cousot, R.: Higher-order abstract interpretation (and application to comportment analysis generalizing strictness, termination, projection and PER analysis of functional languages) (invited paper). In: Proc. of the 1994 IEEE Internat. Conf. on Computer Languages (ICCL 1994), pp. 95–112 (1994)
19. Cousot, P., Cousot, R.: Temporal abstract interpretation. In: Proc. 27th ACM POPL, pp. 12–25 (2000)
20. Cousot, P., Cousot, R., Feret, J., Mauborgne, L., Miné, A., Monniaux, D., Rival, X.: The ASTREÉ analyzer. In: Sagiv, M. (ed.) ESOP 2005. LNCS, vol. 3444, pp. 21–30. Springer, Heidelberg (2005)

21. Cousot, P., Cousot, R., Giacobazzi, R.: Abstract interpretation of resolution-based semantics. Theor. Comput. Sci. 410(46), 4724–4746 (2009)

22. Dams, D., Grumberg, O., Gerth, R.: Abstract interpretation of reactive systems. ACM Trans. Program. Lang. Syst. 16(5), 1512–1542 (1997)

23. Delzanno, G.: An Overview of MSR(C): A CLP-based Framework for the Symbolic Verification of Parameterized Concurrent Systems. ENTCS 76 (2002)

24. Delzanno, G., Podelski, A.: Model Checking in CLP. In: Cleaveland, W.R. (ed.) TACAS 1999. LNCS, vol. 1579, pp. 223–239. Springer, Heidelberg (1999)

25. Dong, Y., Du, X., Ramakrishna, Y.S., Ramakrishnan, C.R., Ramakrishnan, I.V., Smolka, S.A., Sokolsky, O., Stark, E.W., Scott Warren, D.: Fighting Livelock in the i-Protocol: A Comparative Study of Verification Tools. In: Cleaveland, W.R. (ed.) TACAS 1999. LNCS, vol. 1579, pp. 74–88. Springer, Heidelberg (1999)

26. Falaschi, M., Levi, G., Palamidessi, C., Martelli, M.: Declarative modeling of the operational behavior of logic languages. Theor. Comput. Sci. 69(3), 289–318 (1989)

27. Filé, G., Giacobazzi, R., Ranzato, F.: A unifying view of abstract domain design. ACM Comput. Surv. 28(2), 333–336 (1996)

28. Filé, G., Ranzato, F.: Complementation of abstract domains made easy. In: Proc. of the 1996 Joint Internat. Conf. and Symp. on Logic Programming (JICSLP 1996), pp. 348–362 (1996)

29. Fioravanti, F., Pettorossi, A., Proietti, M.: Verification of Sets of Infinite State Processes Using Program Transformation. In: Pettorossi, A. (ed.) LOPSTR 2001. LNCS, vol. 2372, pp. 111–128. Springer, Heidelberg (2002)

30. Fribourg, L., Richardson, J.: Symbolic Verification with Gap-Order Constraints. In: Gallagher, J.P. (ed.) LOPSTR 1996. LNCS, vol. 1207, pp. 20–37. Springer, Heidelberg (1997)

31. Fribourg, L., Olsén, H.: A Decompositional Approach for Computing Least Fixed-Points of Datalog Programs with Z-Counters. Constraints 2(3/4), 305–335 (1997)

32. Gentilini, R., Piazza, C., Policriti, A.: From bisimulation to simulation: coarsest partition problems. J. Automated Reasoning 31(1), 73–103 (2003)

33. Giacobazzi, R., Mastroeni, I.: Compositionality in the puzzle of semantics. In: Proc. of the ACM Symp. on Partial Evaluation and Semantics-Based Program Manipulation (PEPM 2002), pp. 87–97 (2002)

34. Giacobazzi, R., Mastroeni, I.: Transforming semantics by abstract interpretation. Theor. Comput. Sci. 337(1-3), 1–50 (2005)

35. Giacobazzi, R., Mastroeni, I.: Transforming abstract interpretations by abstract interpretation. In: Alpuente, M., Vidal, G. (eds.) SAS 2008. LNCS, vol. 5079, pp. 1–17. Springer, Heidelberg (2008)

36. Giacobazzi, R., Palamidessi, C., Ranzato, F.: Weak relative pseudo-complements of closure operators. Algebra Universalis 36(3), 405–412 (1996)

37. Giacobazzi, R., Quintarelli, E.: Incompleteness, counterexamples, and refinements in abstract model-checking. In: Cousot, P. (ed.) SAS 2001. LNCS, vol. 2126, pp. 356–373. Springer, Heidelberg (2001)

38. Giacobazzi, R., Ranzato, F.: Complementing logic program semantics. In: Hanus, M., Rodríguez-Artalejo, M. (eds.) ALP 1996. LNCS, vol. 1139, pp. 238–253. Springer, Heidelberg (1996)

39. Giacobazzi, R., Ranzato, F.: Refining and compressing abstract domains. In: Degano, P., Gorrieri, R., Marchetti-Spaccamela, A. (eds.) ICALP 1997. LNCS, vol. 1256, pp. 771–781. Springer, Heidelberg (1997)

40. Giacobazzi, R., Ranzato, F.: Optimal domains for disjunctive abstract interpretation. Sci. Comput. Program 32(1-3), 177–210 (1998)

41. Giacobazzi, R., Ranzato, F.: Uniform closures: order-theoretically reconstructing logic program semantics and abstract domain refinements. Information and Computation 145(2), 153–190 (1998)
42. Giacobazzi, R., Ranzato, F.: The reduced relative power operation on abstract domains. Theor. Comput. Sci 216, 159–211 (1999)
43. Giacobazzi, R., Ranzato, F.: Incompleteness of states w.r.t. traces in model checking. Information and Computation 204(3), 376–407 (2006)
44. Giacobazzi, R., Ranzato, F., Scozzari, F.: Making abstract interpretations complete. J. ACM 47(2), 361–416 (2000)
45. Giacobazzi, R., Ranzato, F., Scozzari, F.: Making abstract domains condensing. ACM Transactions on Computational Logic 6(1), 33–60 (2005)
46. Giacobazzi, R., Scozzari, F.: A logical model for relational abstract domains. ACM Trans. Program. Lang. Syst. 20(5), 1067–1109 (1998)
47. van Glabbeek, R.J., Ploeger, B.: Correcting a space-efficient simulation algorithm. In: Gupta, A., Malik, S. (eds.) CAV 2008. LNCS, vol. 5123, pp. 517–529. Springer, Heidelberg (2008)
48. Groote, J.F., Vaandrager, F.: An efficient algorithm for branching bisimulation and stuttering equivalence. In: Paterson, M. (ed.) ICALP 1990. LNCS, vol. 443, pp. 626–638. Springer, Heidelberg (1990)
49. Gupta, G., Pontelli, E.: A constraint-based approach for specification and verification of real-time systems. In: Proc. IEEE Real-Time Systems Symposium 1997, pp. 230–239 (1997)
50. Henzinger, M.R., Henzinger, T.A., Kopke, P.W.: Computing simulations on finite and infinite graphs. In: Proc. 36th FOCS, pp. 453–462 (1995)
51. Henzinger, T.A., Maujumdar, R., Raskin, J.-F.: A classification of symbolic transition systems. ACM Trans. Comput. Log. 6(1), 1–31 (2005)
52. Jensen, T.P.: Disjunctive program analysis for algebraic data types. ACM Trans. Program. Lang. Syst. 19(5), 751–803 (1997)
53. Leuschel, M., Lehmann, H.: Coverability of reset petri nets and other well-structured transition systems by partial deduction. In: Palamidessi, C., Moniz Pereira, L., Lloyd, J.W., Dahl, V., Furbach, U., Kerber, M., Lau, K.-K., Sagiv, Y., Stuckey, P.J. (eds.) CL 2000. LNCS (LNAI), vol. 1861, pp. 101–115. Springer, Heidelberg (2000)
54. Leuschel, M., Lehmann, H.: Solving coverability problems of petri nets by partial deduction. In: Proc. PPDP 2000, pp. 268–279 (2000)
55. Leuschel, M., Massart, T.: Infinite State Model Checking by Abstract Interpretation and Program Specialisation. In: Bossi, A. (ed.) LOPSTR 1999. LNCS, vol. 1817, pp. 62–81. Springer, Heidelberg (2000)
56. Levi, G., Spoto, F.: An experiment in domain refinement: Type domains and type representations for logic programs. In: Palamidessi, C., Meinke, K., Glaser, H. (eds.) ALP 1998 and PLILP 1998. LNCS, vol. 1490, pp. 152–169. Springer, Heidelberg (1998)
57. Levi, G., Spoto, F.: Non pair-sharing and freeness analysis through linear refinement. In: Proc. ACM PEPM, pp. 52–61 (2000)
58. Levi, G., Spoto, F.: Pair-independence and freeness analysis through linear refinement. Information and Computation 182(1), 14–52 (2003)
59. López, P., Pfenning, F., Polakow, J., Watkins, K.: Monadic concurrent linear logic programming. In: Proc. PPDP 2005, pp. 35–46 (2005)
60. Mycroft, A.: Completeness and predicate-based abstract interpretation. In: Proc. of the ACM Symp. on Partial Evaluation and Program Manipulation (PEPM 1993), pp. 179–185 (1993)
61. Nielson, F.: Expected forms of data flow analyses. In: Ganzinger, H., Jones, N.D. (eds.) Programs as Data Objects. LNCS, vol. 217, pp. 172–191. Springer, Heidelberg (1986)
62. Paige, R., Tarjan, R.E.: Three partition refinement algorithms. SIAM Journal on Computing 16(6), 977–982 (1987)

63. Pettorossi, A., Proietti, M., Senni, V.: Transformational Verification of Parameterized Protocols Using Array Formulas. In: Hill, P.M. (ed.) LOPSTR 2005. LNCS, vol. 3901, pp. 23–43. Springer, Heidelberg (2006)

64. Ramakrishnan, C.R.: A Model Checker for Value-Passing Mu-Calculus Using Logic Programming. In: Ramakrishnan, I.V. (ed.) PADL 2001. LNCS, vol. 1990, pp. 1–13. Springer, Heidelberg (2001)

65. Ramakrishna, Y.S., Ramakrishnan, C.R., Ramakrishnan, I.V., Smolka, S.A., Swift, T., Warren, D.S.: Efficient Model Checking Using Tabled Resolution. In: Grumberg, O. (ed.) CAV 1997. LNCS, vol. 1254, pp. 143–154. Springer, Heidelberg (1997)

66. Ranzato, F., Tapparo, F.: A new efficient simulation equivalence algorithm. In: Proc. 22nd IEEE Symp. on Logic in Computer Science (LICS 2007), pp. 171–180 (2007)

67. Ranzato, F., Tapparo, F.: Generalizing the Paige-Tarjan algorithm by abstract interpretation. Information and Computation 206(5), 620–651 (2008)

68. Rosenthal, K.I.: Quantales and their applications. In: Pitman Research Notes in Mathematics. Longman Scientific & Technical, London (1990)

69. Roychoudhury, A., Narayan Kumar, K., Ramakrishnan, C.R., Ramakrishnan, I.V., Smolka, S.A.: Verification of Parameterized Systems Using Logic Program Transformations. In: Schwartzbach, M.I., Graf, S. (eds.) TACAS 2000. LNCS, vol. 1785, pp. 172–187. Springer, Heidelberg (2000)

70. Roychoudhury, A., Ramakrishnan, C.R.: Unfold/Fold Transformations for Automated Verification of Parameterized Concurrent Systems. In: Bruynooghe, M., Lau, K.-K. (eds.) Program Development in Computational Logic. LNCS, vol. 3049, pp. 261–290. Springer, Heidelberg (2004)

71. Scozzari, F.: Logical optimality of groundness analysis. Theor. Comput. Sci. 277(1-2), 149–184 (2002)

72. Singh, A., Ramakrishnan, C.R., Smolka, S.A.: Query-Based Model Checking of Ad Hoc Network Protocols. In: Bravetti, M., Zavattaro, G. (eds.) CONCUR 2009. LNCS, vol. 5710, pp. 603–619. Springer, Heidelberg (2009)

73. Spoto, F.: Optimality and condensing of information flow through linear refinement. Theor. Comput. Sci. 388(1-3), 53–82 (2007)

74. Strachey, C.: The varieties of programming language. In: Proc. of the International Computing Symposium, Cini Foundation, Venice, pp. 222–233. Springer, Heidelberg (1972)

75. Tan, L., Cleaveland, W.R.: Simulation revisited. In: Margaria, T., Yi, W. (eds.) TACAS 2001. LNCS, vol. 2031, pp. 480–495. Springer, Heidelberg (2001)

76. Urbina, L.: Analysis of Hybrid Systems in CLP(R). In: Freuder, E.C. (ed.) CP 1996. LNCS, vol. 1118, pp. 451–467. Springer, Heidelberg (1996)

77. Yang, P., Basu, S., Ramakrishnan, C.R.: Parameterized Verification of π-Calculus Systems. In: Hermanns, H., Palsberg, J. (eds.) TACAS 2006. LNCS, vol. 3920, pp. 42–57. Springer, Heidelberg (2006)

Answer Set Programming

Piero Bonatti[1], Francesco Calimeri[2], Nicola Leone[2], and Francesco Ricca[2]

[1] Dept. of Phisical Sciences - Sec. Informatics, University of Naples "Federico II",
I-80126 Napoli, Italy
bonatti@na.infn.it

[2] Dept. of Mathematics, University of Calabria, I-87036 Rende (CS), Italy
{calimeri,leone,ricca}@mat.unical.it

Abstract. Answer Set Programming (ASP), referred to also as Disjunctive Logic Programming under the stable model semantics (DLP), is a powerful formalism for Knowledge Representation and Reasoning. ASP has been the subject of intensive research studies, and, also thanks to the availability of some efficient ASP systems, has recently gained quite some popularity and is applied also in relevant industrial projects. The Italian logic programming community has been very active in this area, some ASP results achieved in Italy are widely recognized as milestones on the road to the current state of the art. After a formal definition of ASP, this chapter surveys the main contribution given by the Italian community to the ASP field in the last 25 years.

1 Introduction

Answer Set Programming (ASP), [1–5] referred to also as Disjunctive Logic Programming under the stable model semantics (DLP),[1] is a powerful formalism for Knowledge Representation and Reasoning.[2] Bloomed from the work of Gelfond, Lifschitz [2, 3] and Minker [6–9] in the 1980ies, it has enjoyed a continuously increasing interest within the scientific community. One of the main reasons for the success of ASP is the high expressive power of its language: ASP programs, indeed, allow us to express, in a precise mathematical sense, every property of finite structures over a function-free first-order structure that is decidable in nondeterministic polynomial time with an oracle in NP [10, 11] (i.e., ASP captures the complexity class $\Sigma_2^P = \mathrm{NP}^{\mathrm{NP}}$). Thus, ASP allows us to encode also programs which cannot be translated to SAT in polynomial time. Importantly, ASP is fully declarative (the ordering of literals and rules is immaterial), and the ASP encoding of a large variety of problems is very concise, simple, and elegant [1, 12–15].

Example 1. To see an elegant ASP encoding, consider 3-Colorability, a well-known NP-complete problem. Given a graph, the problem is to decide whether there exists an

---

[1] Stable models are also named *answer sets*.

[2] A lot of work has been done by the Italian research community both in the broader field of knowledge representation and non-monotonic reasoning, and in the related field of logic languages for databases. We refer the reader to Chapter 4 and Chapter 9, respectively, for a detailed description of the italian contributions in these specific fields which are closely related and partially overlapping with the ASP contributions.

A. Dovier, E. Pontelli (Eds.): 25 Years of Logic Programming, LNCS 6125, pp. 159–182, 2010.
© Springer-Verlag Berlin Heidelberg 2010

assignment of one out of three colors (say, red, green, or blue) to each node such that adjacent nodes always have different colors. Suppose that the graph is represented by a set of facts F using a unary predicate $node(X)$ and a binary predicate $arc(X, Y)$. Then, the following ASP program (in combination with F) computes all 3-Colorings (as stable models) of that graph.

$$r_1 : \quad color(X, red) \vee color(X, green) \vee color(X, blue) :- node(X).$$
$$r_2 : \quad :- color(X_1, C), color(X_2, C), arc(X_1, X_2).$$

Rule r_1 expresses that each node must either be colored red, green, or blue;[3] due to minimality of the stable models, a node cannot be assigned more than one color. The subsequent integrity constraint checks that no pair of adjacent nodes (connected by an arc) is assigned the same color.

Thus, there is a one-to-one correspondence between the solutions of the 3-Coloring problem and the answer sets of $F \cup \{r_1, r_2\}$. The graph is 3-colorable if and only if $F \cup \{r_1, r_2\}$ has some answer set. □

Unfortunately, the high expressiveness of ASP comes at the price of a high computational cost in the worst case, which makes the implementation of efficient systems a difficult task. Nevertheless, starting from the second half of the 1990ies, and even more in the latest years, a number of efficient ASP systems have been released [16–25], that encouraged a number of applications in many real-world and industrial contexts [26–33, 40]. These applications have confirmed the viability of the ASP exploitation for advanced knowledge-based tasks, and stimulated further research in this field.

The Italian research community produced, in the latest 25 years, a significant contribution in the area, addressing the whole spectrum of issues cited above; this contribution ranged from theoretical results and characterizations [34–39] to practical applications [26–33, 40–45], stepping through language extensions [16, 42, 46–68], evaluation algorithms and optimization techniques [69–78]. Several of the achieved results are widely recognized as milestones on the road to the current state of the art; this is, for instance, the case of the DLV project [16], that produced one of the world leading ASP systems. The Italian community is currently very active on ASP, it contributes in pushing forward the state of the art, as witnessed by the most recent results like, e.g., the ASP extension to deal with infinite domains which is at the frontier of the ASP research [59, 61, 62, 64, 65, 68].

The rest of the Chapter is structured as follows: in Section 2, ASP is formally introduced, syntax and semantics of the language are presented; Section 3 focuses on ASP properties and its theoretical characterizations; Section 4 surveys linguistic extensions; Section 5 reports on ASP with infinite domains; Section 6 first introduces the general architecture of ASP systems, and then surveys algorithms and optimization techniques; Section 7 first describes DLV and number of other ASP-based systems, and then reports on real-world ASP applications; eventually, Section 8 collects a number of further contributions of the Italian ASP community.

---

[3] Variable names start with an upper case letter and constants start with a lower case letter.

2   The ASP Language

In what follows, we provide a formal definition of the syntax and semantics of Answer Set Programming in the spirit of [3].

2.1   Syntax

Following a convention dating back to Prolog, strings starting with uppercase letters denote logical variables, while strings starting with lower case letters denote constants. A *term* is either a variable or a constant. [4] An *atom* is an expression $p(t_1, \ldots, t_n)$, where p is a *predicate* of arity n and $t_1, \ldots, t_n$ are terms. A literal l is either an atom p (*positive* literal) or its negation not p (*negative* literal). A set L of literals is said to be *consistent* if, for every positive literal $l \in L$, its complementary literal not l is not contained in L.

A *disjunctive rule* (*rule*, for short) r is a construct:

$$a_1 \lor \cdots \lor a_n :\text{-} b_1, \cdots, b_k, \text{not } b_{k+1}, \cdots, \text{not } b_m. \tag{1}$$

where $a_1, \cdots, a_n, b_1, \cdots, b_m$ are atoms and $n \geq 0, m \geq k \geq 0$. The disjunction $a_1 \lor \cdots \lor a_n$ is called the *head* of r, while the conjunction $b_1, \ldots, b_k$, not $b_{k+1}, \ldots$, not b_m is referred to as the *body* of r. A rule without head literals (i.e. $n = 0$) is usually referred to as an *integrity constraint*. A rule having precisely one head literal (i.e. $n = 1$) is called a *normal rule*. If the body is empty (i.e. $k = m = 0$), it is called a fact, and in this case the " :- " sign is usually omitted. If r is a rule of form (1), then $H(r) = \{a_1, \ldots, a_n\}$ is the set of literals in the head and $B(r) = B^+(r) \cup B^-(r)$ is the set of the body literals, where $B^+(r)$ (the *positive body*) is $\{b_1, \ldots, b_k\}$ and $B^-(r)$ (*the negative body*) is $\{b_{k+1}, \ldots, b_m\}$. An *ASP program* (also called *Disjunctive Logic Program* or *DLP program*) P is a finite set of rules. A not-free program P (i.e., such that $\forall r \in P, B^-(r) = \emptyset$) is called *positive*, and a v-free program P (i.e., such that $\forall r \in P, |H(r)| \leq 1$) is called *normal logic program*.

In ASP, rules are usually required to be safe; the motivation comes from the field of databases, and for a detailed discussion we refer to [79]. A rule r is *safe* if each variable in r also appears in at least one positive literal in the body of r. An ASP program is safe if each of its rules is safe, and in the following we will only consider safe programs. A term (an atom, a rule, a program, etc.) is called *ground*, if no variable appears in it; a ground program is also called *propositional*.

2.2   Semantics

We next describe the semantics of ASP programs, which is based on the answer set semantics originally defined in [3]. However, different to [3] only consistent answer sets are considered, as it is now standard practice. In ASP the availability of some pre-interpreted predicates is assumed, such as $=, <, >$. However, it would also be possible to define them explicitly as facts, so they are not treated in a special way.

---

[4] Note that, as common in ASP, function symbols are not considered unless explicitly specified (see Section 5).

Herbrand Universe and Herbrand Base. For any program P, the *Herbrand universe*, denoted by U_P, is the set of all constants occurring in P. If no constant occurs in P, U_P consists of one arbitrary constant. The *Herbrand Base* B_P is the set of all ground atoms constructible from predicate symbols appearing in P and constants in U_P.

Ground Instantiation. For any rule r, $Ground(r)$ denotes the set of rules obtained by replacing each variable in r by constants in U_P in all possible ways. For any program P, its ground instantiation is the set $grnd(P) = \bigcup_{r \in P} Ground(r)$. Note that for propositional programs, $P = grnd(P)$ holds.

Answer Sets. For every program P, its answer sets are defined by using its ground instantiation $grnd(P)$ in two steps: first the answer sets of positive disjunctive programs are defined, then the answer sets of general programs are defined by a reduction to positive disjunctive programs and a stability condition. An interpretation I for a program P is a set of ground atoms $I \subseteq B_P$. Let P be a positive program. An interpretation $X \subseteq B_P$ is called *closed under P* if, for every $r \in grnd(P)$, $H(r) \cap X \neq \emptyset$ whenever $B(r) \subseteq X$. An interpretation which is closed under P is also called *model* of P. An interpretation $X \subseteq B_P$ is an *answer set* for a positive program P, if it is minimal (under set inclusion) among all interpretations that are closed under P.

Example 2. The positive program $P_1 = \{a \vee b \vee c.\}$ has the answer sets $\{a\}$, $\{b\}$, and $\{c\}$; they are minimal and correspond to the multiple ways of satisfying the disjunction. Its extension $P_2 = P_1 \cup \{:- a.\}$ has the answer sets $\{b\}$ and $\{c\}$: comparing P_2 with P_1, the additional constraint is not satisfied by interpretation $\{a\}$. Moreover, the positive program $P_3 = P_2 \cup \{b :- c., \; c :- b.\}$ has the single answer set $\{b, c\}$. It is easy to see that, $P_4 = P_3 \cup \{:- c\}$ has no answer set. □

The *reduct* or *Gelfond-Lifschitz transform* [2, 3] of a ground program P w.r.t. a set $X \subseteq B_P$ is the positive ground program P^X, obtained from P by: (i) deleting all rules $r \in P$ for which $B^-(r) \cap X \neq \emptyset$ holds; (ii) deleting the negative body from the remaining rules. An *answer set* of a program P is a set $X \subseteq B_P$ such that X is an answer set of $grnd(P)^X$.

Example 3. For the negative ground program $P_5 = \{a :- \text{not } b.\}$, $A = \{a\}$ is the only answer set, as $P_5^A = \{a.\}$. For example for $B = \{b\}$, $P_5^B = \emptyset$, and so B is not an answer set. □

3   Properties and Theoretical Characterizations

The Italian research community provided relevant contributions to the study of ASP and its theoretical characterizations. In this respect, a relevant bunch of results has been achieved by the work in [34], which has given the theoretical foundation for realization of the ASP system DLV system [16]. There, the authors provide: a declarative characterization of answer sets in terms of unfounded sets; a generalization of the well-founded (W_P) operator to disjunctive logic programs; a fixpoint semantics for function-free programs; an algorithm for answer set computation; an in-depth analysis of the main computational problems related to the concepts. In the this Section, we briefly discuss these contributions.

The definition of unfounded sets for disjunctive logic programs was given as an extension of the analogous concept defined for (disjunction-free) logic programs [80]. As for normal logic programs, unfounded sets single out the atoms that are (definitely) not derivable from a given program w.r.t. a fixed interpretation; thus, according to the closed-world assumption [81], they single out atoms that can be stated to be false. In a disjunctive logic program $\mathcal{P}$, the union of unfounded sets for $\mathcal{P}$ may not be an unfounded set for $\mathcal{P}$; thus, the existence of the greatest unfounded set (i.e., an unfounded set that contains all other unfounded sets) is not guaranteed as in the case of normal programs. The authors proved that for unfounded-free interpretations (i.e., interpretations that do not contain any unfounded atom), the union of different unfounded sets is guaranteed to be an unfounded set even in the disjunctive case; the *greatest unfounded set* of $\mathcal{P}$ w.r.t. I, denoted $GUS_{\mathcal{P}}(I)$, is the union of all unfounded sets.

Several interesting relationships between answer sets and unfounded sets were also discovered, which led to a simple, yet elegant, characterization of answer sets in terms of unfounded sets: the answer sets of a disjunctive program $\mathcal{P}$ coincide with the unfounded-free models of $\mathcal{P}$, and a model of $\mathcal{P}$ is an answer set iff the set of false atoms coincides with the greatest unfounded set.

The authors of [34] defined also a suitable extension of the well-founded operator $\mathcal{W}_{\mathcal{P}}$ of Van Gelder et al. [80] to the disjunctive case; this allowed to achieve another important result: the definition of a fixpoint semantics for disjunctive answer sets in terms of $\mathcal{W}_{\mathcal{P}}$. The set of answer sets of $\mathcal{P}$ coincides with the (total) fixpoints of $\mathcal{W}_{\mathcal{P}}$. By exploiting the theoretical results, the authors designed an algorithm for the computation of the answer set semantics of disjunctive programs. The key idea is that, since answer sets are total interpretations, computing their entire negative portion is superfluous; rather, it is sufficient to restrict the computation to those negative literals that are necessary to derive the positive part. To this end, the notion of *possibly-true literals* is introduced, which plays a crucial role in the computation. The algorithm is based on a controlled search in the space of the interpretations, implemented by a backtracking technique; and the stability of a generated model (answer set candidate) is tested by checking whether it is unfounded-free. This is done by means of a function that runs in polynomial time on *head-cycle-free* (*HCF*) programs [82, 83]. In the general case, the algorithm for the computation of answer sets runs in polynomial space and single exponential time.

4   Language Extensions

The standard language of ASP has been extended in several ways in order to improve its expressiveness. The Italian community provided contributions regarding two of the most relevant extensions of ASP: *Optimization Constructs* and *Aggregates*.

4.1   Optimization Constructs

The basic ASP language can be used to solve complex search problems, but it does not natively provide constructs for specifying optimization problems (i.e. problems where some goal function must be minimized or maximized). In the basic language, constraints represent a condition that *must* be satisfied; for this reason, they are also called *strong* constraints. Contrary to strong constraints, *weak constraints*, introduced in [16, 46], allow one to express desiderata, that is, conditions that *should* be satisfied; their semantics involves minimizing the number of violations, thus allowing to easily encode optimization problems. From a syntactic point of view, a weak constraint is like a strong one, where the implication symbol :– is replaced by :∼ . The informal meaning of a weak constraint :∼ B. is "try to falsify B," or "B should preferably be false.". Additionally, a weight and a priority level for the weak constraint may be specified after the constraint enclosed in brackets (by means of positive integers or variables). If not specified, the default value for weight and priority level is 1. The answer sets are considered which minimize the sum of weights of the violated (unsatisfied) weak constraints in the highest priority level and, among them, those which minimize the sum of weights of the violated weak constraints in the next lower level, and so on.

4.2   Aggregates

There are some simple properties, often arising in real-world applications, which cannot be encoded in a simple and natural manner using ASP [47–50, 84–86]. Especially properties that require the use of arithmetic operators on a set of elements satisfying some conditions (like sum, count, or maximum) require rather cumbersome encodings (often requiring an "external" ordering relation over terms) if one is confined to classic ASP. Similar observations have also been made in related domains, which led to the definition of aggregate functions. Especially in database systems this concept is at present both theoretically and practically fully integrated. When ASP systems started to be used in real applications, the need for aggregates become apparent also here. Hence, ASP has been extended with special atoms handling aggregate functions [47–50, 87, 88]. Intuitively, an aggregate function can be thought of as a (possibly partial) function mapping multisets of constants to a constant. The most common aggregate functions compute the number of terms, the sum of non-negative integers, and minimum/maximum term in a set. Aggregates are especially useful when real-world problems have to be dealt with.

4.3   Other Extensions

In order to meet requirements of different application domains, ASP was extended in other directions; thus, there is a number of interesting languages having the roots on ASP.

For instance, ASP was exploited for defining and implementing the *action language* (i.e., a language conceived for dealing with actions and change) $\mathcal{K}$ [51], while, in [52] a framework for *abduction with penalization* was proposed and implemented as a front-end for the ASP system DLV. Other ASP extensions were conceived to deal with *Ontologies* (i.e. abstract models of a complex domain). In particular, in [42] an ASP-based language for ontology specification and reasoning was proposed, which extends ASP in order to deal with complex real-world entities, like classes, objects, compound objects, axioms, and taxonomies. In [53] an extension of ASP, called *HEX-Programs*, which supports higher-order atoms as well as external atoms was proposed. External atoms allows one to embed external sources of computation in a logic program. Thus, HEX-programs are useful for various tasks, including meta-reasoning, data type manipulations, and reasoning on top of Description Logics (DL) [89] ontologies. *Template predicates* were introduced in [54]; they are special intensional predicates defined by means of generic reusable subprograms, which were conceived for easing coding and improving readability and compactness of programs, and allowing more effective code reusability. An extension of ASP by the introduction of the notion of resource is proposed in [55]. The resulting framework, named RASP, declaratively supports quantitative reasoning on consumption and production of resources. Various forms of preferences, policies, and cost-based criteria can be used to model the processes that produce/consume resources [56].

In [57] standard ASP was enriched by introducing consistency-restoring rules (cr-rules) and preferences, leading to the CR-Prolog language. Basically, in this language, besides standard ASP rules one may specify CR-rules, that are expressions of the form: $r{:}a_1 \vee \ldots \vee a_n {:}{-}{}^+body$ $(n \geq 1)$. The intuitive meaning of CR-rule r is: if *body* is true then one of $a_1, \ldots, a_n$ is "possibly" believed to be true. Importantly, the name of CR-prolog rules can be directly exploited to specify preferences among them. In particular, if the fact $prefer(r_1, r_2)$ is added to a CR-program, then rule r_1 is preferred over rule r_2. This allows one to encode partial orderings among preferred answer sets by explicitly writing preferences among CR-rules.

In [58] Normal Form Nested (NFN) programs, a non-propositional language similar to Nested Logic Programming (NLP) [90] was proposed. NFN programs often allows for more concise ASP representations by permitting a richer syntax in rule heads and bodies. It is worth noting that, NFN programs do allow for variables, whereas NLP are propositional. Since with the presence of variables domain independence is no longer guaranteed, the class of safe NFN programs was defined. Moreover, it was shown that for NFN programs which are also NLPs, the new semantics coincides with the one of [90]; while keeping the standard meaning of answer sets on ASP programs with variables. Finally, an algorithm which translates NFN programs into ASP programs was provided.

In [91] the concept of ordered disjunctions was extended to cardinality constraints. This paved the way to the definition of a policy description language that allows to express preferences among sets of objects and to handle advanced policy description specifications. The work followed some proposals aiming at introducing preferences in policy description languages [92–94].

5   ASP with Infinite Domains

The first ASP languages were based on extensions of Datalog, that is, function-free logic programs.[5] From a syntactic viewpoint, the addition of functions is obtained by generalizing the notion of term: a *term* is either a *simple term* or a *functional term*. A *simple term* (see Section 2) is either a constant or a variable. If $t_1 \ldots t_n$ are terms and f is a function symbol (*functor*) of arity n, then: $f(t_1, \ldots, t_n)$ is a *functional term*. It

---

[5] In this section we use the term *function* to refer to uninterpreted functions (or constructors) as in pure logic programming.

is easy to see that such an extension make U_P, B_P and $grnd(P)$ possibly infinite, and enhances the expressiveness of ASP. Indeed, without function symbols, ASP programs can only reason about finite domains, and have limited data modeling abilities. Such restrictions were motivated by complexity considerations, as answer set reasoning with unrestricted first-order normal programs is Π_1^1-complete, and hence highly undecidable. However, by introducing suitable alternative syntactic restrictions on the usage of functions, it is possible to improve the tradeoff between complexity and expressiveness.

In particular, the introduction of function symbols in ASP languages leads to several benefits [59]: (i) Data encapsulation support, as function symbols are the main logic programming construct for data abstraction [95]; (ii) Enhanced problem solving power, as the class of solvable problems can be extended beyond the second level of the polynomial hierarchy (that is, the class of problems solvable with Disjunctive Datalog with negation); (iii) Support for recursive data structures, such as lists, XML documents, etc. Such data structures are extremely common in modern applications and functions constitute the most natural way of encoding them; (iv) Simulation and extension of description logics [96]; in this context, function symbols are needed to encode existential quantification through skolemization. Such work is of strategic importance given the important role that description logics play in the semantic web.

The first class of computationally well-behaved ASP programs with function symbols, called *finitary programs*, is due to the Italian logic programming community. They were introduced in [60], and soon after were followed by ω-*restricted programs* [97]. The latter address the challenges of ASP with functions only partially. The answer sets of ω-restricted programs are all finite, and recursion over recursive data structures is not allowed—therefore ω-restricted programs address essentially data encapsulation only. Finitary programs constitute a more ambitious effort, capable of supporting ASP programs with infinite and infinitely many answer sets, and a large class of recursive predicates, including the standard list- and tree-manipulation programs [59].

Finitary programs are characterized by two restrictions. To simplify the presentation here we deal only with normal (i.e. disjunction-free) logic programs—see [61, 62] for an account of disjunctive programs. The first restriction applies to recursion, and is expressed in terms of the notion of *dependency graph* of a program P, whose set of nodes is the Herbrand base B_P. The dependency graph contains a directed edge (A, A') if and only if there exists a rule $r \in grnd(P)$ such that $A \in H(r)$ and $A' \in B(r)$. The edge is labelled *positive* if $A' \in B^+(r)$, and *negative* if $A' \in B^-(r)$. Then we say that A *depends* on A' if there exists a path from A to A' in the dependency graph.

Now we are ready to formulate the first restriction: a program P is *finitely recursive* iff every atom in the Herbrand base of P depends only on finitely many other ground atoms. Finitely recursive programs enjoy a number of nice theoretical properties proved in [61]:[6]

- they enjoy an analog of the *compactness* property of first-order logic;
- inconsistency checking and skeptical inference are semidecidable;
- the semantics of a finitely recursive program P can be approximated through a chain of finite programs $P_1 \subseteq P_2 \subseteq \cdots \subseteq P_i \subseteq \cdots \subseteq grnd(P)$.

---

[6] Another contribution of the Italian community; best paper award at ICLP 2007.

The second restriction is based on *odd-cycles*, that are cycles in the dependency graph containing an odd number of negative edges. A normal program is *finitary* iff it is finitely recursive and its dependency graph contains only finitely many odd-cycles.

Finitary programs are very expressive; they comprise a number of useful predicates, including the standard list manipulation predicates, QBF metainterpreters, and programs for reasoning about actions, just to name a few [59]. Moreover, they enjoy very good computational properties [59, 63]. If the set of atoms occurring in an odd-cycle is given, then: (a) ground credulous queries and ground skeptical queries are all decidable; (b) unrestricted ground credulous queries and ground skeptical queries are semidecidable.

Another Italian contribution in this field is the class of *finitely ground* programs [64]. They are characterized by means of an intelligent grounding transformation that turns any given disjunctive program P with functions into an equivalent ground program; P is finitely ground if this transformation yields a finite program. Finitely ground programs—due to the nature of the intelligent grounding—are well-suited for bottom-up evaluation, while finitary programs are naturally well-suited for top-down evaluations. As a consequence finitely ground programs are easier to support in systems like DLV that adopt a bottom-up grounding approach. Finitely ground programs have no restrictions on odd-cycles (and do not need them to be fed to the reasoner as an input). On the other hand, they are required to be safe, which rules out a number of interesting programs, such as list- and tree-manipulation programs. Moreover, like ω-restricted programs, their semantics is always finite, both in terms of the size and the number of answer sets.

In an interesting recent work [65], however, the duality between the two program classes is starting to be reconciled, by showing how given a positive finitely recursive program P and a query Q one can construct—by a magic set transformation—a finitely ground program P' that yields the same answer to Q as P.

The classes of finitary and finitely ground programs, unfortunately, are not decidable. This result motivated further works aimed at characterizing decidable classes of well-behaved programs with function symbols. The fathers of finitely ground programs introduced *finite domain programs*, a subclass of finitely ground programs that can be effectively recognized [64].

This line of research is having an impact on the activity of other groups outside Italy. In [98], an extension of finite domain programs is proposed. In [96, 99, 100], another family of effectively recognizable, well-behaved programs is investigated. This is a very interesting line of investigation, as it covers description logics, and it may eventually lead to interesting nonmonotonic extensions thereof. Moreover, these works adopt a different strategy for achieving inference decidability, based on a tree-model property and on a reasoning method analogous to blocking.

5.1   Calculi and Implementations

Further contributions stemming from the Italian community comprise resolution-based calculi for skeptical and credulous ASP reasoning with function symbols. *Skeptical resolution* [66] consists of five inference rules: resolution, negation as failure, a structural rule for removing successful literals, a rule for detecting contradictions, and a *split* rule

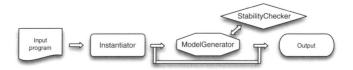

Fig. 1. General architecture of an ASP system

for generating new hypotheses and carrying out reasoning by cases. The skeptical resolution calculus is complete for all finitely recursive programs [61]. Recently, a *credulous resolution* calculus [67] was theoretically studied and experimentally evaluated on a few standard problems with encouraging results that deserve further investigations. The main advantage of resolution calculi is that they need no prior instantiation (grounding) of the input program; instantiation is incremental and on-demand, as in classical resolution. Support for function symbols is also being introduced in DLV for finitely ground programs [68]. We expect it to be soon extended to finitary programs by means of suitable extensions of the magic sets transformation adopted in [65].

5.2 Open Issues

ASP with infinite domains is a lively area which is being further developed by several research groups across the world. The main ongoing investigations concern:

- extending the known decidable classes of well-behaved ASP programs;
- the systematic derivation of new classes of well-behaved programs with functions through the composition of modules belonging to known well-behaved classes [101];
- the development and improvement of reasoning mechanisms for ASP with infinite domains;
- the relationships between finitary and finitely ground programs.

6   Algorithms and Optimization Techniques

The general architecture of an ASP system, depicted in Figure 1, helps in understanding the evaluation flow of the typical computation carried out for computing the answer sets of an ASP program. Upon startup, the input specified by the user is parsed and transformed into the internal data structures of the system.[7]

In general, an input program P contains variables, and the first step of a computation of an ASP system is to eliminate these variables, generating a ground instantiation $grnd(P)$ of P. This variable-elimination process is called *instantiation* of the program (or *grounding*), and is performed by the *Instantiator* module (see Figure 1). A naïve Instantiator would produce the full ground instantiation $grnd(P)$, which is, however, undesirable from a computational point of view, as in general many useless ground rules

---

[7] The input is usually read from text files, but some systems also interface to relational databases for retrieving facts stored in relational tables.

would be generated. An ASP system, therefore, employs a more sophisticated procedure geared towards keeping the instantiated program as small as possible. A necessary condition is, of course, that the instantiated program must have the same answer sets as the original program; however, it should be noted that the Instantiator solves a problem which is in general EXPTIME-hard, the produced ground program being potentially of exponential size with respect to the input program. Optimizations in the Instantiator therefore often have a big impact, as its output is the input for the following modules, which implement computationally hard algorithms. Moreover, if the input program is normal and stratified, the Instantiator module is, in some cases, able to directly compute its answer sets (if they exist).

The subsequent computations, which constitute the non-deterministic part of an ASP system, are then performed on $grnd(P)$ by both the *Model Generator* and the *Model Checker*. Roughly, the former produces some "candidate" answer set, whose stability is subsequently verified by the latter. Model generation is the non-deterministic core of an ASP system, and it is usually implemented as a backtracking search similar to the Davis-Putnam-Logemann-Loveland (DPLL) procedure [102] for SAT solving. Basically, starting from the empty (partial) interpretation, the $ModelGenerator$ module repeatedly assumes truth-values for atoms (branching step), subsequently computing their deterministic consequences (propagation step). This is done until either an answer set candidate is found or an inconsistency is detected. Candidate answer sets are then checked by exploiting the *Model Checker* module; whereas, if an inconsistency is detected, chosen literals have to be undone (backtracking). For disjunctive programs model cheking is as hard as the problem solved by the Model Generator, while it is trivial for non-disjunctive programs. Finally, once an answer set has been found, ASP systems typically print it in text format, and possibly the *Model Generator* resumes in order to look for further solutions.

All the aspects of the evaluation of ASP programs have been subject of analysis by the Italian research community; the obtained results, divided by evaluation task, are surveyed in the following.

Instantiation. The first contributions in this respect date back to 1999, when some optimization techniques, based on a rewriting of the input program, were proposed aiming at reducing the size of the instantiation generated by the grounder [69]. Since computing all the possible instantiations of a rule is, basically, analogous to computing all the answers of a conjunctive query joining the extensions of literals of the rule body, in [70] a new join-ordering technique was proposed, that sensibly improves the instantiation procedures of ASP systems. Some year later, in [71] a new backjumping technique for the instantiation of a rule was proposed which allows for reducing both the size of the generated grounding and the time needed for producing it. All the above mentioned techniques were incorporated in the grounder of the DLV system, and allowed for relevant improvements of the performance of the system. Notably, to our knowledge, the technique in [71] has been successfully exploited also by other two grounders, namely *GrinGo* [103], and GIDL [104].

In the last years, in order to exploit the power of modern multi-core/multiprocessor computers, a number of strategies for the parallelization of the instantiation procedure have been proposed [72, 73]. In particular, three levels of parallelism can be exploited during the instantiation process, namely, components, rules and single rule level. The first two levels were first employed in [72] while the third one was presented in [73]. Also these techniques have been implemented into the DLV grounder, and the resulting parallel instantiator proved to be effective on modern multi-core machines when handling both real-world and classical problem instances [72, 73, 105].

A distributed instantiator working on a Beowulf [106] cluster was presented in [107]; further works appear in [108].

Model Generation. The Italian research community provided relevant contribution regarding all the aspects of model generation. About the propagation step, peculiar properties of ASP programs were exploited in [74, 109], that allow to prune the search space by combining extension of the well-founded operator for disjunctive programs with a number of techniques based on disjunctive ASP program properties. The efficiency of the whole model generation process depends also on two crucial features: a good heuristic (branching rule) to choose the branching literal (i.e., the criterion determining the literal to be assumed true at a given stage of the computation); and a smart recovery procedure for undoing the choices causing inconsistencies. To this end, both look-ahead [75] and look-back [76, 77] techniques and heuristics specifically conceived for enhancing the model generation process were proposed and implemented in the state-of-the-art ASP system DLV [16]. In a lookahead heuristic [75] each possible choice literal is tentatively assumed, its consequences are computed, and some characteristic values on the result are recorded. The look-ahead heuristics of [75] "layers" several criteria based on peculiar properties of ASP, and basically drives the search towards "supported" interpretations (since answer sets are supported interpretations (cfr. [34, 110, 111]). In a look-back heuristics usually choices are made in such a way that the atoms most involved in conflicts are chosen first. Motivated by heuristics implemented in SAT solvers like Chaff [112], a family of new look-back heuristics tailored for disjunctive ASP programs were proposed in [77]. Look-back heuristics are mainly exploited in conjunction with backjumping, where the set of chosen literals that are relevant for an inconsistency are detected, and the system goes back in the search until at least one choice that "entail" the inconsistency is undone. In [76] a *reason calculus* that allows for determining the relevance for an inconsistency was proposed; here the information about the choices ("reasons") whose truth-values have caused truth-values of other deterministically derived atoms is collected and exploited for backjumping.

Native ASP systems exploit backtracking search algorithms that work directly on the ground instantiation of the input program, like the ones described above. An alternative approach to model generation is based on a rewriting into a propositional formula which is then evaluated by a boolean satisfiability solver for finding answer sets. Giunchiglia and Maratea, in collaboration with the members of the Texas Action Group at Austin, led by Prof. Vladimir Lifschitz, designed a SAT-based approach to normal logic programs [21, 113, 114], which is now considered the reference SAT-based work in ASP. A comparison among the techniques employed by ASP systems underlying strengths and weaknesses of each approach was provided in [115, 116].

Techniques for parallel evaluation of ground ASP programs were studied in [117, 118] and, on clusters, in [107, 108]. Furthermore, going beyond the classical methods of computing the answer sets of a logic program, in [119, 120] a method is presented that does not require a preliminary grounding phase.

Model Checking. is a crucial step of the computation of the answer sets. There are two main reason for the importance of the model checking step: the exponential number of possible models (model candidates); and the hardness of stable model checking. Note that, when disjunction is allowed in the head, deciding whether a given model is a stable model of a propositional ASP program is co-NPcomplete [11]. In [78] a new transformation $\mathcal{T}$, which reduces stable model checking to UNSAT, i.e., to deciding whether a given CNF formula is unsatisfiable, is introduced. Thus, the stability of an answer set candidate M of a program P can be verified by calling a SAT solver on the CNF formula obtained by applying $\mathcal{T}$ to P. The transformation is very efficient: it runs in logarithmic space and no new symbol is added. This approach to model checking was implemented in the ASP system DLV [16] and some experiments confirmed its efficacy [78].

7   Systems and Applications

Several ASP systems are available nowadays, and a number of practically relevant real-world applications of ASP have been developed. In the following, we first present DLV [16], a state-of-the-art ASP systems, which is widely used all over the world and is actively developed by Italian researchers; then we mention some relevant systems and application based on ASP.

7.1   The DLV System

The DLV system [16] is widely considered one of the state-of-the-art implementations of answer set programming. The development of DLV started at the end of 1996, within a research project funded by the Austrian Science Funds (FWF) and led by Nicola Leone at the Vienna University of Technology. The first stable release became available in 1997, and at present, DLV is the subject of an international cooperation between the University of Calabria and the Vienna University of Technology. After its first release, the DLV system has been significantly improved over and over in the last years. In particular, the language of DLV was enriched in several ways and currently supports the main ASP extensions: disjunction, aggregates, weak-constraints, and function symbols (see Section 4 and Section 5). Relevant optimization techniques have been incorporated into the DLV engine, including database techniques for efficient instantiation, advanced pruning operators, look-ahead and look-back techniques for model generation, and innovative techniques for answer-set checking (see Section 6). Moreover, in order to deal with data-intensive applications a database oriented version of DLV, called DLV^{DB}, was recently proposed [121, 122]. DLV^{DB} is able to evaluate large amount of data by exploiting an evaluation strategy working mostly onto the database, where input data reside. DLV^{DB} embodies some query-oriented optimization strategies, like magic-sets [44], capable of significantly improving query evaluation performances. As a result, at the time being, DLV is generally recognized to be a state-of-the-art implementation of disjunctive ASP. Importantly, DLV is widely used by researchers all over the world, it is employed in real-world applications (see next Section), and it is competitive from the viewpoint of efficiency with the most advanced systems in the area of Answer Set Programming [13, 123].

7.2   ASP-Based Products

In this section three industrial products strongly based on ASP, and, in particular, on DLV are presented, namely: OntoDLV [41, 42], OLEX [30, 31], H$\imath$LεX [32, 33].

- OntoDLV [41, 42] is a system for ontologies specification and reasoning. The language of OntoDLV is an extension of (disjunctive) ASP with all the main ontology constructs including classes, inheritance, relations, and axioms. Importantly, OntoDLV supports a powerful interoperability mechanism with OWL, allowing the user to retrieve information from external OWL Ontologies and to exploit this data in OntoDLP ontologies and queries. OntoDLV facilitates the development of complex applications in a user-friendly visual environment; it features a rich Application Programming Interface (API) [124], and it is endowed with a robust persistency-layer for saving information transparently on a DBMS, and it seamlessly integrates DLV [16].

- OLEX [30, 31] (OntoLog Enterprise Categorizer System) is a corporate classification system supporting the entire content classification life-cycle, including document storage and organization, ontology construction, pre-processing and classification. OLEX exploits a reasoning-based approach to text classification which synergically combines: (i) ontologies for the formal representation of the domain knowledge; (ii) pre-processing technologies for a symbolic representation of texts and (iii) ASP as categorization rule language. Logic rules, indeed, provides a natural and powerful way to encode how document contents may relate to ontology concepts.

- H$\imath$LεX [32, 33] is an advanced system for ontology-based information extraction from semi-structured and unstructured documents. H$\imath$LεX implements a semantic approach to the information extraction problem able to deal with different document formats (html, pdf, doc, ...). H$\imath$LεX is based on OntoDLP for describing ontologies, and supports a language that is founded on the concept of *ontology descriptor*. A "descriptor" looks like a production rule in a formal attribute grammar, where syntactic items are replaced by ontology elements. Each descriptor allows us to describe: (i) an ontology object in order to recognize it in a document; or (ii) how to "generate" a new object that, in turn, may be added in the original ontology. The obtained specification is rewritten in ASP and evaluated by means of the DLV system.

7.3   Applications

We briefly illustrate here a a number of real-world applications based on DLV or on DLV-based products. They can be grouped in two classes: industrial applications of DLV (developed by the company Exeura s.r.l) and other applications [40].

Industrial Applications. The main commercial applications exploiting DLV are the following:

- *Team Building in the Gioia-Tauro Seaport*. A system based on DLV has been developed to automatically produce an optimal allocation of the available personnel of the international seaport of Gioia Tauro [125]. The system currently employed by the trans-shipment company ICO BLG can build new teams satisfying a number of constraints or complete the allocation automatically when the roles of some key employees are fixed manually.

• *E-Tourism.* IDUM [26] is an intelligent e-tourism system. IDUM system helps both employees and customers of a travel agency in finding the best possible travel solution in a short time. In IDUM an ontology modeling the tourism scenario was developed by using OntoDLV, and is automatically filled by processing the offers received by a travel agent with H*i*L*ε*X. IDUM mimics the behavior of the typical employee of a travel agency by running a set of specifically devised logic programs that reason on the information contained in the tourism ontology. The result is a system that combines the speed of computers with the knowledge of a travel agent.

• *Automatic Itinerary Search.* In this application, a web portal conceived for better exploiting the whole transportation system of the Italian region Calabria, including both public and private companies. The system is very precise; it tells you where and what time to catch your bus/train, where to get off and transfer, how long your trip will take, walking directions etc. A set of specifically devises ASP programs are used to build the required itineraries.

• *e-Government.* In this field, an application of the OLEX system was developed, in which legal acts and decrees issued by public authorities are classified. The system was validated with the help of the employees of the Calabrian Region administration, and it performed very well by obtaining an f-measure of 92% and a mean precision of 96% in real-world documents.

• *e-Medicine.* OLEX was employed for developing a system able to classify auto-matically case histories and documents containing clinical diagnoses. The system was commissioned, with the goal of conducting epidemiological analyses, by the ULSS n.8 (which is, a local authority for health services) of the area of Asolo, in the Italian region Veneto. The system has been deployed and is currently employed by the personnel of the ULSS of Asolo.

Other Applications. The European Commission funded a project on Information Integration, which produced a sophisticated and efficient data integration system, called INFOMIX, which uses DLV at its computational core [28]. The powerful mechanisms for database interoperability, together with magic sets [43, 44] and other database optimization techniques, which are implemented in DLV, make DLV very well-suited for handling information integration tasks. And DLV (in INFOMIX) was succesfully employed to develop in a real-life integration system for the information system of the University of Rome "La Sapienza" The DLV system has been experimented also with an application for Census Data Repair [29], in which errors in census data are identified and eventually repaired.

DLV has been employed at CERN, the European Laboratory for Particle Physics, for an advanced deductive database application that involves complex knowledge manipulation on large-sized databases.

The Polish company Rodan Systems S.A. has exploited DLV in a tool for the detection of price manipulations and unauthorized use of confidential information, which is used by the Polish Securities and Exchange Commission.

In the area of self-healing Web Services, moreover, DLV is exploited for implementing the computation of minimum cardinality diagnoses [45].

In [126] MASEL, A Multi Agent System for E-Learning and Skill Management has been proposed. In MASEL personalized learning paths are automatically composed by exploiting suitable ASP programs run on the DLV system. A prototype tool implementing MASEL using JADE (Java Agent DEvelopment Framework) was developed.

In [127] a complete on-line exam taking portal has been described, called EXAM. The system allows teachers and students to be assisted in the whole process of assessment test building, exam taking, and test correction. The system exploits ASP for automatically generating assessment tests based on user defined constraints: a teacher is made able to build up an assessment test template; her preferences are then translated into a logic specification executable by DLV.

The cooperation between the University of Milan-Bicocca and the University of Potsdam led to the implementation of intelligent monitoring systems based on gringo/clasp [22], where the ASP reasoning module is crucial (see, for instance, [128, 129]).

Italian researchers have exploited ASP capabilities also for diagnosis [130] and e-ealth [131] applications.

8   Further Contributions

This Section briefly mentions several other contributions to the ASP field due to the work of Italian researchers.

In [132, 133], an integrated information retrieval agent based on an ASP inference engine, named GSA_2, was presented. The GSA_2 approach is general and reusable, and the result constitutes a good example of real implementation of agents based on logics.

The first purely syntactic characterization of answer sets in the context of logic programming was introduced in [35]. In the same work, it was pointed out explicitly that answer sets are supersets of the well-founded model (wfm) and can thus be in principle computed after a simplification w.r.t. the wfm (this property was independently discovered in [134]). In [36], the authors introduced a graphical representation of ASP programs, called Extended Dependency Graph (EDG). EDG is defined on a simplified form of programs called *kernel*. In [37, 38], kernel programs were exploited for defining an algorithm for answer set computation, as answer sets can be characterized as *admissible colorings* of the EDG. Moreover, the kernel normal form and other normal forms of ASP programs were studied in [39]. In [135], some features that graph representations of ASP programs should exhibit, especially isomorphism between a program and the corresponding graph, were identified. It turns out that isomorphism is possible only if the graph representation formalism is able to distinguish the cycles occurring in the program, and the different connections among them. Investigating the program structure is also important for understanding the effects of updates of given program on the existence, the number and the content of answer sets. In particular, a graph representation can be useful to understand what happens after asserting lemmas [136] and/or adding new rules [137]. The work [138] showed that representations like the EDG (or others that have been proposed in the literature), which are oriented to atoms and rules, can be usefully condensed into more compact representations, called *Cycle Graph*, which is oriented to components. In the Cycle Graph, vertices are not atoms or rules, but significant subprograms. The Cycle Graph allows the relationship between the syntax of programs and the existence of answer sets to be investigated, and thus can be the basis of software engineering methodologies for answer set programming. In [139] inconsistency and incompleteness in data integration are handled by introducing an "helper model" acting as a mediator between the global conceptual data model and the data sources.

ASP was exploited as a core inference engine for a system for qualitative management of probabilistic uncertainty [140–142]. The system supports basic reasoning tasks by mechanizing various notions of comparative preference notions that represent plausible models of cognitive unconscious humans mental processes.

ASP was integrated with arithmetic and finite domain constraint solvers in [143]. The benefits, besides enhanced expressiveness, comprise reduced memory requirements because the part of a program involving constraints needs not be instantiated. Consequently, it was possible to extend significantly the size of the problems solved by an ASP planner for Space Shuttle operations (see also [144]).

The mutual interdependence of ASP-based agents has been investigated [145–148] at Università Mediterranea of Reggio Calabria. In [145], agreements possibly reached by a collection of agents are represented. In [146, 147], a community of agents where individual conclusions rely on others ones is modeled by nested social predicates. This language is refined in [148] by adding social aggregates and a form of reasoning where models include also "unfounded" interpretations in case they are mutually supported by multiple agents. Finally, a form of preferences under uncertainty is modeled under ASP in [149].

References

1. Baral, C.: Knowledge Representation, Reasoning and Declarative Problem Solving. CUP (2003)
2. Gelfond, M., Lifschitz, V.: The Stable Model Semantics for Logic Programming. In: ICLP/SLP 1988, pp. 1070–1080. MIT Press, Cambridge (1988)
3. Gelfond, M., Lifschitz, V.: Classical Negation in Logic Programs and Disjunctive Databases. NGC 9, 365–385 (1991)
4. Lifschitz, V.: Answer Set Planning. In: ICLP 1999, pp. 23–37 (1999)
5. Marek, V.W., Truszczyński, M.: Stable Models and an Alternative Logic Programming Paradigm. In: The Logic Programming Paradigm – A 25-Year Perspective, pp. 375–398 (1999)
6. Minker, J. (ed.): Foundations of Deductive Databases and Logic Programming, Washington, DC (1988)
7. Minker, J., Rajasekar, A.: A Fixpoint Semantics for Disjunctive Logic Programs. JLP 9(1), 45–74 (1990)
8. Lobo, J., Minker, J., Rajasekar, A.: Foundations of Disjunctive Logic Programming. The MIT Press, Cambridge (1992)
9. Fernández, J.A., Minker, J.: Semantics of Disjunctive Deductive Databases. In: Hull, R., Biskup, J. (eds.) ICDT 1992. LNCS, vol. 646, pp. 21–50. Springer, Heidelberg (1992)
10. Eiter, T., Gottlob, G., Mannila, H.: Disjunctive Datalog. ACM TODS 22(3), 364–418 (1997)

11. Dantsin, E., Eiter, T., Gottlob, G., Voronkov, A.: Complexity and Expressive Power of Logic Programming. ACM Computing Surveys 33(3), 374–425 (2001)
12. Eiter, T., Faber, W., Leone, N., Pfeifer, G.: Declarative Problem-Solving Using the DLV System. In: Logic-Based Artificial Intelligence, pp. 79–103. Kluwer, Dordrecht (2000)
13. Gebser, M., Liu, L., Namasivayam, G., Neumann, A., Schaub, T., Truszczyński, M.: The First Answer Set Programming System Competition. In: Baral, C., Brewka, G., Schlipf, J. (eds.) LPNMR 2007. LNCS (LNAI), vol. 4483, pp. 3–17. Springer, Heidelberg (2007)
14. Zhao, Y.: The Second Answer Set Programming Competition homepage (2009x), http://www.cs.kuleuven.be/~dtai/ASP-competition
15. Dovier, A., Formisano, A., Pontelli, E.: An Empirical Study Of Constraint Logic Programming And Answer Set Programming Solutions Of Combinatorial Problems. J. Exp. Theor. Artif. Intell. 21(2), 79–121 (2009)
16. Leone, N., Pfeifer, G., Faber, W., Eiter, T., Gottlob, G., Perri, S., Scarcello, F.: The DLV System for Knowledge Representation and Reasoning. ACM TOCL 7(3), 499–562 (2006)
17. Simons, P.: Smodels Homepage (since (1996), http://www.tcs.hut.fi/Software/smodels/
18. Simons, P., Niemelä, I., Soininen, T.: Extending and Implementing the Stable Model Semantics. AI 138, 181–234 (2002)
19. Zhao, Y.: ASSAT homepage (since 2002), http://assat.cs.ust.hk/
20. Lin, F., Zhao, Y.: ASSAT: Computing Answer Sets of a Logic Program by SAT Solvers. In: AAAI 2002, Edmonton, Alberta, Canada. AAAI Press / MIT Press (2002)
21. Babovich, Y., Maratea, M.: Cmodels-2: SAT-based Answer Sets Solver Enhanced to Nontight Programs (2003), http://www.cs.utexas.edu/users/tag/cmodels.html
22. Gebser, M., Kaufmann, B., Neumann, A., Schaub, T.: Conflict-Driven Answer Set Solving. In: IJCAI 2007, pp. 386–392 (2007)
23. Janhunen, T., Niemelä, I., Seipel, D., Simons, P., You, J.H.: Unfolding Partiality and Disjunctions in Stable Model Semantics. ACM TOCL 7(1), 1–37 (2006)
24. Lierler, Y.: Disjunctive Answer Set Programming via Satisfiability. In: Baral, C., Greco, G., Leone, N., Terracina, G. (eds.) LPNMR 2005. LNCS (LNAI), vol. 3662, pp. 447–451. Springer, Heidelberg (2005)
25. Drescher, C., Gebser, M., Grote, T., Kaufmann, B., König, A., Ostrowski, M., Schaub, T.: Conflict-Driven Disjunctive Answer Set Solving. In: Proceedings of the Eleventh International Conference on Principles of Knowledge Representation and Reasoning (KR 2008), Sydney, Australia, pp. 422–432. AAAI Press, Menlo Park (2008)
26. Ielpa, S.M., Iiritano, S., Leone, N., Ricca, F.: An ASP-Based System for e-Tourism. In: Erdem, E., Lin, F., Schaub, T. (eds.) LPNMR 2009. LNCS, vol. 5753, pp. 368–381. Springer, Heidelberg (2009)
27. Leone, N., Ricca, F., Terracina, G.: An ASP-Based Data Integration System. In: Erdem, E., Lin, F., Schaub, T. (eds.) LPNMR 2009. LNCS, vol. 5753, pp. 528–534. Springer, Heidelberg (2009)
28. Leone, N., Gottlob, G., Rosati, R., Eiter, T., Faber, W., Fink, M., Greco, G., Ianni, G., Kałka, E., Lembo, D., Lenzerini, M., Lio, V., Nowicki, B., Ruzzi, M., Staniszkis, W., Terracina, G.: The INFOMIX System for Advanced Integration of Incomplete and Inconsistent Data. In: SIGMOD 2005, Baltimore, Maryland, USA, pp. 915–917. ACM Press, New York (2005)
29. Franconi, E., Palma, A.L., Leone, N., Perri, S.: Census Data Repair: A Challenging Application of Disjunctive Logic Programming. In: Nieuwenhuis, R., Voronkov, A. (eds.) LPAR 2001. LNCS (LNAI), vol. 2250, pp. 561–578. Springer, Heidelberg (2001)
30. Cumbo, C., Iiritano, S., Rullo, P.: OLEX – A Reasoning-Based Text Classifier. In: Alferes, J.J., Leite, J. (eds.) JELIA 2004. LNCS (LNAI), vol. 3229, pp. 722–725. Springer, Heidelberg (2004)

31. Rullo, P., Cumbo, C., Policicchio, V.L.: Learning Rules With Negation For Text Categorization. In: ACM Symposium on Applied Computing (SAC 2007), Seoul, Korea, 11-15, pp. 409–416. ACM, New York (2007)

32. Ruffolo, M., Manna, M.: HiLeX: A System for Semantic Information Extraction from Web Documents. In: ICEIS. Lecture Notes in Business Information Processing, vol. (3), pp. 194–209. Springer, Heidelberg (2008)

33. Ruffolo, M., Leone, N., Manna, M., Saccà, D., Zavatto, A.: Exploiting ASP for Semantic Information Extraction. In: Proceedings ASP 2005 - Answer Set Programming: Advances in Theory and Implementation, Bath, UK, pp. 248–262 (2005)

34. Leone, N., Rullo, P., Scarcello, F.: Disjunctive Stable Models: Unfounded Sets, Fixpoint Semantics and Computation. Information and Computation 135(2), 69–112 (1997)

35. Costantini, S.: Contributions to the stable model semantics of logic programs with negation. Theoretical Computer Science 149 (1995); preliminary version in Proc. of LPNMR93

36. Brignoli, G., Costantini, S., D'Antona, O., Provetti, A.: Characterizing and Computing Stable Models of Logic Programs: the Non–stratified Case. In: Proc. of the 1999 Conference on Information Technology, Bhubaneswar, India (1999)

37. Bertoni, A., Grossi, G., Provetti, A., Kreinovich, V., Tari, L.: The Prospect for Answer Set Computation by a Genetic Model. In: AAAI Spring Symposium ASP 2001, pp. 1–5. AAAI Press, Menlo Park (2001)

38. Grossi, G., Marchi, M., Pontelli, E., Provetti, A.: Improving the AdjSolver Algorithm for ASP Kernel Programs. In: ASP 2007, 4th International Workshop on Answer Set Programming at ICLP 2007 (2007)

39. Costantini, S., Provetti, A.: Normal Forms for Answer Sets Programming. J. on TPLP 5(6) (2005)

40. Grasso, G., Iiritano, S., Leone, N., Ricca, F.: Some DLV Applications for Knowledge Management. In: Erdem, E., Lin, F., Schaub, T. (eds.) LPNMR 2009. LNCS, vol. 5753, pp. 591–597. Springer, Heidelberg (2009)

41. Ricca, F., Gallucci, L., Schindlauer, R., Dell'Armi, T., Grasso, G., Leone, N.: OntoDLV: an ASP-based System for Enterprise Ontologies. Journal of Logic and Computation (2009)

42. Ricca, F., Leone, N.: Disjunctive Logic Programming With Types And Objects: The Dlv^+ System. Journal of Applied Logics 5(3), 545–573 (2007)

43. Cumbo, C., Faber, W., Greco, G., Leone, N.: Enhancing the Magic-Set Method for Disjunctive Datalog Programs. In: Demoen, B., Lifschitz, V. (eds.) ICLP 2004. LNCS, vol. 3132, pp. 371–385. Springer, Heidelberg (2004)

44. Faber, W., Greco, G., Leone, N.: Magic Sets and their Application to Data Integration. JCSS 73(4), 584–609 (2007)

45. Friedrich, G., Ivanchenko, V.: Diagnosis From First Principles For Workflow Executions. Tech. Rep.,
http://proserver3-iwas.uni-klu.ac.at/download_area/
Technical-Reports/technical_report_2008_02.pdf

46. Buccafurri, F., Leone, N., Rullo, P.: Enhancing Disjunctive Datalog by Constraints. IEEE TKDE 12(5), 845–860 (2000)

47. Calimeri, F., Faber, W., Leone, N., Perri, S.: Declarative and Computational Properties of Logic Programs with Aggregates. In: IJCAI 2005, pp. 406–411 (2005)

48. Dell'Armi, T., Faber, W., Ielpa, G., Leone, N., Pfeifer, G.: Aggregate Functions in Disjunctive Logic Programming: Semantics, Complexity, and Implementation in DLV. In: IJCAI 2003, Acapulco, Mexico, pp. 847–852 (2003)

49. Faber, W., Leone, N.: On the Complexity of Answer Set Programming with Aggregates. In: Baral, C., Brewka, G., Schlipf, J. (eds.) LPNMR 2007. LNCS (LNAI), vol. 4483, pp. 97–109. Springer, Heidelberg (2007)

50. Faber, W., Leone, N., Pfeifer, G.: Recursive Aggregates in Disjunctive Logic Programs: Semantics and Complexity. In: Alferes, J.J., Leite, J. (eds.) JELIA 2004. LNCS (LNAI), vol. 3229, pp. 200–212. Springer, Heidelberg (2004)

51. Eiter, T., Faber, W., Leone, N., Pfeifer, G., Polleres, A.: A Logic Programming Approach to Knowledge-State Planning: Semantics and Complexity. ACM TOCL 5(2), 206–263 (2004)

52. Perri, S., Scarcello, F., Leone, N.: Abductive Logic Programs with Penalization: Semantics, Complexity and Implementation. TPLP 5(1–2), 123–159 (2005)

53. Eiter, T., Ianni, G., Schindlauer, R., Tompits, H.: A Uniform Integration of Higher-Order Reasoning and External Evaluations in Answer Set Programming. In: IJCAI 2005, Edinburgh, UK, pp. 90–96 (2005)

54. Calimeri, F., Ianni, G., Ielpa, G., Pietramala, A., Santoro, M.C.: A System with Template Answer Set Programs. In: Alferes, J.J., Leite, J. (eds.) JELIA 2004. LNCS (LNAI), vol. 3229, pp. 693–697. Springer, Heidelberg (2004)

55. Costantini, S., Formisano, A.: Answer Set Programming with Resources. Journal of Logic and Computation (to appear, 2009), www.dipmat.unipg.it/~formis/papers/report200816.ps.gz Draft available as Report-16/2008 of Dip. di Matematica e Informatica, Univ. di Perugia

56. Costantini, S., Formisano, A.: Modeling Preferences And Conditional Preferences On Resource Consumption And Production In Asp. Journal of of Algorithms in Cognition, Informatics and Logic 64(1), 3–15 (2009)

57. Balduccini, M., Gelfond, M.: Logic Programs with Consistency-Restoring Rules. In: International Symposium on Logical Formalization of Commonsense Reasoning. AAAI 2003 Spring Symposium Series (2003)

58. Bria, A., Faber, W., Leone, N.: Normal Form Nested Programs. FI (2009) (accepted for publication)

59. Bonatti, P.A.: Reasoning with Infinite Stable Models. Artif. Intell. 156(1), 75–111 (2004)

60. Bonatti, P.: Reasoning with Infinite Stable Models. In: Proceedings of the Seventeenth International Joint Conference on Artificial Intelligence, IJCAI 2001, pp. 603–610 (2001)

61. Baselice, S., Bonatti, P.A., Criscuolo, G.: On Finitely Recursive Programs. TPLP 9(2), 213–238 (2009)

62. Bonatti, P.A.: Reasoning with infinite stable models II: Disjunctive programs. In: Stuckey, P.J. (ed.) ICLP 2002. LNCS, vol. 2401, pp. 333–346. Springer, Heidelberg (2002)

63. Bonatti, P.A.: Erratum to: Reasoning with infinite stable models [artificial intelligence 156(1) (2004) 75–111]. Artif. Intell. 172(15), 1833–1835 (2008)

64. Calimeri, F., Cozza, S., Ianni, G., Leone, N.: Computable Functions in ASP: Theory and Implementation. In: [150], pp.407–424

65. Calimeri, F., Cozza, S., Ianni, G., Leone, N.: Magic Sets for the Bottom-Up Evaluation of Finitely Recursive Programs. In: [151], 71–86

66. Bonatti, P.A.: Resolution for Skeptical Stable Model Semantics. J. Autom. Reasoning 27(4), 391–421 (2001)

67. Bonatti, P.A., Pontelli, E., Son, T.C.: Credulous Resolution for Answer Set Programming. In: AAAI, pp. 418–423. AAAI Press, Menlo Park (2008)

68. Calimeri, F., Cozza, S., Ianni, G., Leone, N.: An ASP System with Functions, Lists, and Sets. In: [151], 483–489

69. Faber, W., Leone, N., Mateis, C., Pfeifer, G.: Using Database Optimization Techniques for Nonmonotonic Reasoning. In: DDLP 1999, Prolog Association of Japan, pp. 135–139 (1999)

70. Leone, N., Perri, S., Scarcello, F.: Improving ASP Instantiators by Join-Ordering Methods. In: Eiter, T., Faber, W., Truszczyński, M. (eds.) LPNMR 2001. LNCS (LNAI), vol. 2173, pp. 280–294. Springer, Heidelberg (2001)

71. Perri, S., Scarcello, F., Catalano, G., Leone, N.: Enhancing DLV Instantiator by Backjumping Techniques. AMAI 51(2-4), 195–228 (2007)
72. Calimeri, F., Perri, S., Ricca, F.: Experimenting with Parallelism for the Instantiation of ASP Programs. Journal of Algorithms in Cognition, Informatics and Logics 63(1-3), 34–54 (2008)
73. Vescio, S., Perri, S., Ricca, F.: Efficient Parallel ASP Instantiation via Dynamic Rewriting. In: Proceedings of the First Workshop on Answer Set Programming and Other Computing Paradigms (ASPOCP 2008), Udine, Italy (2008)
74. Calimeri, F., Faber, W., Leone, N., Pfeifer, G.: Pruning Operators for Disjunctive Logic Programming Systems. FI 71(2-3), 183–214 (2006)
75. Faber, W., Leone, N., Pfeifer, G., Ricca, F.: On look-ahead heuristics in disjunctive logic programming. AMAI 51(2-4), 229–266 (2007)
76. Ricca, F., Faber, W., Leone, N.: A Backjumping Technique for Disjunctive Logic Programming. AI Communications 19(2), 155–172 (2006)
77. Maratea, M., Ricca, F., Faber, W., Leone, N.: Look-Back Techniques and Heuristics in DLV: Implementation, Evaluation and Comparison to QBF Solvers. Journal of Algorithms in Cognition, Informatics and Logics 63(1-3), 70–89 (2008)
78. Koch, C., Leone, N., Pfeifer, G.: Enhancing Disjunctive Logic Programming Systems by SAT Checkers. AI 15(1-2), 177–212 (2003)
79. Abiteboul, S., Hull, R., Vianu, V.: Foundations of Databases. Addison-Wesley, Reading (1995)
80. Van Gelder, A., Ross, K.A., Schlipf, J.S.: The Well-Founded Semantics for General Logic Programs. J. ACM 38(3), 620–650 (1991)
81. Reiter, R.: On Closed World Data Bases. In: Logic and Data Bases, pp. 55–76. Plenum Press, New York (1978)
82. Ben-Eliyahu, R., Dechter, R.: Propositional Semantics for Disjunctive Logic Programs. AMAI 12, 53–87 (1994)
83. Ben-Eliyahu, R., Palopoli, L.: Reasoning with Minimal Models: Efficient Algorithms and Applications. In: Proceedings Fourth International Conference on Principles of Knowledge Representation and Reasoning (KR 1994), pp. 39–50 (1994)
84. Denecker, M., Pelov, N., Bruynooghe, M.: Ultimate Well-Founded and Stable Model Semantics for Logic Programs with Aggregates. In: Codognet, P. (ed.) ICLP 2001. LNCS, vol. 2237, p. 212. Springer, Heidelberg (2001)
85. Hella, L., Libkin, L., Nurmonen, J., Wong, L.: Logics with Aggregate Operators. J. ACM 48(4), 880–907 (2001)
86. Pelov, N., Denecker, M., Bruynooghe, M.: Well-founded and Stable Semantics of Logic Programs with Aggregates. TPLP 7(3), 301–353 (2007)
87. Elkabani, I., Pontelli, E., Son, T.C.: Smodels^A - A System for Computing Answer Sets of Logic Programs with Aggregates. In: Baral, C., Greco, G., Leone, N., Terracina, G. (eds.) LPNMR 2005. LNCS (LNAI), vol. 3662, pp. 427–431. Springer, Heidelberg (2005)
88. Son, T.C., Pontelli, E.: A Constructive Semantic Characterization of Aggregates in ASP. TPLP 7, 355–375 (2007)
89. Baader, F., Calvanese, D., McGuinness, D.L., Nardi, D., Patel-Schneider, P.F. (eds.): The Description Logic Handbook: Theory, Implementation, and Applications. CUP (2003)
90. Lifschitz, V., Tang, L.R., Turner, H.: Nested Expressions in Logic Programs. AMAI 25(3-4), 369–389 (1999)
91. Mileo, A., Schaub, T.: Qualitative Constraint Enforcement in Advanced Policy Specification. In: Mellouli, K. (ed.) ECSQARU 2007. LNCS (LNAI), vol. 4724, pp. 695–706. Springer, Heidelberg (2007)

92. Bertino, E., Mileo, A., Provetti, A.: PDL with Preferences. In: POLICY, pp. 213–222. IEEE Computer Society, Los Alamitos (2005)
93. Marchi, M., Mileo, A., Provetti, A.: Specification and Execution of Declarative Policies for Grid Service Selection. In (LJ) Zhang, L.-J., Jeckle, M. (eds.) ECOWS 2004. LNCS, vol. 3250, pp. 102–115. Springer, Heidelberg (2004)
94. Bertino, E., Mileo, A., Provetti, A.: PDL with Maximum Consistency Monitors. In: Zhong, N., Raś, Z.W., Tsumoto, S., Suzuki, E. (eds.) ISMIS 2003. LNCS (LNAI), vol. 2871, pp. 65–74. Springer, Heidelberg (2003)
95. Sterling, L., Shapiro, E.: The Art of Prolog, 2nd edn. MIT Press, Cambridge (1994)
96. Šimkus, M., Eiter, T.: FDNC: Decidable Non-monotonic Disjunctive Logic Programs with Function Symbols. In: Dershowitz, N., Voronkov, A. (eds.) LPAR 2007. LNCS (LNAI), vol. 4790, pp. 514–530. Springer, Heidelberg (2007)
97. Syrjänen, T.: Omega-Restricted Logic Programs. In: Eiter, T., Faber, W., Truszczyński, M. (eds.) LPNMR 2001. LNCS (LNAI), vol. 2173, pp. 267–279. Springer, Heidelberg (2001)
98. Lierler, Y., Lifschitz, V.: One More Decidable Class of Finitely Ground Programs. In: [152], pp. 489–493
99. Eiter, T., Ortiz, M., Šimkus, M.: Reasoning Using Knots. In: Cervesato, I., Veith, H., Voronkov, A. (eds.) LPAR 2008. LNCS (LNAI), vol. 5330, pp. 377–390. Springer, Heidelberg (2008)
100. Simkus, M.: Fusion of Logic Programming and Description Logics. In: [152], pp. 551–552
101. Baselice, S., Bonatti, P.A.: Composing Normal Programs with Function Symbols. In: [150], pp. 425–439
102. Davis, M., Logemann, G., Loveland, D.: A Machine Program for Theorem Proving. Communications of the ACM 5, 394–397 (1962)
103. Gebser, M., Schaub, T., Thiele, S.: GrinGo: A New Grounder for Answer Set Programming. In: Baral, C., Brewka, G., Schlipf, J. (eds.) LPNMR 2007. LNCS (LNAI), vol. 4483, pp. 266–271. Springer, Heidelberg (2007)
104. Wittocx, J., Mariën, M., Denecker, M.: GidL: A Grounder for FO+. In: Proceedings of the Twelfth International Workshop on Non-Monotonic Reasoning, pp. 189–198 (2008)
105. Perri, S., Ricca, F., Sirianni, M.: A Parallel ASP Instantiator Based on DLV. In: DAMP, pp. 73–82. ACM, New York (2010)
106. Beowulf.org: The Beowulf Cluster Site, http://www.beowulf.org.
107. Balduccini, M., Pontelli, E., Elkhatib, O., Le, H.: Issues in Parallel Execution of Non-Monotonic Reasoning Systems. Parallel Computing 31(6), 608–647 (2005)
108. Grossi, G., Marchi, M., Pontelli, E., Provetti, A.: Experimental Analysis of Graph-based Answer Set Computation over Parallel and Distributed Architectures. J. of Logic and Computation 19(4), 697–715 (2009)
109. Faber, W., Leone, N., Pfeifer, G.: Pushing Goal Derivation in DLP Computations. In: Gelfond, M., Leone, N., Pfeifer, G. (eds.) LPNMR 1999. LNCS (LNAI), vol. 1730, pp. 177–191. Springer, Heidelberg (1999)
110. Marek, V.W., Subrahmanian, V.: The Relationship between Logic Program Semantics and Non-Monotonic Reasoning. In: ICLP 1989, pp. 600–617. MIT Press, Cambridge (1989)
111. Baral, C., Gelfond, M.: Logic Programming and Knowledge Representation. JLP (19/20), 73–148 (1994)
112. Moskewicz, M.W., Madigan, C.F., Zhao, Y., Zhang, L., Malik, S.: Chaff: Engineering an Efficient SAT Solver. In: DAC 2001, pp. 530–535 (2001)
113. Giunchiglia, E., Lierler, Y., Maratea, M.: Answer Set Programming Based on Propositional Satisfiability. Journal of Automated Reasoning 36(4), 345–377 (2006)
114. Giunchiglia, E., Lierler, Y., Maratea, M.: A SAT-based Polynomial Space Algorithm for Answer Set Programming. In: Proceedings of the 10th International Workshop on Non-Monotonic Reasoning (NMR 2004), pp. 189–196 (2004)

115. Giunchiglia, E., Maratea, M.: On the relation between answer set and SAT procedures (or, between CMODELS and SMODELS). In: Gabbrielli, M., Gupta, G. (eds.) ICLP 2005. LNCS, vol. 3668, pp. 37–51. Springer, Heidelberg (2005)

116. Giunchiglia, E., Leone, N., Maratea, M.: On the Relation among Answer Set Solvers. AMAI 53(1-4), 169–204 (2008)

117. Pontelli, E., El-Khatib, O.: Exploiting Vertical Parallelism from Answer Set Programs. In: Proceedings of the 1st Intl. ASP 2001 Workshop on Answer Set Programming, Towards Efficient and Scalable Knowledge Representation and Reasoning, Stanford, pp. 174–180 (2001)

118. Le, H.V., Pontelli, E.: Dynamic Scheduling in Parallel Answer Set Programming Solvers. In: SpringSim (2), SCS/ACM, pp. 367–374 (2007)

119. Dal Palù, A., Dovier, A., Pontelli, E., Rossi, G.: Answer Set Programming with Constraints Using Lazy Grounding. In: [152], pp. 115–129

120. Dal Palù, A., Dovier, A., Pontelli, E., Rossi, G.: GASP: Answer Set Programming with Lazy Grounding. FI 96(3), 297–322 (2009)

121. Terracina, G., Leone, N., Lio, V., Panetta, C.: Experimenting with Recursive Queries in Database and Logic Programming Systems. TPLP 8, 129–165 (2008)

122. Terracina, G., De Francesco, E., Panetta, C., Leone, N.: Enhancing a DLP System for Advanced Database Applications. In: Calvanese, D., Lausen, G. (eds.) RR 2008. LNCS, vol. 5341, pp. 119–134. Springer, Heidelberg (2008)

123. Denecker, M., Vennekens, J., Bond, S., Gebser, M., Truszczyński, M.: The Second Answer Set Programming Competition. In: Erdem, E., Lin, F., Schaub, T. (eds.) LPNMR 2009. LNCS, vol. 5753, pp. 637–654. Springer, Heidelberg (2009)

124. Gallucci, L., Ricca, F.: Visual Querying and Application Programming Interface for an ASP-based Ontology Language. In: Proceedings of the Workshop on Software Engineering for Answer Set Programming (SEA 2007), pp. 56–70 (2007)

125. Grasso, G., Iiritano, S., Leone, N., Lio, V., Ricca, F., Scalise, F.: An ASP-Based System for Team-Building in the Gioia-Tauro Seaport. In: Peña, R. (ed.) PADL 2010. LNCS, vol. 5937, pp. 40–42. Springer, Heidelberg (2010)

126. Garro, A., Palopoli, L., Ricca, F.: Exploiting Agents in E-Learning and Skills Management Context. AI Communications 19(2), 137–154 (2006)

127. Ianni, G., Ricca, F., Panetta, C.: Specification of Assessment-Test Criteria through ASP Specification. In: Answer Set Programming: Advances in Theory and Implementation, Bath, UK, Research Press International, P.O. Box 144, Bristol BS 1YA, pp. 293–302 (2005)

128. Mileo, A., Merico, D., Bisiani, R.: Non-monotonic Reasoning Supporting Wireless Sensor Networks for Intelligent Monitoring: The SINDI System. In: [151], pp. 585–590

129. Mileo, A., Merico, D., Bisiani, R.: A Logic Programming Approach to Home Monitoring for Risk Prevention in Assisted Living. In: [150], pp. 145–159

130. Balduccini, M., Gelfond, M.: Diagnostic reasoning with A-Prolog. TPLP 3, 425–461 (2003)

131. Bisiani, R., Merico, D., Mileo, A., Pinardi, S.: A Logical Approach to Home Healthcare with Intelligent Sensor-Network Support. The Computer Journal (2009); bxn074

132. Ianni, G., Calimeri, F., Lio, V., Galizia, S.: Reasoning about the Semantic Web using Answer Set Programming. In: APPIA-GULP-PRODE, pp. 324–336 (2003)

133. Ianni, G., Ricca, F., Calimeri, F., Lio, V., Galizia, S.: An agent system reasoning about the web and the user. In: WWW (Alternate Track Papers & Posters), pp. 492–493 (2004)

134. Subrahmanian, V., Nau, D., Vago, C.: WFS + Branch and Bound = Stable Models. IEEE TKDE 7(3), 362–377 (1995)

135. Costantini, S.: Comparing Different Graph Representations of Logic Programs under the Answer Set Semantics. In: Proc. of the AAAI Spring Symposium Answer Set Programming: Towards Efficient and Scalable Knowledge Representation and Reasoning, CA (2001)

136. Costantini, S., Lanzarone, G.A., Magliocco, G.: Asserting Lemmas in the Stable Model Semantics. In: Logic Programming – Proc. of the 1996 Joint International Conference, USA (1996)

137. Costantini, S., Intrigila, B., Provetti, A.: Coherence of Updates in Answer Set Programming. In: IJCAI 2003 Workshop on Nonmonotonic Reasoning, Action and Change, NRAC 2003, pp. 66–72 (2003)

138. Costantini, S.: On the Existence of Stable Models of Non-Stratified Logic Programs. J. on TPLP 6(1-2) (2006)

139. Costantini, S., Formisano, A., Omodeo, E.G.: Mappings Between Domain Models in Answer Set Programming. In: Answer Set Programming, Advances in Theory and Implementation, Proc. of the 2nd Intl. ASP 2003. CEUR Workshop Proc., vol. 78 (2003)

140. Capotorti, A., Formisano, A.: Comparative Uncertainty: Theory and Automation. Mathematical Structures in Computer Science 18(1) (2008)

141. Capotorti, A., Formisano, A., Murador, G.: Qualitative Uncertainty Orderings Revised. Electronic Notes in Theoretical Computer Science 169, 43–59 (2007)

142. Capotorti, A., Formisano, A.: Management of Uncertainty Orderings Through ASP. In: Modern Information Processing: From Theory to Applications. Elsevier, Amsterdam (2004) ISBN: 0-444-52075-9

143. Baselice, S., Bonatti, P.A., Gelfond, M.: Towards an Integration of Answer Set and Constraint Solving. In: Gabbrielli, M., Gupta, G. (eds.) ICLP 2005. LNCS, vol. 3668, pp. 52–66. Springer, Heidelberg (2005)

144. Nogueira, M., Balduccini, M., Gelfond, M., Watson, R., Barry, M.: An A-Prolog Decision Support System for the Space Shuttle. In: Ramakrishnan, I.V. (ed.) PADL 2001. LNCS, vol. 1990, pp. 169–183. Springer, Heidelberg (2001)

145. Buccafurri, F., Gottlob, G.: Multiagent compromises, joint fixpoints, and stable models. In: Kakas, A.C., Sadri, F. (eds.) Computational Logic: Logic Programming and Beyond. LNCS (LNAI), vol. 2407, pp. 561–585. Springer, Heidelberg (2002)

146. Buccafurri, F., Caminiti, G.: A Social Semantics for Multi-agent Systems. In: Baral, C., Greco, G., Leone, N., Terracina, G. (eds.) LPNMR 2005. LNCS (LNAI), vol. 3662, pp. 317–329. Springer, Heidelberg (2005)

147. Buccafurri, F., Caminiti, G.: Logic Programming with Social Features. TPLP 8(5–6), 643–690 (2008)

148. Buccafurri, F., Caminiti, G., Laurendi, R.: A Logic Language with Stable Model Semantics for Social Reasoning. In: Garcia de la Banda, M., Pontelli, E. (eds.) ICLP 2008. LNCS, vol. 5366, pp. 718–723. Springer, Heidelberg (2008)

149. Buccafurri, F., Caminiti, G., Rosaci, D.: Logic Programs with Multiple Chances. In: ECAI, pp. 347–351 (2006)

150. Garcia de la Banda, M., Pontelli, E. (eds.): ICLP 2008. LNCS, vol. 5366. Springer, Heidelberg (2008)

151. Erdem, E., Lin, F., Schaub, T. (eds.): LPNMR 2009. LNCS, vol. 5753, pp. 14–18. Springer, Heidelberg (2009)

152. Hill, P.M., Warren, D.S. (eds.): Logic Programming. LNCS, vol. 5649, pp. 14–17. Springer, Heidelberg (2009)

Logic Programming Languages for Databases and the Web

Sergio Greco[1] and Francesca A. Lisi[2]

[1] Dipartimento di Elettronica, Informatica e Sistemistica, Università della Calabria
Via P. Bucci, 41C - Arcavacata di Rende (CS), Italy
greco@deis.unical.it
[2] Dipartimento di Informatica, Università degli Studi di Bari "Aldo Moro"
Via E. Orabona, 4 - 70125 Bari, Italy
lisi@di.uniba.it

Abstract. This chapter contains a reference selection of Italian contributions in the intersection of Logic Programming (LP) with databases and the (Semantic) Web. More precisely, we will survey the main contributions on deductive databases such as the coupling of Prolog systems and database systems, evaluation and optimization techniques, Datalog extensions for expressing nondeterministic and aggregate queries, and active rules and their relation to deductive rules. Also we will illustrate solutions employing LP for querying the Web, manipulating Web pages, representing knowledge in the Semantic Web and learning Semantic Web ontologies and rules.

1 Introduction

Deductive databases started more than 30 years ago and this area has been characterized by intensive research for the past years. It stemmed from earlier work on logic and databases [37,39] that was reviewed in an excellent paper by Gallaire et al. [38]. Deductive databases extend the power of relational systems in several ways [96] by allowing:

- the capability to express, by means of logical rules, recursive queries and efficient algorithms for their evaluation against stored data;
- support for the use of nonmonotonic features such as negation;
- the expansion of the underlying data domain to include structured objects;
- extensions beyond first-order logic for the declarative specification of database operations as updates;
- the development of optimization methods that guarantee the translation of the declarative specifications into efficient access plans and their termination when executed.

Although deductive databases have not found widespread adoptions outside academia, some of their concepts are used in many fields where databases and information systems are used. Over the years the research in different areas where logic-based languages are used for modeling information system features and

A. Dovier, E. Pontelli (Eds.): 25 Years of Logic Programming, LNCS 6125, pp. 183–203, 2010.

managing large datasets has benefited of the results of deductive databases (e.g. integration of advanced features in SQL standards, nonmonotonic reasoning, artificial intelligence and others).

The World Wide Web (WWW, or simply Web) is nowadays the most famous information system. Its success is witnessed by its enormous size and rate of growth. However, the success itself has given raise to a status of the WWW where more sophisticated techniques are urgently needed to properly handle such information overload. Recent years have seen a tremendous interest for Web technologies that can employ some form of logical reasoning. In particular, the ambitious plan for an evolution of the WWW, the so-called Semantic Web [10], has shown that it is of primary importance to find an appropriate interaction between the Web infrastructure, and solutions coming from the LP research area. Interest in the (Semantic) Web application context is testified by initiatives such as the ALPSWS series of international workshops on *Applications of Logic Programming to the (Semantic) Web and Web Services*[1] started in 2006 and traditionally co-located with the *International Conference on Logic Programming*, and the recent special issue of the TPLP journal [72].

We point out that several topics considered in this chapter have also been investigated in the areas of Non-Monotonic Reasoning (NMR) and Answer Set Programming (ASP) and, for more information, we refer readers to [46,12]. The connection between the areas of deductive databases and (Semantic) Web is strong as the Web can be modeled as a (huge) database and the research in both fields is mainly devoted to extend Datalog to have enough expressivity and ensure efficiency in querying and managing relational databases and the (Semantic) Web [15]. This chapter contains a reference selection of Italian contributions in the intersection of LP with databases (section 2) and the (Semantic) Web (section 3).

2   Deductive Databases and Logic Programming

In this section we will discuss some of the research aspects in deductive databases with a particular attention to some fields which have been of particular interest for the Italian deductive database community. In the next subsection we will present some basic definitions on Datalog [22,98]. Next, we will discuss the coupling of Prolog systems and database systems (subsection 2.2), and evaluation and optimization techniques (subsection 2.3). Subsequently, we will present some Datalog extensions regarding the possibility to express nondeterministic and aggregate queries (subsections 2.4 and 2.5). Finally, we will discuss active rules and their relation to deductive rules (subsection 2.6).

2.1   Datalog

A Datalog program is a logic program without function symbols. The restriction imposed by Datalog allows to have finite models which can be efficiently computed by means of a standard bottom-up evaluation. Extensions allowing

---

[1] http://www.kr.tuwien.ac.at/events/alpsws2008/

complex, finite objects have also been considered, but for the sake of simplicity, we restrict ourselves to only consider simple terms.

The semantics of a *positive* (i.e., negation-free) program P is given by the minimum model that coincides with the least fixpoint $T_P^\infty(\emptyset)$ of the *immediate transformation* T_P [68]. The semantics of a logic program with negation is given by the stable model semantics [42,90]. Stable models are said to be total (2-valued) if atoms can be either *true* or *false*, with the standard order $false < true$, or partial (3-valued) if the truth value of atoms can be *true*, *false* or undefined, with the order $false < undefined < true$. On the set of partial stable models it is possible to define an order relation so that they define a semi-lattice [90]. Minimal and stable model semantics have also been extended for programs with disjunctive heads [84]. It is worth observing that although different alternative semantics have been proposed so far (e.g. well-founded semantics, deterministic models, minimal founded semantics [41,93,49,35]), total stable model semantics has been widely accepted by the nonmonotonic reasoning community and for stratified programs these semantics coincide. In particular, general programs with negation may have 0, one or several stable models, *positive* (i.e., negation-free) standard programs have unique (total) stable models, corresponding to the minimum model, and *stratified* programs (i.e. programs where recursion does not "pass" through negated atoms) have a unique (total) stable model which is called *perfect* model.

Generally, the logic language Datalog is denoted by DATALOG¬, whereas restrictions allowing only stratified negation or positive rules only are denoted, respectively, by DATALOG¬ˢ and DATALOG; the extension allowing disjunctive heads is denoted by DATALOG¬∨ [27]. Predicates are usually partitioned into *extensional* (or EDB) and *intensional* (or IDB) predicates. EDB predicates are associated with facts denoting the input databases[2], whereas IDB predicates are associated with rules denoting a set of (possibly recursive) views. The input database will be denoted by D, the program will be denoted by P and it is assumed that $|P| \ll |D|$ (the size of D is much greater than the size of P). It is also assumed that the rules of our programs are *safe* [98], i.e. variables appearing in the head or in negated literals in the body of rules are range restricted, i.e., they take values from the database.

The unique stable model of a stratified program P, applied to a database D, can be computed in polynomial time in the size of D (i.e., the number of symbols in D). For general programs we have that i) the existence of a stable model is not guaranteed, ii) finding a stable model is $\mathcal{NP}$-hard, and iii) deciding whether a program admits some stable model is $\mathcal{NP}$-complete. For programs with disjunctive heads the complexity is even higher (in the general case, in the second level of the polynomial hierarchy). The notion of *data complexity* is defined naturally by viewing the program P as a function computed on the database (which is thus viewed as the input variable). Another important notion is that of *genericity* which means that the database is unchanged if all constants are consistently renamed [3].

---

[2] For each tuple t belonging to a relation r of the input database there is a fact $r(t)$.

A *query* Q is a pair $\langle g, P \rangle$ where P is a program and g is a predicate symbol in P denoting the output relation; The *answer* of Q on D, denoted by $Q(D)$, is the set of relations on g denoted as $A_g = \{M(g)|M$ is a stable model of $P \cup D\}$. Two queries Q_1 and Q_2 are equivalent ($Q_1 \equiv Q_2$) if for each database D the answers of Q_1 and Q_2 on D are the same. The query is called *deterministic* or *non-deterministic* according to whether the mapping is single-valued or multi-valued.

Since it is assumed to deal with large databases, only tractable (i.e. polynomial time) queries are considered. Therefore, in the rest of this chapter we only consider stratified queries and tractable extensions. For queries using unstratified negation and/or disjunctive heads we address the reader to "nonmonotonic formalisms" which are discussed in [46,12].

2.2   Coupling Relational Databases with LP Systems

Integrating LP and relational databases has been recognized to be very promising since both LP and relational databases are related by their common ancestry of mathematical logic [38]. The combination of advanced query processing facilities, typical of expert systems, and efficient access techniques of relational database systems has been very promising. In particular, Prolog systems would greatly benefit from both the ability to store large amounts of information in secondary memory and the optimization techniques built into database systems.

The coupling of a Prolog front-end with a database back-end has been a very promising vehicle for developing database and knowledge-based applications and has received a lot of attention in the field's early years. In practice, systems linking Prolog and a relational database system simply tack on a software interface between a pre-existing Prolog implementation and a pre-existing relational database system. In other words, the two systems are loosely coupled. The interface allows Prolog to query the database when needed, either via the automatic translation of Prolog goals into SQL or else by directly embedding SQL statements into the Prolog code.

In designing the interface between a relational database and a Prolog interpreter, persistence and efficiency were the major concerns. Persistence is obtained by the capability of storing not only data but rules in the database as well. Thus, after a session is over and a new session starts, the user does not need to re-assert the knowledge asserted in the past. Considering efficiency, the objective of minimizing the interaction between the two systems is achieved by means of an optimizing translation mechanism.

A method for loading into the memory-resident database of Prolog facts permanently stored in secondary storage was proposed in [23,22]. The rationale of the method is to save queries accessing the database by never repeating the same query. This is carried out by storing in the main memory, in a compact and efficient way, information about the past interaction with the database. An underlying assumption of this approach is the availability of large core memories on the machine running Prolog [47].

2.3   Query Evaluation and Optimization

In the computation of a query Q over a database D, two main approaches have been proposed in the literature: the *top-down* computation (used by Prolog) and the *bottom-up* computation (used by deductive database languages). The latter is based on the fixpoint operator T_P and on "database implementations" such as the naive algorithm which transforms rules into relational algebra expressions which are evaluated repeatedly until a fixpoint is reached; the semi-naive algorithm improves the naive algorithm avoiding to re-evaluate relational expressions on the same sets of facts [98]. With the top-down approach, only rules and atoms relevant to the query are considered, but termination and duplicated computation are problematic issues. The bottom-up strategy always terminates instead, but it may compute irrelevant atoms. Therefore, optimization techniques combining top-down and bottom-up strategies have been proposed to try to compute only atoms which may be "relevant" for the query in a bottom-up fashion. The key idea here consists of rewriting the rules with respect to the query in order to answer it without actually referring to irrelevant facts. The well-known *magic-set* technique is based on rewriting the rules in P (for a given query $Q = \langle g, P \rangle$) into a set P^α such that, let $Q = \langle g, P^\alpha \rangle$, Q and Q^α are query-equivalent, i.e. the sets of g-facts computed by Q and Q^α are the same [9,97]. General rewriting techniques can be applied to all queries, but their efficiency is limited, while specialized techniques can be very efficient, but have limited applicability. An interesting class of queries is the one known as chain queries, i.e. queries where bindings are propagated from arguments in the head to arguments in the tail of the rules, in a chain-like fashion. For these queries, which are rather frequent in practice, insisting on general optimization methods is not convenient, while specialized methods for subclasses thereof have been proposed, but do not fully exploit bindings. The *counting* method specialized for bound chain queries was proposed to further improve the efficiency of queries [91,53]. However, although proposed in the context of general queries, it preserves the original efficiency only for a subset of chain queries whose recursive rules are linear. The so-called *pushdown* method, exploiting the relationship between chain queries, context-free languages and pushdown automata, was later proposed [51,52]. It rewrites queries into a form that is more suitable for the bottom-up evaluation, i.e. translates a chain query into a factorized left-linear program implementing the pushdown automaton recognizing the language associated with the query. A nice property here is that it reduces to the counting method in all the cases where the latter method behaves efficiently and introduces a unified framework for the treatment of special cases, such as the factorization of right-, left-, mixed-linear programs, as well as the linearization of non-linear programs [97].

2.4   Choice and Non-determinism in Datalog

Stable model semantics introduces a sort of non-determinism in the sense that programs may have more than one "intended" model [42]. Non-determinism offers a solution to overcome the limitations in expressive power of deterministic languages [4,5]. For instance, non-determinism can be used to capture the class of polynomial-time queries on unordered domains [5,50]. The problem with stable model semantics is that the expressive power can blow up without control, so that polynomial time resolution is no longer guaranteed. Thus, it is possible that polynomial time queries are computed in exponential time, that is, it is possible to get exponential time resolution. In order to guarantee polynomial time computability and the existence of stable models, nondeterministic constructs and semantics have been proposed.

Given a query $Q = \langle g, P \rangle$ and a database D, the *answer* to Q on D is a relation defined as follows:

1. under *non-deterministic semantics*: $M(g)$, for some stable model M for $P \cup D$ and $\emptyset$ if no stable model exists;
2. under *possibility semantics* with ground query goal: $M(g)$, if there exists a stable model M such that $M(g) \neq \emptyset$ and $\emptyset$ otherwise — thus, the answer to a query can be either "true" or "false";
3. under *certainty semantics*: $\bigcap M_i(g)$ for all stable models M_i.

Moreover, the mappings defined by queries are multi-valued under non-deterministic semantics and single-valued under possible and certain semantics, i.e., possible and certain semantics are deterministic. In practice, to answer a query under non-deterministic semantics it is sufficient to find any relation in $Q(D)$; this corresponds to determining any stable model. As discussed above, full negation under stable model semantics cannot be used in practical database languages because the complexity is not guaranteed to be polynomial also for queries expressing polynomial problems.

A controlled usage of stable model semantics has been proposed in [44,50,92], where the *choice* construct, first introduced in [60], has been given a stable model semantics. The *choice* construct is used to enforce functional constraints in rules. Thus, an atom of the form, $choice((X), (Y))$, where X and Y denote vectors of variables, in a rule r denotes that any consequence derived from r must respect the functional dependency $X \rightarrow Y$.

A fixpoint algorithm for Datalog programs with choice constructs (called Choice Fixpoint Procedure) has been proposed in [44,45], where it has also been shown that the time complexity of computing, nondeterministically, a stable model, and consequently a nondeterministic answer, is polynomial. This procedure has been extended to programs with stratified negation in [50] where it has also been shown that given a database D and a stratified program with choice P, the problem of deciding if there exists a stable model M (non-deterministic semantics) for $P \cup D$ is polynomial time. In the same paper it has been demonstrated that the class of nondeterministic polynomial problems is captured by Datalog with stratified negation and choice. Therefore, *choice* is a powerful don't-care form of non-determinism which allows one to express some problems for which domain ordering is needed but is not available [5,43].

2.5   Aggregates in Datalog

Early research on deductive databases strived to support a declarative high-level formulation of problem solution without surrendering the performance obtainable by careful programming in an imperative language. In this respect, an interesting challenge is posed by optimization problems, such as finding the minimum spanning tree in a graph or the knapsack problem, that are encountered in several applications.

Datalog, enriched with extrema (*least/most*) and *choice* constructs, can express and efficiently solve optimization problems requiring a greedy search strategy [40,54]. Moreover, many optimization problems can be solved efficiently using a *dynamic programming* technique that is based on the division of the problem into subproblems: the original problem is divided into simpler subproblems that are solved separately; their solutions are then used to solve the original problem. Therefore, Datalog extensions allowing to express classical problems whose efficient solutions are based on greedy and dynamic programming methods have been proposed as well [48].

These extensions are based on the definition of built-in aggregate predicates which enhance Datalog representational capabilities, making it possible to naturally express many well-known algorithms that have wide applicability. The extension of Datalog with classical aggregates (*least*, *most*, *count* and *sum*) has been investigated by considering two main aspects: the definition of suitable semantics for programs with aggregates and the efficiency of the evaluation.

The main novelty of the proposed approach is that only stratified aggregation and a semantics allowing to define linear orders on the input domain are considered. Moreover, the paper also considers in some cases unstratified negation to guarantee efficiency and termination. Another important novelty is that the paper introduces a new aggregate, called *summation*, which combined with *least* and *most* permits us to express and efficiently compute optimization problems such as dynamic programming and integer programming problems. More specifically, the global class of integer programming problems can be easily expressed in the proposed framework and extended programs can be efficiently computed by using a dynamic programming evaluation technique.

The possibility of transforming queries with *least* and *most* predicates into equivalent queries that can be computed more efficiently has been investigated in [36]. Recently there have been further proposals to extend the well-founded and stable model semantics with unstratified aggregates [16,80,95]. Moreover, as pointed out in [77], unstratified aggregates are not necessary if ordered domains and arithmetics are available.

2.6   Deductive and Active Databases

The field of active databases is based on logics and combines techniques from databases, expert systems and artificial intelligence. The main peculiarity of this technology is the support for automatic 'triggering' of rules in response to events. Automatic triggering of rules can be useful in different areas such as integrity constraint maintenance, update of materialized views, knowledge bases and expert systems [100].

Active rules follow the so called *Event-Condition-Action* (ECA) paradigm; rules autonomously react to events occurring on the data, by evaluating a data dependent condition and executing a reaction whenever the condition is true. Active rules consist of three parts: *Event* (which causes the rule to be triggered), *Condition* (which is checked when the rule is triggered) and *Action* (which is executed when the rule is triggered and the condition is true). Thus, according to the semantics of a single active rule, the rule reacts to a given event, tests a condition, and performs a given action.

Understanding the behavior of active rules, especially in the case of rules which interact with one another, is very difficult, and often the actions performed are not the expected ones. The semantics of active rules are usually given in terms of execution models, specifying how and when rules will be applied, but execution models are not completely satisfactory since their behavior is not always clear and could result in nonterminating computations. Most commercial active rule systems operate at a relatively low-level of abstraction and are heavily influenced by implementation-dependent procedural features. A further problem of active databases is that, as shown in [81], most of the operational semantics proposed in the literature have very high complexity and expressivity (PSPACE or even higher complexity).

Different solutions using deductive database semantics to provide a clear semantics to active rules have been proposed. Here we recall the solution proposed in [61,11,33] where declarative semantics are associated to active rules, and in [101,75], where active rules are modeled by means of deductive rules with an attribute denoting the state of the computation. The advantage of associating a declarative semantics to active rules is that confluence and termination are guaranteed and complexity is much lower.

In some sense, active and deductive rules can be seen as opposite ends of a spectrum of database rule languages [99]. Deductive rules provide a high-level powerful framework for specifying intensional relations. In contrast, active rules are more low-level and often need explicit control on rule execution. The problem of providing a homogeneous framework for integrating, in a database environment, active rules, which allow the specification of actions to be executed whenever certain events take place, and deductive rules, which allow the specification of deductions in a logic programming style has been investigated in [79,61].

Since active rules are often used to make databases consistent, *active integrity constraints (AICs)*, an extension of integrity constraints for consistent database maintenance [21], have been recently proposed [17]. An active integrity constraint is a special constraint whose body contains a conjunction of literals which must be false and whose head contains a disjunction of update actions representing actions (insertions and deletions of tuples) to be performed if the constraint is not satisfied (that is its body is true). The AICs work in a domino-like manner as the satisfaction of one AIC may trigger the violation and therefore the activation of another one. The advantage of AICs is that they have declarative semantics (i.e. they can be rewritten into logic rules), lower complexity than active rules and can be used to compute consistent answers, even if the source database is inconsistent [18]. An alternative semantics for AICs is proposed in [19], whereas its relationships to Revision Programming is investigated in [20].

3 From Databases to the (Semantic) Web

The Web has caused a revolution in how we represent, retrieve, and process information. Its growth has given us a universally accessible database but in the form of a largely unorganized collection of documents. This is changing, thanks to the simultaneous emergence of new ways of representing data: from within the Web community, the eXtensible Markup Language (XML)[3]; and from within the database community, *semistructured data*. The convergence of these two approaches has rendered them nearly identical, thus promoting a concerted effort to develop effective techniques for retrieving and processing both kinds of data [2]. In spite of the success of XML as data interchange format, it has turned out very soon that XML has severe limits in conveying data semantics.

The Semantic Web is an evolving extension of the Web in which the semantics of information and services on the Web is defined, making it possible for the Web to understand and satisfy the requests of people and machines to use the Web content [10]. It derives from W3C (World Wide Web Consortium) director Sir Tim Berners-Lee's vision of the Web as a universal medium for data, information, and knowledge exchange. At its core, the Semantic Web comprises a set of design principles, collaborative working groups, and a variety of enabling technologies. Some elements of the Semantic Web are expressed as prospective future possibilities that are yet to be implemented or realized. The Semantic Web architecture is a stack of layers, on top of XML, each of which equipped with one or more mark-up languages, notably the Resource Description Framework (RDF)[4], the RDF Schema (RDFS)[5] and the Web Ontology Language (OWL)[6] all of which are intended to provide a formal description of concepts, terms, and relationships within a given knowledge domain. The use of formal specifications, also called *ontologies*, fairly overcomes the aforementioned limits of XML.

In this section, we will survey solutions employing LP for querying the Web (subsection 3.1), for manipulating Web pages (subsection 3.2), for representing knowledge in the Semantic Web (subsection 3.3) and for learning Semantic Web ontologies and rules (subsection 3.4).

3.1 LP-Based Query Languages for the Web

The Web can be seen as a vast heterogeneous collection of databases, which must be queried in order to extract information. In fact, in many ways the Web is not similar to a database system: it has no uniform structure, no integrity

---

[3] http://www.w3.org/XML/

[4] http://www.w3.org/RDF/

[5] http://www.w3.org/TR/rdf-schema/

[6] http://www.w3.org/2004/OWL/

constraints, no support for transaction processing, no management capabilities, no standard query language, or data model. Perhaps the most popular data model for the Web is the labelled graph, where nodes represent Web pages (or internal components of pages) and arcs correspond to links. Labels on the arcs can be viewed as attribute names for the nodes. The lack of structure in Web pages has motivated the use of semistructured data techniques, which also facilitate the exchange of information between heterogeneous sources. Abiteboul [1] suggests the following features for a semistructured data query language: standard relational database operations (using an SQL viewpoint), navigational capabilities in the hypertext/Web style, information retrieval influenced search using patterns, temporal operations, and the ability to mix data and schema (type) elements together in queries. Many languages support regular path expressions over the graph for stating navigational queries along arcs. The inclusion of wild cards allows arbitrarily deep data and cyclic structures to be searched, although restrictions must be applied to prevent looping.

Queries can be posed to Web pages with XML or RDF content. XML is a notation for describing labelled ordered trees with references. Specifying a query language for XML has been an active area of research, much of it coordinated by the XML Query Working Group of the W3C. The suggested features for such a language are almost identical to those for querying semistructured data. It is hardly surprising that most proposals adopt models which view XML as an edge-labelled directed graph, and use semistructured data query languages. The main difference is that the elements in an XML document are sometimes ordered. The XPath language[7] is based on a tree representation of the XML document, and provides the ability to navigate around the tree, selecting nodes by a variety of criteria. Conversely, XQuery[8] is a query and functional programming language that is designed to query collections of XML data. XQuery provides the means to extract and manipulate data from XML documents or any data source that can be viewed as XML, such as relational databases. Therefore it finally supports the needed interaction between the Web world and the database world. Ultimately, collections of XML files will be accessed like databases. XPath 2.0 is in fact a subset of XQuery 1.0.

RDF is an application of XML aimed at facilitating the interoperability of meta-data across heterogeneous hosts. With RDF, the most suitable approach is to focus on the underlying data model. Even though XQuery could be used to query RDF descriptions in their XML encoded form, a single RDF data model could not be correctly determined with a single query due to the fact that RDF allows several XML syntax encodings for the same data model. Conceived to address this issue, **Metalog** is a LP language where facts and rules are translated and stored as RDF statements [73,71]. Facts are treated as RDF triples, while rule syntax is supported with additional RDFS statements for LP elements such as head, body, if and variable. A query language for RDF, called SPARQL[9],

---

[7] http://www.w3.org/TR/xpath20/
[8] http://www.w3.org/TR/xquery/
[9] http://www.w3.org/TR/rdf-sparql-query/

has been recommended by the RDF Data Access Working Group of the W3C in 2008. Also it has been proved that SPARQL and non-recursive safe DATALOG⌐ have equivalent expressive power, and hence, by classical results, SPARQL is equivalent from an expressive point of view to Relational Algebra [6]. A LP-based rule system for querying persistent RDFS data is suggested in [58] as an alternative to SPARQL engines.

3.2   LP for Web Computation

The ability to support the execution of logic and constraint programs on parallel and distributed architectures have prompted LP researchers to consider some natural generalization of these programming paradigms to suit the needs of some specific application areas among which the Web.

The concurrent constraint-based LP language **W-ACE** has explicit support for the Web computation [83]. Some of its novel ideas include representing Web pages as LP trees and the use of constraints to manipulate tree components and the relationship between trees. W-ACE also contains modal operators for reasoning about groups of pages, and composition operators very similar to those in LogicWeb [69].

In [82] the author studies the use of distributed logic programming models to provide a natural concurrent framework for Web programming. A concurrent logic-based framework (called **WEB-KLIC**) has already been developed and is currently publicly distributed as part of the ICOT Free Software Project[10]. A relevant component of this part of the project includes the design of constraint domains for representing HTML and XML documents. Also, a primary goal has been the improvement of its CGI facilities (i.e., for server-side computation).

3.3   LP for Knowledge Representation in the Semantic Web

The advent of the Semantic Web has given a tremendous impulse on research in Knowledge Representation (KR) due to the key role played by ontologies in the Semantic Web architecture. Indeed the design of OWL has been based on KR formalisms known as *Description Logics* (DLs) [7], more precisely on the $\mathcal{SH}$ family of the so-called very expressive DLs [56]. DLs are a family of decidable First Order Logic (FOL) fragments that allow for the specification of knowledge in terms of classes (*concepts*), binary relations between classes (*roles*), and instances (*individuals*). Complex concepts can be defined from atomic concepts and roles by means of constructors such as atomic negation ($\neg$), concept conjunction ($\sqcap$), value restriction ($\forall$), and limited existential restriction ($\exists$) - just to mention the basic ones. A DL KB can state both is-a relations between concepts (*axioms*) and instance-of relations between individuals (resp. couples of individuals) and concepts (resp. roles) (*assertions*). Concepts and axioms form the so-called TBox whereas individuals and assertions form the so-called ABox. An $\mathcal{SH}$ KB encompasses also a RBox, i.e. axioms defining hierarchies over roles.

---

[10] http://www.jipdec.or.jp/icot/ARCHIVE/Museum/IFS/

The semantics of DLs can be defined through a mapping to FOL. Thus, coherently with the *Open World Assumption* (OWA) that holds in FOL semantics, a DL KB represents all its models. The main reasoning task for a DL KB is the *consistency check* that is performed by applying decision procedures mostly based on tableau calculus. Summing up, when a DL-based ontology language is adopted, an ontology is nothing else than a TBox eventually coupled with a RBox. If the ontology is populated, it corresponds to a whole DL KB, i.e. encompassing also an ABox.

The Semantic Web architecture poses several challenges to KR like (i) the scalability of ontology reasoning, and (ii) the integration of rules and ontologies. It turns out that LP can help facing these challenges, as explained in the following subsections, though Italian research has focused more on the latter challenge.

DL reasoning with LP

A second round of standardization at W3C has very recently delivered OWL 2^{11} which now includes several profiles (or fragments) that can be more simply and/or efficiently implemented than the former OWL proposal. E.g., the OWL 2 RL profile is aimed at applications that require scalable reasoning without sacrificing too much expressive power. It is designed to accommodate both OWL 2 applications that can trade the full expressivity of the language for efficiency, and RDFS applications that need some added expressivity from OWL 2. This is achieved by defining a syntactic subset of OWL 2 which is amenable to implementation using rule-based technologies such as LP. The design of OWL 2 RL has been inspired by Description Logic Programs [55] which are at the intersection of DLs and Datalog. Yet the influence of LP tradition on the implementation of DL systems is also testified by, e.g., KAON2^{12} and DLog^{13}.

Contrary to most currently available DL reasoners, **KAON2** does not implement the tableaux calculus [57]. Rather, it implements novel algorithms which reduce an $\mathcal{SHIQ}$ KB to a disjunctive Datalog program. These algorithms allow applying well-known deductive database techniques, such as magic sets or join-order optimizations, to DL reasoning, thus making answering queries in KAON2 one or more orders of magnitude faster than in existing systems.

DLog is a DL ABox reasoner that uses resolution [70]. It performs query-driven execution whereby the terminological part of the DL KB is converted into a Prolog program using a specialisation of the PTTP Theorem Proving approach and the assertional facts are accessed dynamically from a database. DLog 2 will ensure scalability by specialising well-established LP techniques for parallel computation and efficiency by using an ad-hoc abstract machine.

Rule Systems combining LP and DLs

Rules are currently in the focus within the Semantic Web architecture, and consequently interest and activity in this area has grown rapidly over recent years. They would allow the integration, transformation and derivation of data

---

[11] http://www.w3.org/TR/owl2-overview/
[12] http://kaon2.semanticweb.org/
[13] http://www.dlog-reasoner.org/

from numerous sources in a distributed, scalable, and transparent manner. The rules landscape features design aspects of rule markup; engineering of engines, translators, and other tools; standardization efforts, such as the recent Rules Interchange Format (RIF)[14] activity at W3C; and applications. Rules complement and extend *ontologies* on the Semantic Web. They can be used in combination with ontologies, or as a means to specify ontologies. Rules are also frequently applied over ontologies, to draw inferences, express constraints, specify policies, react to events, discover new knowledge, transform data, etc. Rule markup languages enrich Web ontologies by supporting publishing rules on the Web, exchange rules between different systems and tools, share guidelines and policies, merge and maintain rulebases, and more.

The debate around a RIF is still ongoing. Because of the great variety in rule languages and rule engine technologies, this format will consist of a core language to be used along with a set of standard and non-standard extensions. These extensions need not all be combinable into a single unified language. As for the expressive power, two directions are followed: monotonic extensions towards full FOL and non-monotonic extensions based on the LP tradition, i.e. on *Clausal Logics* (CLs). In particular, the latter will most likely be the so-called *hybrid systems* that integrate DLs and (fragments of) CLs. These KR systems are constituted by two or more subsystems dealing with distinct portions of a single KB by performing specific reasoning procedures [34]. The motivation for investigating and developing such systems is to improve on two basic features of KR formalisms, namely *representational adequacy* and *deductive power*, by preserving the other crucial feature, i.e. *decidability*. Indeed DLs and CLs are FOL fragments incomparable as for the expressiveness [13] and the semantics [85] but combinable at different degrees of integration: tight, loose, full.

The semantic integration is *tight* when a model of the hybrid KB is defined as the union of two models, one for the DL part and one for the CL part, which share the same domain. In particular, combining DLs with CLs in a tight manner can easily yield to undecidability if the interaction scheme between the DL and the CL part of a hybrid KB does not fulfill the condition of *safeness*, i.e. does not solve the semantic mismatch between DLs and CLs [86]. E.g., the hybrid KR system CARIN is *unsafe* [63] because the interaction scheme is left unrestricted. Conversely, $\mathcal{AL}$-**log** [24] is a *safe* hybrid KR system that integrates Datalog with the DL $\mathcal{ALC}$ [94]. In particular, variables occurring in the body of rules may be constrained with $\mathcal{ALC}$ concept assertions to be used as "typing constraints". This makes rules applicable only to explicitly named objects. As in CARIN, query answering is decided using the constrained SLD-resolution which however in $\mathcal{AL}$-log is decidable and runs in single non-deterministic exponential time. The hybrid KR framework of $\mathcal{DL}+$**log** [87] allows for the *weakly-safe* integration of DATALOG$^{\neg\vee}$ with any DL. The condition of weak safeness allows to overcome the main representational limits of the approaches based on the DL-safeness condition, e.g. the possibility of expressing a *union of conjunctive queries* (UCQ), by keeping the integration scheme still decidable. Apart from the FOL semantics,

---

[14] http://www.w3.org/2005/rules/

$\mathcal{DL}$+log has a NM semantics obtained by extending the stable model semantics. According to it, DL-predicates are still interpreted under OWA, while Datalog predicates are interpreted under CWA. The problem statement of satisfiability for finite $\mathcal{DL}$+log KBs relies on the problem known as the *Boolean CQ/UCQ containment problem* in DLs. It is shown that the decidability of reasoning in $\mathcal{DL}$+log, thus of ground query answering, depends on the decidability of the Boolean CQ/UCQ containment problem in $\mathcal{DL}$.

The semantic integration is *loose* when the DL part and the CL part are separate components connected through a minimal interface for exchanging knowledge. An example of one such kind of coupling is the integration scheme for ASP and DLs illustrated in [28]. It derives from the previous work of the same authors on the extension of ASP with higher-order reasoning and external evaluations [29] which has been implemented into the system DLVHEX[15].

The semantic integration is *full* when there is no separation between vocabularies of the two parts of the hybrid KB. In [76], the authors introduce a so-called faithful integration scheme of LP with DLs using the logic of Minimal Knowledge and Negation as Failure (MKNF).

A complete picture of the computational properties of systems combining DL ontologies and Datalog rules can be found in [88]. An updated survey of the literature on hybrid DL-CL systems [26] is suggested for further reading.

3.4   LP for Learning Semantic Web Ontologies and Rules

The advent of the Semantic Web has also raised a knowledge acquisition bottleneck problem for ontologies and rules. Some promising solutions to this problem come from that Machine Learning approach known under the name of Inductive Logic Programming (ILP).

Rooted into LP, the methodological apparatus of ILP inherits the inferential mechanisms for induction from Concept Learning, the most prominent of which is *generalization* [78]. In Concept Learning, thus in ILP, generalization is traditionally viewed as search through a partially ordered space of inductive hypotheses [74]. According to this vision, an inductive hypothesis is a clausal theory and the induction of a single clause requires (i) structuring, (ii) searching and (iii) bounding the space of clauses. In order to achieve (i), a *generality order* is imposed on clauses for determining which one, between two clauses, is more general than the other. Since partial orders are considered, uncomparable pairs of clauses are admitted. Once structured, the space of hypotheses can be searched (ii) by means of refinement operators. A *refinement operator* is a function which computes a set of specializations or generalizations of a clause according to whether a top-down or a bottom-up search is performed. The two kinds of refinement operators have been therefore called *downward* and *upward*, respectively. The definition of refinement operators presupposes the investigation of the properties of the various orderings and is usually coupled with the specification of a *declarative bias* for bounding the space of clauses (iii). This

---

[15] http://www.kr.tuwien.ac.at/research/systems/dlvhex/

concerns anything which constrains the search for theories, e.g. a *language bias* specifies syntactic constraints on the clauses in the search space.

A distinguishing feature of ILP with respect to other forms of Concept Learning is the use of background knowledge (BK), i.e. prior knowledge of the domain of interest, during the induction process. Therefore, an ILP algorithm generalizes from individual instances/observations in the presence of BK, finding valid hypotheses. *Validity* depends on the underlying *setting*. At present, there exist several settings in ILP that vary according to: (i) the *scope of induction* (prediction vs description) and (ii) the *representation of observations* (ground definite clauses vs ground unit clauses). *Prediction* aims at inducing hypotheses with discriminant power as required in tasks such as classification where observations encompass both positive and negative examples. *Description* is more suitable for finding regularities in a data set. This corresponds to learning from positive examples only. Aspect (ii) affects the notion of *coverage*, i.e. the condition under which a hypothesis explains an observation. In *learning from entailment*, hypotheses are clausal theories, observations are ground definite clauses, and a hypothesis covers an observation if the hypothesis logically entails the observation. In *learning from interpretations*, hypotheses are clausal theories, observations are Herbrand interpretations (ground unit clauses) and a hypothesis covers an observation if the observation is a model for the hypothesis.

ILP has been historically concerned with Concept Learning from examples and BK within the representation framework of Horn CL and with the aim of prediction. More recently ILP has considered the problems of learning in different FOL fragments such as DLs and hybrid DL-CL formalisms. This bunch of ILP research is relevant to the Semantic Web application domain.

Inducing DL Concept Descriptions

An ILP characterization of the problem has been proposed by [8,62]. Contributions from the Italian LP community are on the formal treatment of learning in DLs, e.g.: Supervised learning in $\mathcal{ALC}$ [30]; Unsupervised learning (concept formation) in $\mathcal{ALC}$ [32]; Supervised learning in OWL DL [31].

Inducing Hybrid DL-CL Rules

Only three ILP frameworks have been proposed that adopt a hybrid DL-CL representation for both hypotheses and background knowledge: [89] chooses CARIN-$\mathcal{ALN}$, [64] resorts to $\mathcal{AL}$-log, and [65] builds upon $\mathcal{SHIQ}$+log. They can be considered as attempts at accommodating ontologies in ILP by having ontologies as BK. Indeed both proposals extend previous work in ILP, notably the order of generalized subsumption [14], to hybrid DL-CL KR frameworks [66].

The framework proposed in [89] focuses on discriminant induction and adopts the ILP setting of learning from interpretations. Hypotheses are represented as CARIN-$\mathcal{ALN}$ non-recursive rules with a Horn literal in the head that plays the role of target concept. The coverage relation adapts the usual one in the ILP setting of learning from interpretations to the case of hybrid CARIN-$\mathcal{ALN}$ BK. Procedures for testing both the coverage relation and the generality relation are based on the existential entailment algorithm of CARIN. In [59], the author studies the learnability of CARIN-$\mathcal{ALN}$ and provides a pre-processing method which enables traditional ILP systems to learn CARIN-$\mathcal{ALN}$ rules.

In [64], hypotheses are represented as $\mathcal{AL}$-log rules. As opposite to [89], this framework is general, meaning that it is valid whatever the scope of induction (prediction/description) is. Therefore the literal in the head of hypotheses represents a concept to be either discriminated from others or characterized. The generality order for one such hypothesis language can be checked with a decidable procedure based on constrained SLD-resolution. Coverage relations for both ILP settings of learning from interpretations and learning from entailment have been defined on the basis of query answering in $\mathcal{AL}$-log. As opposite to [89], the framework has been implemented into an ILP system that supports a variant of a very popular data mining task - frequent pattern discovery - where rich prior conceptual knowledge is taken into account during the discovery process in order to find patterns at multiple levels of description granularity [67].

The ILP framework presented in [65] represents hypotheses as $\mathcal{DL}$+log rules restricted to the DL $\mathcal{SHIQ}$ and positive Datalog. Analogously to [64], it encompasses both scopes of induction but, differently from [64], it assumes the ILP setting of learning from interpretations only. Both the coverage relation and the generality relation boil down to query answering in $\mathcal{DL}$+log. Refinement operators are defined to search the hypothesis space either top-down or bottom-up. Compared to [89] and [64], this framework shows an added value which can be summarized as follows. First, it relies on a more expressive DL, i.e. $\mathcal{SHIQ}$. Second, it allows for inducing definitions for new DL concepts, i.e. rules with a $\mathcal{SHIQ}$ literal in the head. Third, it adopts a more expressive integration scheme of DLs and CLs, i.e. the weakly-safe one.

References

1. Abiteboul, S.: Querying semi-structured data. In: Afrati, F.N., Kolaitis, P.G. (eds.) ICDT 1997. LNCS, vol. 1186, pp. 1–18. Springer, Heidelberg (1996)
2. Abiteboul, S., Buneman, P., Suciu, D.: Data on the Web: From Relations to Semistructured Data and XML. Morgan Kaufmann, San Francisco (2000)
3. Abiteboul, S., Hull, R., Vianu, V.: Foundations of Databases. Addison-Wesley, Reading (1995)
4. Abiteboul, S., Simon, E.: Fundamental Properties of Deterministic and Nondeterministic Extensions of Datalog. Theoretical Compututer Science 78(1), 137–158 (1991)
5. Abiteboul, S., Vianu, V.: Non-Determinism in Logic-Based Languages. Annals of Mathematics and Artificial Intelligence 3(2-4), 151–186 (1991)
6. Angles, R., Gutierrez, C.: The Expressive Power of SPARQL. In: Sheth, A.P., Staab, S., Dean, M., Paolucci, M., Maynard, D., Finin, T., Thirunarayan, K. (eds.) ISWC 2008. LNCS, vol. 5318, pp. 114–129. Springer, Heidelberg (2008)
7. Baader, F., Calvanese, D., McGuinness, D., Nardi, D., Patel-Schneider, P.F. (eds.): The Description Logic Handbook: Theory, Implementation and Applications, 2nd edn. Cambridge University Press, Cambridge (2007)
8. Badea, L., Nienhuys-Cheng, S.-W.: A Refinement Operator for Description Logics. In: Cussens, J., Frisch, A.M. (eds.) ILP 2000. LNCS (LNAI), vol. 1866, pp. 40–59. Springer, Heidelberg (2000)

9. Beeri, C., Ramakrishnan, R.: On the Power of Magic. Journal of Logic Programming 10(1-4), 255–299 (1991)
10. Berners-Lee, T., Hendler, J., Lassila, O.: The Semantic Web. Scientific American (May 2001)
11. Bidoit, N., Maabout, S.: A Model Theoretic Approach to Update Rule Programs. In: Afrati, F.N., Kolaitis, P.G. (eds.) ICDT 1997. LNCS, vol. 1186, pp. 173–187. Springer, Heidelberg (1996)
12. Bonatti, P., Calimeri, F., Leone, N., Ricca, F.: Answer Set Programming. In: Dovier, Pontelli [25], ch. 8, vol. 6125, pp. 159–178 (2010)
13. Borgida, A.: On the Relative Expressiveness of Description Logics and Predicate Logics. Artificial Intelligence 82(1-2), 353–367 (1996)
14. Buntine, W.: Generalized Subsumption and Its Applications to Induction and Redundancy. Artificial Intelligence 36(2), 149–176 (1988)
15. Calì, A., Gottlob, G., Lukasiewicz, T.: Tractable Query Answering over Ontologies with Datalog+/-. In: Description Logics (2009)
16. Calimeri, F., Faber, W., Leone, N., Perri, S.: Declarative and Computational Properties of Logic Programs with Aggregates. In: IJCAI, pp. 406–411 (2005)
17. Caroprese, L., Greco, S., Sirangelo, C., Zumpano, E.: Declarative Semantics of Production Rules for Integrity Maintenance. In: Etalle, S., Truszczyński, M. (eds.) ICLP 2006. LNCS, vol. 4079, pp. 26–40. Springer, Heidelberg (2006)
18. Caroprese, L., Greco, S., Zumpano, E.: Active Integrity Constraints for Database Consistency Maintenance. IEEE Transactions on Knowledge and Data Engineering 21(7), 1042–1058 (2009)
19. Caroprese, L., Truszczyński, M.: Declarative Semantics for Active Integrity Constraints. In: Garcia de la Banda, M., Pontelli, E. (eds.) ICLP 2008. LNCS, vol. 5366, pp. 269–283. Springer, Heidelberg (2008)
20. Caroprese, L., Truszczyński, M.: Declarative Semantics for Revision Programming and Connections to Active Integrity Constraints. In: Hölldobler, S., Lutz, C., Wansing, H. (eds.) JELIA 2008. LNCS (LNAI), vol. 5293, pp. 100–112. Springer, Heidelberg (2008)
21. Ceri, S., Fraternali, P., Paraboschi, S., Tanca, L.: Automatic Generation of Production Rules for Integrity Maintenance. ACM Transactions on Database Systems 19(3), 367–422 (1994)
22. Ceri, S., Gottlob, G., Tanca, L.: Logic Programming and Databases. Springer, Heidelberg (1990)
23. Ceri, S., Gottlob, G., Wiederhold, G.: Efficient Database Access from Prolog. IEEE Transaction on Software Engineering 15(2), 153–164 (1989)
24. Donini, F.M., Lenzerini, M., Nardi, D., Schaerf, A.: $\mathcal{AL}$-log: Integrating Datalog and Description Logics. J. of Intelligent Information Systems 10(3), 227–252 (1998)
25. Dovier, A., Pontelli, E. (eds.): 25 Years of Logic Programming in Italy. LNCS, vol. 6125. Springer, Heidelberg (2010)
26. Drabent, W., Eiter, T., Ianni, G.B., Krennwallner, T., Lukasiewicz, T., Maluszynski, J.: Hybrid Reasoning with Rules and Ontologies. In: REWERSE, pp. 1–49 (2009)
27. Eiter, T., Gottlob, G., Mannila, H.: Disjunctive Datalog. ACM Transactions on Database Systems 22(3), 364–418 (1997)
28. Eiter, T., Ianni, G., Lukasiewicz, T., Schindlauer, R., Tompits, H.: Combining answer set programming with description logics for the Semantic Web. Artificial Intelligence 172(12-13), 1495–1539 (2008)

29. Eiter, T., Ianni, G.B., Schindlauer, R., Tompits, H.: A Uniform Integration of Higher-Order Reasoning and External Evaluations in Answer-Set Programming. In: IJCAI, pp. 90–96 (2005)

30. Esposito, F., Fanizzi, N., Iannone, L., Palmisano, I., Semeraro, G.: Knowledge-Intensive Induction of Terminologies from Metadata. In: McIlraith, S.A., Plexousakis, D., van Harmelen, F. (eds.) ISWC 2004. LNCS, vol. 3298, pp. 441–455. Springer, Heidelberg (2004)

31. Fanizzi, N., d'Amato, C., Esposito, F.: DL-FOIL Concept Learning in Description Logics. In: Železný, F., Lavrač, N. (eds.) ILP 2008. LNCS (LNAI), vol. 5194, pp. 107–121. Springer, Heidelberg (2008)

32. Fanizzi, N., Iannone, L., Palmisano, I., Semeraro, G.: Concept Formation in Expressive Description Logics. In: Boulicaut, J.-F., Esposito, F., Giannotti, F., Pedreschi, D. (eds.) ECML 2004. LNCS (LNAI), vol. 3201, pp. 99–110. Springer, Heidelberg (2004)

33. Flesca, S., Greco, S.: Declarative semantics for active rules. Theory and Practice of Logic Programming 1(1), 43–69 (2001)

34. Frisch, A.M., Cohn, A.G.: Thoughts and Afterthoughts on the 1988 Workshop on Principles of Hybrid Reasoning. AI Magazine 11(5), 84–87 (1991)

35. Furfaro, F., Greco, G., Greco, S.: Minimal founded semantics for disjunctive logic programs and deductive databases. Theory and Practice of Logic Programming 4(1-2), 75–93 (2004)

36. Furfaro, F., Greco, S., Ganguly, S., Zaniolo, C.: Pushing extrema aggregates to optimize logic queries. Information Systems 27(5), 321–343 (2002)

37. Gallaire, H., Minker, J. (eds.): Logic and Data Bases. Plenum Press, New York (1978)

38. Gallaire, H., Minker, J., Nicolas, J.M.: Logic and databases: A deductive approach. ACM Computing Surveys 16(2), 153–185 (1984)

39. Gallaire, H., Nicolas, J.M., Minker, J. (eds.): Advances in Data Base Theory, vol. 2. Plenum Press, New York (1984)

40. Ganguly, S., Greco, S., Zaniolo, C.: Extrema Predicates in Deductive Databases. Journal of Computer and Systems Science 51(2), 244–259 (1995)

41. Van Gelder, A., Ross, K.A., Schlipf, J.S.: The Well-Founded Semantics for General Logic Programs. J. ACM 38(3), 620–650 (1991)

42. Gelfond, M., Lifschitz, V.: The Stable Model Semantics for Logic Programming. In: ICLP/SLP, pp. 1070–1080 (1988)

43. Giannotti, F., Greco, S., Saccà, D., Zaniolo, C.: Programming with Non-Determinism in Deductive Databases. Annals of Mathematics and Artificial Intelligence 19(1-2), 97–125 (1997)

44. Giannotti, F., Pedreschi, D., Saccà, D., Zaniolo, C.: Non-Determinism in Deductive Databases. In: Delobel, C., Masunaga, Y., Kifer, M. (eds.) DOOD 1991. LNCS, vol. 566, pp. 129–146. Springer, Heidelberg (1991)

45. Giannotti, F., Pedreschi, D., Zaniolo, C.: Semantics and Expressive Power of Nondeterministic Constructs in Deductive Databases. Journal of Computer and Systems Science 62(1), 15–42 (2001)

46. Giordano, L., Toni, F.: Knowledge representation and non-monotonic reasoning. In: Dovier, Pontelli [25], ch. 5, vol. 6125, pp. 86–110 (2010)

47. Gozzi, F., Lugli, M., Ceri, S.: An overview of PRIMO: a portable interface between PROLOG and relational databases. Information Systems 15(5), 543–553 (1990)

48. Greco, S.: Dynamic Programming in Datalog with Aggregates. IEEE Transactions on Knowledge and Data Engineering 11(2), 265–283 (1999)

49. Greco, S., Saccà, D.: Complexity and Expressive Power of Deterministic Semantics for DATALOG. Information and Computation 153(1), 81–98 (1999)
50. Greco, S., Saccà, D., Zaniolo, C.: Datalog Queries with Stratified Negation and Choice: from p to d^P. In: Y. Vardi, M., Gottlob, G. (eds.) ICDT 1995. LNCS, vol. 893, pp. 82–96. Springer, Heidelberg (1995)
51. Greco, S., Saccà, D., Zaniolo, C.: The PushDown Method to Optimize Chain Logic Programs. In: Fülöp, Z., Gecseg, F. (eds.) ICALP 1995. LNCS, vol. 944, pp. 523–534. Springer, Heidelberg (1995)
52. Greco, S., Saccà, D., Zaniolo, C.: Grammars and Automata to Optimize Chain Logic Queries. Int. Journal Foundations of Computer Science 10(3), 349 (1999)
53. Greco, S., Zaniolo, C.: Optimization of Linear Logic Programs Using Counting Methods. In: Pirotte, A., Delobel, C., Gottlob, G. (eds.) EDBT 1992. LNCS, vol. 580, pp. 72–87. Springer, Heidelberg (1992)
54. Greco, S., Zaniolo, C.: Greedy algorithms in Datalog. Theory and Practice of Logic Programming 1(4), 381–407 (2001)
55. Grosof, B.N., Horrocks, I., Volz, R., Decker, S.: Description logic programs: combining logic programs with description logic. In: WWW, pp. 48–57 (2003)
56. Horrocks, I., Patel-Schneider, P.F., van Harmelen, F.: From $\mathcal{SHIQ}$ and RDF to OWL: The Making of a Web Ontology Language. Journal of Web Semantics 1(1), 7–26 (2003)
57. Hustadt, U., Motik, B., Sattler, U.: Deciding expressive description logics in the framework of resolution. Information and Computation 206(5), 579–601 (2008)
58. Ianni, G.B., Krennwallner, T., Martello, A., Polleres, A.: A Rule System for Querying Persistent RDFS Data. In: Aroyo, L., Traverso, P., Ciravegna, F., Cimiano, P., Heath, T., Hyvönen, E., Mizoguchi, R., Oren, E., Sabou, M., Simperl, E. (eds.) ESWC 2009. LNCS, vol. 5554, pp. 857–862. Springer, Heidelberg (2009)
59. Kietz, J.-U.: Learnability of Description Logic Programs. In: Matwin, S., Sammut, C. (eds.) ILP 2002. LNCS (LNAI), vol. 2583, pp. 117–132. Springer, Heidelberg (2003)
60. Krishnamurthy, R., Naqvi, S.A.: Non-Deterministic Choice in Datalog. In: JCDKB, pp. 416–424 (1988)
61. Lausen, G., Ludäscher, B., May, W.: On Logical Foundations of Active Databases. In: Logics for Databases and Information Systems, pp. 389–422 (1998)
62. Lehmann, J., Hitzler, P.: Foundations of Refinement Operators for Description Logics. In: Blockeel, H., Ramon, J., Shavlik, J., Tadepalli, P. (eds.) ILP 2007. LNCS (LNAI), vol. 4894, pp. 161–174. Springer, Heidelberg (2008)
63. Levy, A.Y., Rousset, M.-C.: Combining Horn rules and description logics in CARIN. Artificial Intelligence 104, 165–209 (1998)
64. Lisi, F.A.: Building Rules on Top of Ontologies for the Semantic Web with Inductive Logic Programming. Theory and Practice of Logic Programming 8(03), 271–300 (2008)
65. Lisi, F.A., Esposito, F.: Foundations of Onto-Relational Learning. In: Železný, F., Lavrač, N. (eds.) ILP 2008. LNCS (LNAI), vol. 5194, pp. 158–175. Springer, Heidelberg (2008)
66. Lisi, F.A., Esposito, F.: On Ontologies as Prior Conceptual Knowledge in Inductive Logic Programming. In: Knowledge Discovery Enhanced with Semantic and Social Information, pp. 3–18. Springer, Heidelberg (2009)
67. Lisi, F.A., Malerba, D.: Inducing Multi-Level Association Rules from Multiple Relations. Machine Learning 55, 175–210 (2004)

68. Lloyd, J.W.: Foundations of Logic Programming, 2nd edn. Springer, Heidelberg (1987)
69. Loke, S.W., Davison, A.: LogicWeb: Enhancing the Web with Logic Programming. Journal of Logic Programming 36(3), 195–240 (1998)
70. Lukácsy, G., Szeredi, P.: Efficient Description Logic Reasoning in Prolog: The DLog system. Theory and Practice of Logic Programming 9(3), 343–414 (2009)
71. Marchiori, M.: Towards a people's web: Metalog. In: Web Intelligence, pp. 320–326. IEEE Computer Society Press, Los Alamitos (2004)
72. Marchiori, M.: Introduction to the Special Issue on Logic Programming and the Web. Theory and Practice of Logic Programming 8(3), 247–248 (2008)
73. Marchiori, M., Saarela, J.: Query + Metadata + Logic = Metalog. In: W3C Workshop on Query Languages (1998)
74. Mitchell, T.M.: Generalization as Search. Artificial Intelligence 18, 203–226 (1982)
75. Motakis, I., Zaniolo, C.: Temporal Aggregation in Active Database Rules. In: SIGMOD Conference, pp. 440–451 (1997)
76. Motik, B., Rosati, R.: A Faithful Integration of Description Logics with Logic Programming. In: IJCAI, pp. 477–482 (2007)
77. Mumick, I.S., Shmueli, O.: How Expressive is Statified Aggregation? Annals of Mathematics and Artificial Intelligence 15(3-4), 407–434 (1995)
78. Nienhuys-Cheng, S.-H., de Wolf, R.: Foundations of Inductive Logic Programming. Springer, Heidelberg (1997)
79. Palopoli, L., Torlone, R.: Generalized Production Rules as a Basis for Integrating Active and Deductive Databases. IEEE Transactions on Knowledge and Data Engineering 9(6), 848–862 (1997)
80. Pelov, N., Denecker, M., Bruynooghe, M.: Well-founded and stable semantics of logic programs with aggregates. Theory and Practice of Logic Programming 7(3), 301–353 (2007)
81. Picouet, P., Vianu, V.: Semantics and Expressiveness Issues in Active Databases. Journal of Computer and Systems Science 57(3), 325–355 (1998)
82. Pontelli, E.: Concurrent Web-Programming in CLP(WEB). In: HICSS (2000)
83. Pontelli, E., Gupta, G.: W-ACE: A Logic Language for Intelligent Internet Programming. In: IEEE ICTAI, pp. 2–10 (1997)
84. Przymusinski, T.C.: Semantics of Disjunctive Logic Programs and Deductive Databases. In: Delobel, C., Masunaga, Y., Kifer, M. (eds.) DOOD 1991. LNCS, vol. 566, pp. 85–107. Springer, Heidelberg (1991)
85. Rosati, R.: On the decidability and complexity of integrating ontologies and rules. Journal of Web Semantics 3(1), 61–73 (2005)
86. Rosati, R.: Semantic and Computational Advantages of the Safe Integration of Ontologies and Rules. In: Fages, F., Soliman, S. (eds.) PPSWR 2005. LNCS, vol. 3703, pp. 50–64. Springer, Heidelberg (2005)
87. Rosati, R.: $\mathcal{DL}$+log: Tight Integration of Description Logics and Disjunctive Datalog. In: KR, pp. 68–78 (2006)
88. Rosati, R.: On Combining Description Logic Ontologies and Nonrecursive Datalog Rules. In: Calvanese, D., Lausen, G. (eds.) RR 2008. LNCS, vol. 5341, pp. 13–27. Springer, Heidelberg (2008)
89. Rouveirol, C., Ventos, V.: Towards Learning in CARIN-$\mathcal{ALN}$. In: Cussens, J., Frisch, A.M. (eds.) ILP 2000. LNCS (LNAI), vol. 1866, pp. 191–208. Springer, Heidelberg (2000)
90. Saccà, D.: The Expressive Powers of Stable Models for Bound and Unbound DATALOG Queries. Journal of Computer System Sciences 54(3), 441–464 (1997)

91. Saccà, D., Zaniolo, C.: The Generalized Counting Method for Recursive Logic Queries. Theoretical Computer Science 62(1-2), 187–220 (1988)
92. Saccà, D., Zaniolo, C.: Stable Models and Non-Determinism in Logic Programs with Negation. In: PODS, pp. 205–217 (1990)
93. Saccà, D., Zaniolo, C.: Deterministic and Non-Deterministic Stable Models. Journal of Logic and Computation 7(5), 555–579 (1997)
94. Schmidt-Schauss, M., Smolka, G.: Attributive Concept Descriptions with Complements. Artificial Intelligence 48(1), 1–26 (1991)
95. Son, T.C., Pontelli, E., Elkabani, I.: An unfolding-based semantics for logic programming with aggregates. CoRR, abs/cs/0605038 (2006)
96. Tsur, S.: Deductive Databases in Action. In: PODS, pp. 142–153 (1991)
97. Ullman, J.D.: Principles of Database and Knowledge-Base Systems, vol. II. Computer Science Press (1989)
98. Ullman, J.D.: Principles of Database and Knowledge-Base Systems, vol. I. Computer Science Press (1988)
99. Widom, J.: Deductive and Active Databases: Two Paradigms or Ends of a Spectrum? In: Rules in Database Systems, pp. 306–315 (1993)
100. Widom, J., Ceri, S. (eds.): Active Database Systems: Triggers and Rules For Advanced Database Processing. Morgan Kaufmann, San Francisco (1996)
101. Zaniolo, C.: The Nonmonotonic Semantics of Active Rules in Deductive Databases. In: Bry, F., Ramamohanarao, K. (eds.) DOOD 1997. LNCS, vol. 1341, pp. 265–282. Springer, Heidelberg (1997)

Agents, Multi-Agent Systems and Declarative Programming: What, When, Where, Why, Who, How?

Matteo Baldoni[1], Cristina Baroglio[1], Viviana Mascardi[2],
Andrea Omicini[3], and Paolo Torroni[3]

[1] Dipartimento di Informatica, Università degli Studi di Torino,
c.so Svizzera, 185 - I-10149, Torino, Italy
{baldoni,baroglio}@di.unito.it
[2] DISI, Università degli Studi di Genova,
Via Dodecaneso 35 - I-16146, Genova, Italy
viviana.mascardi@unige.it
[3] DEIS, ALMA MATER STUDIORUM–Università di Bologna
V.le Risorgimento, 2 - I-40136, Bologna, Italy
{paolo.torroni,andrea.omicini}@unibo.it

Abstract. This chapter tackles the relation between declarative languages and multi-agent systems by following the dictates of the five Ws (and one H) that characterize investigations. The aim is to present this research field, which has a long-term tradition, and discuss about its future. The first question to answer is *"What? What are declarative agents and multi-agent systems?"*. Therefore, we will introduce the history of declarative agent systems up to the state of the art by answering the question *"When? When did research on them begin?"*. We will, then, move to the question *"Where? Where can it take place?"*: in which kind of real applications and for which kind of problems declarative agents and MAS have already proven useful? Connected to where is *"Why? Why should it happen?"*. We will discuss the benefits of adopting the abstractions offered by declarative approaches for developing communication, interaction, cooperation mechanisms. We will compare with other technologies, mainly service-based and object-oriented ones. *"Who? Who can be involved?"*: in order to exploit this kind of technology what sort of background does a specialist have to acquire? We address this question by looking at the Italian landscape of Computer Science research and education. Finally, with the question *"How? How can it happen?"* we will shortly report some examples of existing declarative languages and frameworks for the specification, verification, implementation and prototyping of agents and MAS.

1 What? Declarative Agent Systems

The notion of *declarative agent system* should be taken as a conventional one, to be used in order to focus on an essential theme in agent-oriented computing, rather than to clearly delimit the boundaries of a well-defined research subfield. In fact, given the ever-lasting relationship between agents and MAS, on the one hand, and declarative approaches, languages and technologies, on the other, declarative agent systems are not so easily distinguishable from the notion of MAS themselves. For instance, by adopting

A. Dovier, E. Pontelli (Eds.): 25 Years of Logic Programming, LNCS 6125, pp. 204–230, 2010.

the *strong* notion of agency as promoted by [181], mentalistic notions, like beliefs and desires, with an obvious declarative taste are at the core of the very notion of agent. More specifically, they are at the core of the notion of intelligent agent.

Of course, when adopting weaker notions of agent, such as weak agents in [181], or the autonomy-grounded definition of agent in the A&A metamodel [133], declarative approaches and techniques are no longer strictly required, at a first glance. Whenever MAS are adopted to build up non-trivial systems, however, declarative technologies are typically the only viable approach, mainly due to the high-level of abstraction over complexity they promote [134].

Declarative Agents. Historically, one landmark of declarative agent system is represented by Shoham's AgentSpeak [156], the pioneering framework for agent-oriented programming (AOP) promoting a mentalistic view of agents based on components such as beliefs, decisions, capabilities, and obligations, and where the mental state of agents is described formally in an extension of standard epistemic logics. According to [182]

AOP may be regarded as a kind of "post-declarative" programming. ... In AOP, the idea is that, as in declarative programming, we state our goals, and let the built-in control mechanism figure out what to do in order to achieve them. In this case, however, the control mechanism implements some model of rational agency ... Hopefully, this computational model corresponds to our own intuitive understanding of (say) beliefs and desires, and so we need no special training to use it. Ideally, as AOP programmers, we need not be too concerned with how an agent achieves its goals.

Even more so, languages ranging from AgentSpeak(L) [141] to Jason [26], based on the AOP framework, clearly show how declarative and procedural techniques cannot be but combined when building intelligent agents. In particular, when declarative knowledge, required to represent the mentalistic structures of an intelligent agent, needs to be properly combined with procedural knowledge so as to result in an effective process of practical reasoning. More practically, Prolog-like syntax and operators for beliefs, rules, goals, and plans in Jason provide a clear example of how declarative and logic-based technologies are the most suitable approach for the engineering of intelligent agents, covering some of the essential intra-agent aspects of MAS.

Declarative Multiagent Systems. Individual aspects, however, are far from covering all the issues of declarative agent systems. In fact, a huge space for declarative and logic-based approaches is represented by agent societies: intuitively, the social level is where the complexity of MAS typically grows [134]. Handling a MAS composed by hundreds or thousands of agents, as an open system where both known and unknown agents coexist and interact in an unpredictable way, is obviously more than a challenge to MAS engineers. For this very reason, the social level is the one where declarative models and technologies are likely to provide the most relevant contribution: for instance, by allowing system properties to be assessed at design time, and then enforced at run time by suitable declarative technologies, independently of the MAS dynamics, and of the MAS environment as well.

Agent communication languages (ACL) have represented the first and most immediate representatives for the use of declarative technologies in the construction of agent societies. In particular, it is well-known that languages as KQML [108] and FIPA ACL [77] represent the first and the current standards, respectively, for inter-agent communication. However, while it is simple to understand how speech-act communication can be based on a declarative approach, ACL are only the first example of declarative techniques adopted for the construction of agent societies.

Even though it was not meant to address the problem of MAS coordination, one of the milestones of declarative technologies for MAS is represented by the work on Shared Prolog [31], a Linda-based language for the coordination of Prolog agents. There, for the first time, a logic-based language was used to coordinate a number of concurrent agents, thus exploiting a declarative technology in order to govern interaction within an agent society. Subsequently, other declarative and logic-based languages were conceived and developed for the construction of agent societies based on coordination abstractions, such as the coordination language **ReSpecT**: there, both the messages and the social rules of are specified in terms of first-order logic tuples hosted in distributed logic-based tuple centres [132].

A step further is represented by the notion of social integrity constraints [7], which formalizes social concepts such as violation, fulfilment, social expectation within a logic-based framework, concepts that can be enforced at run-time through a suitably-defined logic-based infrastructure. At the infrastructural level, declarative technologies are essential in the definition of the concept of MAS institution. This is the case of Basic Institutions, formally defined in [39], and founded on the social interpretation of agent communicative acts, and of Logic-based Electronic Institutions [172], first-order logic tools aimed at the specification of open agent organizations.

Declarative Agent Systems

Overall, it is apparent that declarative languages and techniques are essential in the design and development of modern MAS, where they are typically used to address most critical aspects. Both intra- and inter-agent issues, in fact, are more and more faced by adopting declarative approaches, the most relevant of which are presented in the remainder of this chapter. So, in the end, it would be quite artificial to draw a line between declarative and non-declarative agent systems: more easily, it is typically the case that one should devise those portions of (nearly) any MAS that exploit declarative and logic-based technologies.

2   When?

The history of declarative agent systems partially coincides with that of intentional systems in Artificial Intelligence: the notion of an intelligent agent as an entity which appears to be the subject of beliefs, desires, commitments, and other mental attitudes [156] is well known and accepted by many researchers. The philosopher Dennett coined the term *intentional system* to denote systems of this kind [50]. In that period (after STRIPS), Artificial Intelligence posed a great emphasis on the use of formal representations, often associated with deductive forms of reasoning [180], and logic programming developed very fast, producing languages that allow for writing executable specifications [160].

Since intelligent software agents must be programmed using languages that can be compiled or interpreted, as any other piece of software, the need for programming languages that could fill the gap between the logical theory and the practical issues concerned with software agents' development arose very soon. Computational logic emerged as a natural tool for developing approaches and solutions, in regards to many aspects. First of all, for the formalization of state-related information (*knowledge, beliefs, goals, environment*). Moreover, for the formalization of *behavior*, and therefore, of the skill of reasoning in order to find new information, to take decisions, to build plans. Generally, to produce proper reactions to the environment and to other agents.

The first real implementation of an intentional system was SRI's Procedural Reasoning System, PRS [82,97], developed to represent and use an expert's procedural knowledge for accomplishing goals and tasks, based on the research on procedural reasoning carried out at the Artificial Intelligence Center, SRI International. Procedural knowledge amounts to descriptions of collections of structured actions for use in specific situations. PRS supported the definition of real-time, continuously-active, intelligent systems that make use of procedural knowledge, such as diagnostic programs and system controllers. In order to formalize intentional systems, different logics were developed, among which the theory of intentions [38], the Belief-Desire-Intention (BDI) logic [142], and that of Knowledge-Abilities-Results-Opportunity [168]. The success of these first implementations gave new impulse to the use of logic approaches for representing and giving a semantics to agents and to their behaviors. A noteworthy example in this respect is Wooldridge's Ph.D. thesis [178], which paved the way to research on agent theories, architectures and languages [129].

Shoham's paper *Agent-Oriented Programming* [156] describes one of the first attempts to define a programming language based on intentional notions. The mental categories upon which Agent-0 is based are *belief* and *obligation* (or *commitment*). *Decision* (or *choice*) is treated as obligation to oneself. Relevant is also dMARS [54], implemented at the Australian AI Institute under the direction of Georgeff, which was a kind of second generation PRS, implemented in C++ and used for commercial agent development projects [83].

These first attemps bear the ambition of developing an approach that more fully draws from the experience of computational logic. A first proposal in this direction is METATEM [66]. So in the '90s, there was, on the one hand, a strain towards the engineering of agents and agent systems in order to meet the requirements of commercialization. To this respect it is important to mention AAII, spun out of Agentis International[1] to address the commercialization of the developed technology, and Agent Oriented Software (AOS), formed by a number of ex-AAII staff to pursue agent technology developing JACK Intelligent Agents [35].

On the other hand, METATEM proved the importance of computational logic for the feasibility of the verification of properties, like interoperability, of complex (agent) systems. New themes started to be tackled. In order to reason about systems of agents it is in fact necessary to represent also the other agents' beliefs and goals, and to represent in a declarative way the rules that govern their interactions and the communication. It is also important to introduce processes of negotiation and to deal with possibly conflicting

---

[1] http://www.agentissoftware.com/

sets of goals. This led to the proposal of languages like Golog [114] and ConGolog [48], of approaches for the representation of interaction protocols like those of Singh [157], and proposals like AgentSpeak(L) [141] which aimed to help the understanding of the relation between practical implementations of the BDI architecture such as PRS and the formalization of the ideas behind the BDI architecture using modal logics [143]. It is important to notice how, in the same years, the revamp of programming languages exploiting garbage collection, such as Java, and the greater efficiency of hardware due to the technological advancements brought a renovated attention onto declarative programming languages due to their ability of dealing with the openness, the dynamicity, and the flexibility that characterize complex systems.

3   Where? Applications

The exploitation of declarative agent systems for industrial projects and applications has a long and successful history dating back to the early and mid nineties. In the following, we provide some meaningful examples coming from different application domains— some of which developed and tested in real, safety-critical scenarios. In the overall, they show that an agent-oriented solution adopting declarative techniques can be fruitfully exploited to satisfy concrete industrial needs, and demonstrate as well the success of declarative agent technologies and systems outside the boundaries of academia.

Among the oldest applications of declarative agents we may mention a re-implementation of TEAM-CPS [175] where agents used the PRODIGY planning system [122] for local network planning, and the Agent-Orientated Programming framework for communication and control. In 1997, Leckie et al. [110] developed a prototype agent-based system for performance monitoring and fault diagnosis in a telecommunications network, where agents were implemented using C5 [148], based on the OPS5 rule language [75], and communicated using KQML.

ARCHON (ARchitecture for Cooperative Heterogeneous ON-line systems [99]) was Europe's largest ever project in the area of Distributed Artificial Intelligence. It was employed for monitoring and controlling the cycle of generating, transporting and distributing electrical energy to industrial and domestic customers, for the Iberdrola company, one of the world's leading private energy groups. ARCHON's Planning and Coordination Module was implemented as a rule-based system.

In [152], Schroeder et al. describe a declarative and reactive diagnostic agent based on extended logic programming. Both the inference engine used for computing diagnoses and the reactive layer that implements a meta-interpreter for the agent were implemented in Prolog extended with communication facilities.

The IMPACT agent framework [12] integrates concepts from deontic logic and was used to develop real applications. They include combat information management where IMPACT was used to provide yellow pages matchmaking services and aerospace applications where IMPACT technology led to the development of a multiagent solution to the "controlled flight into terrain" problem.

Moving to nowadays, [154] describes Space Shuttle Ground Processing with Monitoring Agents. JESS, the Java Expert System Shell [78], is used to realize a system that helps the monitoring of all the processes, instrumentation and data flows of the Kennedy Space Center's Launch Processing System. The system, called NESTA, helps to monitor and above all to discover problems concerning the "ground process", i.e. the set of operations carried out in the weeks before the Space Shuttle's launch. NESTA autonomously and continuously monitors shuttle telemetry data and automatically alerts NASA shuttle engineers if it discovers predefined situations.

Daimler A.G. is exploiting BDI-agent features to develop a "goal-context" modeling technique for describing and executing agile business processes. The goal-context approach aims at (i) having a modular process model that describes the process' steps separate from the process' goals and contexts; and (ii) having different modeling levels, for the different parts of the process model. The goal-context approach was used for the engineering change management process of Daimler, and Jadex [29] was employed for developing a running prototype [34]. The change management real application is currently being implemented using the Whitestein agent platform [147,33]. Go4Flex[2] is a follow-up of these activities, where the goal-context approach will be applied to another area at Daimler.

Other applications of Jadex include a prototype developed for a company to use semantic Web Technologies for improving the search [174], and two simulations based on real (company) data. The first one dealt with logistics in a big packet delivery company [146]. In the second scenario Jadex was used to build a patient scheduling system evaluated using statistical data from over 3000 patient cases collected at the Klinikum Kulmbach hospital [186].

In a recent project that involved DISI, the Computer Science Department of Genoa University, and Ansaldo Segnalamento Ferroviario, the Italian leader in design and construction of signalling and automation systems for conventional and high speed railway lines, a MAS prototype was developed which monitors processes running in a railway signalling plant, detects functioning anomalies, and provides support to the early notification of problems [30]. The prototype was implemented and tested using DCaseLP [117]. Due to the intrinsic rule-based nature of monitoring agents, Prolog proved extremely suitable for their implementation.

In the past, DCaseLP and its ancestor, CaseLP, were used for many industrial research projects: the Kicker project, based on a previous "freight train traffic" application [42], was developed within the framework of the EuROPE-TRIS Project as a result of an industrial collaboration with the Information Systems Division of Italian Railways (Ferrovie dello Stato s.p.a.), and dealt with the train dispatching problem. CaseLP was employed for the design and development of a working prototype of a vehicle monitoring system, which was carried out in collaboration with Elsag s.p.a. [11]. Finally, a prototype of a multimedia, multichannel, personalized news provider [49] was developed using CaseLP in collaboration with Ksolutions s.p.a. as part of the ClickWorld project, a national, Ministry-funded research project.

2APL has been employed for virtual training systems in the TNO research organization in the Netherlands. In [88] 2APL is used as an example to illustrate how a virtual training system can be modeled, whereas in [89] some experiments are reported in which 2APL agents are used to generate explanations in virtual training systems.

---

[2] http://jadex.informatik.uni-hamburg.de/go4flex/

Descriptions of industrial applications of commercial BDI style agent systems include [61] and [21] which cover the JACK and Agentis agent platforms, respectively. Whitestein's Living Systems technology[3] has been applied in scenarios spanning from telecommunications to logistics, supply chain management, and manufacturing.

4 Why? Benefits

The reasons of the success of the agents and declarative programming binomial is that declarative approaches are particularly suitable to handle the complexity of agent systems. Agent systems are dynamic, in the sense that at runtime agents can enter and exit the system and they can be modified at any moment. The interacting are heterogeneous, they have their own goals and they may need to define agreements for cooperating. Declarative languages abstract away from the execution mechanisms and, by merging semantics and computation, they allow the study of a solution and of its properties in the world of concepts. Although in the industry there seems to be a minor interest towards declarative languages, many concepts introduced by declarative approaches are adopted by more widely spread languages and tools. *Bytecode*, *scripting*, *assertions*, *pattern matching*, *destructuring*, *correctness* are a few examples. Agent and multi-agent systems based on declarative approaches supply very effective mechanisms for communication, interaction, and cooperation that can be used to implement choreographies, interaction protocols and orchestrations.

These are particularly useful in addressing computing problems which share with multi-agent systems the properties of openness, dynamicity, and flexibility, involving a large number of heterogeneous components that are physically distributed and that interoperate. Some examples are Web Services [184,9], Mashups [100], SOA [151], Sensor Networks [106], Middleware [98], Distributed components [138]. These developments require the specification of proper interfaces that make components accessible through standard protocols and make it possible to develop new applications by combining and integrating existing components. To this aim, components should bear some public information about themselves, their structure, the way in which they are supposed to be used, and so forth. This information should be represented according to some conventional formalism which relies on well-founded models, upon which it is possible to define access and usage mechanisms. In the following we briefly highlight some areas in which the declarative approach is clearly emerging as predominant w.r.t. other approaches.

Exemplary is the case of the Semantic Web, where declarative languages are becoming very important in the Semantic Web, and where the focus started to shift from the *ontology layer* to the *logic layer*, with a consequent need of expressing rules and of applying various forms of reasoning [2], an interest also witnessed by the RULE Markup Language initiative, by the creation of a W3C working group to define a Rule Interchange Format [3].

The Internet itself provides interesting hooks to the declarative languages community. For instance, OASIS, with the language BPEL4WS [130] (a de facto standard for

---

[3] http://www.whitestein.com/autonomic-technology-platform/
overview

the specification of single services, allowing the representation of a local view of the interaction that should take place, i.e. the interaction from the point of view of the process), has emphasized the need of a language that can be used both as an execution language for specifying the actual behavior of a participant in a business interaction, and noticeably as a *modeling* language, for specifying the interaction at an abstract level. The need of an *abstract representation* that can be reasoned about emerged even more notably for the *composition* of services. Here it is crucial to have tools that allow the verification at design time of properties regarding the behavioral aspects of the composed system. Although proposals have been made for composition rules and models, like BPMN [176] and WS-CDL [102], a comprehensive solution is currently lacking (BPEL4WS is not enough) and research is moving towards considering declarative approaches [167,137,123].

For instance, the work by Zaremski and Wing on software components matching, based on a logic representation of their preconditions and effects [185], inspired most of the work on semantic matchmaking for Web Service discovery. Semantic Web approaches commonly describe services in terms of inputs, outputs, preconditions and effects [1,153]. Inputs and outputs are usually expressed by ontological terms, while preconditidons and effects are often expressed by means of logic representations. Amongst the works on semantic matchmaking, Paolucci et al. [135] propose four degrees of match (exact, plugin, subsumes, and fail). These matches tackle representations, in which services are described by means of inputs and outputs; specifically, matches are computed on the ontological relations of the outputs of an advertisement for a service and a query. The advantage of these kinds of match is that a service description does not need to exactly correspond to the request: this flexibility fosters the re-use of Web Services. The work by Zaremski and Wing also influenced the Web Service Modeling Ontology proposal [62], an organizational framework for Semantic Web Services.

On the other hand, often services are not sought to be used individually but rather to be used *jointly* for executing tasks that none of them alone can accomplish. Semantic annotations of the kind "inputs, outputs, preconditions, and effects" are not sufficient in this case: it becomes useful to introduce a notion of *goal* [139,171,14], which can be used to guide both the selection and the composition of services. The introduction of goals strengthens the need of adopting declarative agents. Agents, in fact, include the ability of dealing with goals and performing goal-driven forms of reasoning; agents also feature autonomy and proactivity, which help when dealing with open environments, allowing for instance a greater *fault tolerance* and an easy approach to *exception handling* [119,140,28].

Besides being used as modeling languages and for reasoning in a goal-driven manner, declarative approaches are starting to gain attention also as a means for designing *behaviors*, replacing more traditional (in the area of Business Process Management) procedural approaches and languages, like Petri nets [144] and PI-calculus [121]. The reason (see [137]) is that systems which allow users to maneuver within the process model or change it while working are considered as the most suitable for dynamic process management. Traditional approaches, having an imperative nature, appear to be too rigid as they strictly prescribe how to work, often forcing an overspecification, which as a side effect compromises dinamicity. Opposed to the imperative approaches, Pesic and van der Aalst [137] have proposed *ConDec*, a language for modeling and enacting dynamic business processes. A ConDec model mainly consists of a set of activities and a set of relationships that constrain the way activities can be executed, and are referred to as *constraints*. Constraints can be interpreted as *policies/business rules*, and may reflect different kinds of knowledge, e.g., external regulations and norms, internal policies and best practices, service/choreography goals. Differently than in the prescriptive approaches, where what is not explicitly modeled is forbidden, ConDec models are open: activities can be freely executed, unless they are subject to constraints. This choice has two implications. First, a ConDec model accommodates many different possible executions, improving flexibility. Second, the language provides abstractions to explicitly capture not only what is mandatory, but also what is forbidden. In this way, the set of possible executions does not need to be expressed extensionally and models remain compact. Agent research too explored similar approaches to obtain openness, flexibility, and heterogeneity. Yolum and Singh [183] propose to adopt the notion of commitment to provide a declarative semantics to the interaction protocols: an agent (the debtor) makes a commitment to another agent (the creditor) to bring about a certain property. Commitments capture and handle mutual obligations which relate interacting agents, giving a meaning to the exchanged messages in terms of their impact on commitments. The adoption of commitments allows a greater flexibility in two respects: the interacting parties can be heterogeneous in their implementations as long as they have the ability of understanding the social commitments and of reasoning about them; their executions do not have to attain to a rigidly encoded behavior, but just not to violate the commitments. The commitment approach has been studied also by others such as [76,86].

A social approach, closer to Logic Programming, has been developed within the SOCS EU Project (see also Section "How?") where global interactions protocols are specified by means of the SCIFF language [6] and its Abductive Logic Programming (ALP) semantics [101]. Protocols are specified only by considering the external observable behavior of interacting entities, and by the concept of expectation about desired events and interaction. Events and expectations are linked by way of forward rules. The SCIFF language comes with an associated proof procedure, used to verify at runtime (or a posteriori, by analyzing a log of the interaction) whether interacting agents conform to the interaction protocols defined. The SCIFF approach is starting to attract the attention of researchers working on Web Services because of the its potential as a tool for verifying the interoperability and for giving an executable semantics to languages like ConDec [124]. The SCIFF framework has also been used to implement commitments [164] via a reactive version of the event calculus [37]. A discussion of commitments and expectations together is proposed in [166].

Another issue in which agents' declarative approaches proved their usefulness and that also gained attention in other fields is trust negotiation. Trust negotiation [18,22,93] is an approach to security and privacy preserving interactions in open networked environment. In such scenarios peers often interact without any previous relationship and without sharing any common security domain. As a consequence, traditional authentication is sometimes undesirable and frequently impossible. Access control policies and privacy policies are based on user properties. Such properties can be encoded in various ways, including digital credentials, unsigned declarations, and reputation measures

[23]. Some proposals for declarative languages that allow the representation of different kinds of policies (e.g. XACML [131] and P3P [173]) have been made by standardization committees and for reasoning about them [24].

Bordering with Trust Negotiation is Argumentation theory [58], where logic models for debate and negotiation are used for modeling agent reasoning and dialogue. The possibility of structuring rational discussion aimed at reaching mutually acceptable conclusions, and the potential for intuitive, modular and tractable implementations are promising tools in all those fields where there is the need of testing the validity of certain kinds of evidence.

5   Who? Required Background

In 1987, while GULP was founded and IJCAI was held in Italy, the main Italian ICT and consumer electronics event, SMAU, was just discovering AI [19]. The heterogeneous mix of AI promoters included small enterprises of academic roots, such as Delphi, a University of Pisa's spin-off then based in Viareggio, and big actors such as IBM, and included many more in between. Back then AI mainly meant Expert Systems, and the use of Prolog inference engines and the adoption of declarative technologies in general was considered a very promising approach. Nixdorf Italia, involved in Esprit-2 research projects and in the development of air fleet optimization tools for Alitalia, was using a development environment written in Prolog, called Twaice [120]. IBM, Unisys, Pirelli Informatica and Datitalia Processing, among others, were all promoting expert systems for configuration and diagnosis which made use of knowledge bases and declarative rules. IBM was pushing expert systems technologies by announcing a series of AI courses.

As discussed in [149] the interest for declarative solutions seemed to fade in the years that preceded 2000. In spite of that, declarative programming started to being taught at Italian universities and a growing number of AI-related courses put a significant emphasis on Prolog and rule-based languages. In 2007, GULP ran a survey to evaluate the extent of computational logic teaching at Italian universities. It turns out that nowadays declarative programming is being taught in 20 Italian universities at around 50 courses, at various levels in computer science and engineering curricula. Some of these courses have been running for as long as 20 years. They are sometimes elective courses attended by small classes. In many cases, however, they are fundamental courses (programming methodologies, AI, logics) attended by large classes with as many as 150 students. In 80% of the cases, the syllabus includes practical lab sessions that teach students how to use SWI Prolog, SICStus, ECLIPSe or other Prolog engines, ASP solvers such as DLV, SAT solvers and model checkers. Every year, around 1500 university students over the country attend on average 20 hours of lectures on computational logic topics, 80% of which focus on logic programming. This is an immense heritage. Many graduates who join the labour market master the basics of declarative languages and technologies.

In more recent times, a number of applications of logic programming have been developed, mainly by academic actors, and most of them were never fully fledged [43]. However, even if the majority of Italian software companies chose not to endorse the declarative paradigm, most of Italian programmers and software engineers do have the necessary background to start working with rules, knowledge bases and inference engines. The effects of this situation are cultural rather than practical. Declarative technologies do not play a major role in implementing systems, but they nevertheless influence the way many programmers and software engineers conceive the systems they implement. Or, at least, they have the potential to do so.

Agent technologies can bring this potential to the surface and help exploit it. As discussed in Section "What", declarative agent systems are a collection of paradigms and ways of thinking about software systems, rather than a unique, well-defined engineering solution. Although younger than logic programming as a discipline, autonomous agents and multi-agent systems also started to being taught at Italian universities. These are sometimes a part of software engineering and AI courses, but also live as stand-alone courses.[4]

Who can be involved in declarative agent and MAS technologies then, and what kind of background is required to do so? The answer to the second question is easy: today's graduates already have such a background, or they can easily acquire it since it is already a part of academic curricula. Then, who can or should be involved?

In our opinion, the ability to think in terms of declarative agent and MAS technologies should be mastered by all software engineers who need to develop systems of some complexity. With a warning. The research effort in this domain is considerable and steady, and we hope to see a constant improvement in the theory and in the tools. However, in approaching the world of agents, today's software engineers should neither seek for revolutionary solutions nor expect to find out that all they have being using so far has become obsolete. That would be a wrong approach. Instead, they should consider declarative agents as a way of thinking that should guide many separate aspects of a system's design.

The ideas of goals, capabilities, action, interaction, delegation, commitment, trust, artifact and so on could be exported to so many concrete software engineering problems. This does not necessarily mean that one should use Tropos, Gaia or West2East for requirements elicitation and system design, DCaseLP, Jade, KGP or DyLog for implementing the components, CArtAgO for the middleware and SCIFF for monitoring their execution. But we suggest that these be considered as sources of inspiration, because a deep understanding of such technologies will help producing software solutions that are more correct and thus safer, and at the same time more scalable, easier to maintain and monitor, and more suited to integration and interoperation.

6   How? Tools and Languages

In this section we briefly survey (without the presumption of being exhaustive) the tools and the languages that exploit declarative approaches. We structure the presentation in two parts. The former presents the most noticeable BDI-style proposals, while the latter presents approaches based on computational logic. For each of the main proposals, we

---

[4] The Universities of Palermo, Genova, L'Aquila, Torino, Bologna, Firenze, Pavia, Roma La Sapienza, Trento, Bari, Modena e Reggio Emilia, Pisa, Milano's Politecnico and many other ones offer such courses.

describe the same four facets, so to facilitate a comparison: *Semantics, Implementation, Extensions*, and *Purpose of use*. The last part of the presentation is mainly devoted to Italian research.

6.1   BDI-Style Tools and Languages

AgentSpeak(L) [141] takes as its starting point PRS and its dMARS implementation. It is based on a restricted first-order language with events and actions. Beliefs, desires and intentions of the agent are not represented as modal formulas, but they are ascribed to agents, in an implicit way, at design time. The current state of the agent can be viewed as its current belief base; states that the agent wants to bring about can be viewed as desires; and the adoption of programs to satisfy such stimuli can be viewed as intentions.

Semantics: At run-time, an agent consists of a set of beliefs, a set of plans, a set of intentions, a set of events, a set of actions, and a set of selection functions. The operational semantics is driven by the rules for selecting plans, adopting them as intentions, and executing the adopted intentions [55].

Implementation: There are many implementations of the AgentSpeak(L) language, among which: *(a)* SIM_Speak [115] (the first AgentSpeak(L) interpreter), which runs on Sloman's SIM_AGENT toolkit, a testbed for cognitively rich agent architectures [158]; *(b)* Jason [27] that implements, in Java, the operational semantics of an extended version of AgentSpeak(L) (http://jason.sourceforge.net)

Extensions: The community working on AgentSpeak(L) is, and has been in the past, very active. Many extensions exist, among which: cooperation through plan exchange [10]; ontological reasoning [126]; belief revision [8]; team formation [96]; combination with the Semantic Web [107].

Purpose of use: The main application of AgentSpeak(L) is in formal verification. Bordini et al. [25] developed model-checking techniques that apply directly to multi-agent programs written in AgentSpeak(L). AgentSpeak(L) multi-agent systems are translated into either Promela or Java models, and then, respectively, SPIN or JPF are used as model checkers.

3APL – "An Abstract Agent Programming Language" [90] – supports the design and construction of intelligent agents for the development of complex systems through the concepts beliefs and procedural goals (also often termed plans). In turn, these can be used to describe and understand the computational system in a natural way. Beliefs represent the issues the agent must deal with, while goals allow the agent both to focus on what it must achieve and to represent the way in which it can achieve it. The practical reasoning rules provide the agent with planning capabilities to find an appropriate plan to achieve a goal, capabilities to create new goals to deal with a particular situation, and capabilities to use the rules to revise a plan.

Semantics: 3APL semantics was originally specified by means of Plotkin-style transition semantics [91] and has been re-specified in Z later on [53]. In [45], the specification of a programming language for implementing the deliberation cycle of cognitive agents is shown, and 3APL has been used as the object language.

Implementation: Both a Java version and an Haskell version of 3APL can be downloaded from `http://www.cs.uu.nl/3apl/`. More recently, a simplified version has been implemented in the Maude term rewriting language [170].

Extensions: The newest incarnation of 3APL is 2APL (A Practical Agent Programming Language) [44]. It can be downloaded from `http://www.cs.uu.nl/2apl/`

Purpose of use: The 2APL platform which provides a set of tools designed to support the implementation, execution, and testing of multi-agent systems. Its application in the field of virtual training has been discussed in Section 3.

Among the other proposals, it is worthwhile to mention Agent-0 by Shoham [156], which exploits a declarative approach and is the first proposal of an agent-oriented approach to programming. For Shoham, a complete AOP system will include three primary components: *(a)* A restricted formal language with clear syntax and semantics for describing mental states, the mental state will be defined uniquely by several modalities, such as belief and commitments; *(b)* An interpreted programming language in which to define and program agents, with primitive commands such as *REQUEST* and *INFORM*; *(c)* An "agentification process" to treat existing hardware devices or software applications like agents. Agent-0 is targetted towards the second component. Two prototype interpreters were developed: one implemented in Common Lisp, and another developed by Hewlett Packard as part of a joint project to incorporate AOP in the New Wave^{TM} architecture. Agent-0 has two extensions, PLACA [162] and Agent-K [46].

Another interesting tool is Jadex [29], which brings together BDI-style reasoning and FIPA-compliant communication [64] and extends the traditional BDI-model (e.g. with explicit goals). Jadex agents have beliefs, goals, that are implicit or explicit descriptions of states to be achieved, and plans. The Jadex research project is conducted by the Distributed Systems and Information Systems Group at the University of Hamburg. The developed software framework is available under GNUs LGPL license[5]. It allows for programming intelligent software agents in XML and Java and can be deployed on different kinds of middleware such as JADE, a software framework implemented in Java that facilitates development of interoperable intelligent multi-agent systems and that is distributed under an Open Source License [20].

Finally, Dribble [169] is a propositional language that constitutes a synthesis between the declarative features of the language GOAL [92], and the procedural features of 3APL. GOAL agents do not provide planning features, but they do offer the possibility to use declarative goals to select actions. The language Dribble thus incorporates beliefs and goals as well as planning features. Also worthwhile to mention MYWORLD [179], in which agents are directly programmed in terms of beliefs and intentions; ViP [105], a visual programming language for plan execution systems with a formal semantics based upon an agent process algebra; CAN [177], a conceptual notation for agents with procedural and declarative goals; NUIN [52], a Java framework for building BDI agents, with strong emphasis on Semantic Web aspects; SPARK [127], that builds on PRS and supports the construction of large-scale, practical agent systems; and JAM [95] that combines ideas drawn from the BDI theories, the PRS system and its UMPRS and PRS-CL implementations, the SRI International's ACT plan interlingua [128], and

---

[5] `http://sourceforge.net/projects/jadex/`

the Structured Circuit Semantics (SCS) representation [111]. It also addresses mobility aspects from Agent Tcl [87], Agents for Remote Action (ARA) [136], Aglets [109] and others. A survey of languages for programming BDI-style agents can be found in [116].

6.2   Computational Logic-Based Tools and Languages

The IMPACT Agent Language [12] is a relevant example of use of deontic logic to specify agents.

Semantics: The paper [60] provides a series of successively more refined semantics for action programs that compute the set of all action status atoms that are true with respect to an agent program P, the current state S and the set IC of underlying integrity constraints on agent states.

Implementations: The implementation of an IMPACT agent program consists of two major parts, both implemented in Java: *(a)* the IMPACT Agent Development Environment which is used by the developer to build and compile agents, and *(b)* the run-time part that allows the agent to autonomously update its "reasonable status set" and execute actions as its state changes.

Extensions: Many extensions to the IMPACT framework are discussed in [161] which analyses meta agent programs to reason about other agents based on the beliefs they hold; temporal agent programs to specify temporal aspects of actions and states; probabilistic agent programs to deal with uncertainty; and secure agent programs to provide agents with security mechanisms. Agents able to recover from an integrity constraints violation and able to continue to process some requests while continuing to recover are discussed in [59]. The integration of planning algorithms in the IMPACT framework is discussed in [56].

Purpose of use: IMPACT's purpose is to allow the integration of heterogeneous information sources and software packages for solving real problems.

Golog [114] is a logic-programming language based on situation calculus, that allows for reasoning on both atomic and complex actions. ConGolog [48] is the concurrent extension of Golog, and it includes facilities for prioritizing the concurrent execution, interrupting the execution when certain conditions become true, and dealing with exogenous actions. Golog is an alternative to traditional plan synthesis, since it allows forms of procedural planning.

Semantics: The semantics of ConGolog and Golog is based on situation calculus and is in the style of transition semantics.

Implementations: Interpreters have been developed in SWI-Prolog and for ECLIPSE as well (http://www.cs.toronto.edu/cogrobo/main/systems/).

Extensions: Many extensions exist: Legolog (*LEGO MINDSTORM in (Con)Golog* [113]), IndiGolog (*Incremental Deterministic (Con)Golog* [47]), CASL (*Cognitive Agent Specification Language* [155]), and an extension of ConGolog with sensing actions [145]. More recently, Golog has been exploited to represent flexible templates of Web service composition and integrate user preferences in the composition process [159]. In [79], the compilation of ConGolog into Basic Action Theories for planning is discussed.

Purpose of use: Golog and ConGolog allow the design of flexible controllers for agents living in complex scenarios. IndiGolog provides a practical framework for real robots that must sense the environment and react to changes occurring in it. CASL is an environment based on ConGolog which provides a verification environment.

Concurrent METATEM [66] is a programming language for distributed artificial intelligence, based on first-order linear temporal logic [65]. A Concurrent METATEM system contains a number of concurrently executing agents which are able to communicate through message passing. Each agent executes a first-order temporal logic specification of its desired behavior.

Semantics: METATEM semantics is the one defined for first-order linear temporal logic.

Implementations: Two implementations have been produced. The first is a prototype interpreter for propositional METATEM implemented in Scheme. A more robust Prolog-based interpreter for a restricted first-order version of METATEM has been used as a transaction programming language for temporal databases [63].

Extensions: Single Concurrent METATEM agents have been extended with deliberation and beliefs [68] and with resource-bounded reasoning [71]. Compilation techniques for MASs specified in Concurrent METATEM are analyzed in [103]. Concurrent METATEM has been proposed as a coordination language in [104]. The definition of groups of agents in Concurrent METATEM is discussed in [69,73]. The research on single Concurrent METATEM agents converged with the research on Concurrent METATEM MASs in the paper [72] where "confidence" is added to both single and multiple agents. The development of teams of agents is discussed in [94].

Purpose of use: In [67] a range of sample applications of Concurrent METATEM utilizing both the core features of the language and some of its extensions are discussed. They include bidding, problem solving, process control, fault tolerance. Concurrent METATEM has the potential of specifying and verifying applications in all of the areas above [74], but it is not suitable for the development of real systems.

SCIFF is a framework, developed within the EU-funded SOCS project,[6] thought to specify and verify interaction in open agent societies [6].

The SCIFF language is equipped with a semantics based on Abductive Logic Programming (ALP) [101]. Interaction is modeled by way of rules (*Social integrity constraints*), which associate the current state of affairs, including all the relevant events detected so far, with a number of alternative possible future worlds, characterized in terms of what is expected or not expected of them. SCIFF's operational component is an ALP proof procedure for reasoning with expectations in dynamic environments. The SOCS approach to the specification and verification of agent societies [4], is *open*, aimed at minimally restricting the operation of system components, and it is inspired by the deontic notions of prohibitions and permission.

---

[6] Societies Of ComputeeS (SOCS, IST-2001-32530): a computational logic model for the description, analysis and verification of global and open societies of heterogeneous computees. http://lia.deis.unibo.it/research/SOCS/

Semantics: The semantics of SCIFF is given as a mapping to ALP, augmented with a notion of consistency of expectations. SCIFF is sound and complete under realistic domain assumptions [6].

Implementations: SCIFF is implemented using Constraint Handling Rules [80]. It runs on SICStus Prolog and on SWI Prolog. SCIFF is also embedded in SOCS-SI, a Java tool for runtime monitoring and verification of agent interaction [5].

Extensions: Recent extensions of SCIFF are an efficient implementation of the Event Calculus for Commitment tracking [165,37], extensions for static verification of declarative models [125], its integration with Tropos [32], its extension for constraint optimization [81], and a number of extensions for several application domains described in the SCIFF[7] and CLIMB[8] Web sites.

Purpose of use: SCIFF is used for interaction specification and verification. Its main application domains, beside multi-agent systems, are business processes, Web service choreographies, and medical guidelines.

DCaseLP is a multi-language development environment for Multi-Agent Systems. It provides tools and languages for modelling and implementing a MAS prototype following a set of steps which guide the developer from the late requirement analysis to the prototype implementation. The languages and tools that DCaseLP integrates are UML and an XML-based language for the analysis and design stages, Java, JESS and TuProlog [51] for the implementation stage, and JADE for the execution stage. Software libraries for translating UML class diagrams into code and for integrating JESS and TuProlog agents into the JADE platform are also provided.

Semantics: No unifying formal semantics of the agents and the MAS, despite the language they are modeled or implemented in, have been defined.

Implementations: DCaseLP is implemented on top of JADE and provides libraries for seamless integration of agents implemented in TuProlog or JESS and enriched with FIPA-compliant communication capabilities. It can be downloaded from the web site[9].

Extensions: A translator from UML sequence diagrams to Prolog agent skeletons that can be embedded into DCaseLP has been developed and integrated within the computer-aided Agent-Oriented Software Engineering West2East framework [36].

Purpose of use: DCaseLP main purpose is fast prototyping of agent systems. Its applications in industrial research projects have been discussed in Section 3.

Dynamics in Logic [17,13] is a programming language for reasoning about actions, that can be used for specifying agents and for executing agent specifications. The authors adopt a modal action theory, in which actions are represented by modalities. The adoption of Dynamic Logic or a modal logic to deal with the problem of reasoning about actions and change is motivated by the fact that modal logic allows a very natural

---

[7] http://lia.deis.unibo.it/research/sciff/
[8] http://lia.deis.unibo.it/research/climb/
[9] http://www.disi.unige.it/person/MascardiV/Software/DCaseLP.html

representation of actions as state transitions, through the accessibility relation of Kripke structures. Since the intentional notions (or attitudes), which are used to describe agents, are usually represented as modalities, the proposed modal action theory is well suited to incorporate them. The language can represent incomplete belief states and can deal with sensing actions as well as with complex actions.

Semantics. The logical characterization of *Dynamics in Logic* is provided in two steps: *(a)* a multimodal logic interpretation of a dynamic domain description which describes the monotonic part of the language is introduced; *(b)* an abductive semantics to account for non-monotonic behavior of the language is provided [17]. The language relies on such abductive semantics to provide a nonmonotonic solution to the frame problem; when there are no ramifications, it has been proved to be equivalent to the language $\mathcal{A}$.

Implementation. A goal-directed proof procedure for reasoning about complex actions (including sensing actions), which can be considered as an interpreter of the language, is supplied. This procedure can be extended for constructing linear and conditional plans to achieve a given goal from an incompletely specified initial state. The interpreter was implemented in Sicsuts Prolog; it is a straightforward implementation of its operational semantics and is available on request.

Extensions. In [13] the language was extended to represent beliefs of other agents in order to reason about conversations. A communication kit including a primitive set of speech acts, a set of special "get message" actions, was included, allowing for the specification of conversation protocols. Other proposals with a causality operator are presented in [84,85].

Purpose of use. The language *Dynamics on Logic* is suitable for building agents acting, interacting and planning in dynamic environments. A web agent system called *WLog* [15], supplying adaptive services in a web-based application context, has been developed to demonstrate the language potential in developing adaptative web applications as software agents. More recently, the language has been used also for giving a declarative interpretation to web services [14,16].

DALI language and agent architecture [41]. DALI is an agent programming language encompassing basic patterns for reactivity, proactivity, internal thinking, and memory. A DALI agent is a logic program that contains reactive rules, aimed at interacting with an external environment. The reactive and proactive behavior of a DALI agent is triggered by several kinds of events: external, internal, present and past events.

Semantics: The declarative and procedural semantics of DALI is defined as an evolutionary semantics in order to cope with the evolution of an agent corresponding to the perception of events. The semantics has been generalized so as to include the communication architecture by resorting to the general framework RCL (Reflective Computational Logic) based on the concept of reflection principle.

Implementations: The DALI interpreter has been implemented in SICStus Prolog, and includes a FIPA-compliant communication library. The DALI interpreter is in principle able to interoperate with other FIPA-compliant platforms; interoperability with JADE is already guaranteed. DALI agents can be distributed on the web, as the implementation of the communication primitives is based on TCP/IP.

Purpose of use: DALI is suitable to implement reactive agents, embedded in an interactive environment. Cultural heritage applications have been proposed, where DALI agents discover the users' movements via a Galileo satellite signal and proactively learn and enhance user profiles to competently assist users during their visits [40].

Finally, for more information on computational logics and MAS, we forward the interested reader to a number of comprehensive surveys already available in the literature, among which [150,163,118,70].

Acknowledgments

The authors acknowledge Alexander Pokahr and Lars Braubach for their support in describing Jadex applications, and Maaike Harbers and Mehdi Dastani for the help in describing 2APL ones. Viviana Mascardi is partially supported by the Iniziativa Software CINI-FinMeccanica Project.

References

1. OWL-S: Semantic markup for web services, http://www.w3.org/Submission/2004/SUBM-OWL-S-20041122/
2. Reasoning on the web with rules and semantics, network of excellence, http://rewerse.net
3. Rule interchange format. W3C, http://www.w3.org/2005/rules/wiki/RIF_Working_Group
4. Alberti, M., Chesani, F., Gavanelli, M., Lamma, E., Mello, P., Torroni, P.: The SOCS computational logic approach to the specification and verification of agent societies. In: Priami, C., Quaglia, P. (eds.) GC 2004. LNCS, vol. 3267, pp. 314–339. Springer, Heidelberg (2005)
5. Alberti, M., Chesani, F., Gavanelli, M., Lamma, E., Mello, P., Torroni, P.: Compliance verification of agent interaction: a logic-based tool. Applied Artificial Intelligence 20(2-4), 133–157 (2006)
6. Alberti, M., Chesani, F., Gavanelli, M., Lamma, E., Mello, P., Torroni, P.: Verifiable agent interaction in abductive logic programming: The sciff framework. ACM Trans. Comput. Logic 9(4), 1–43 (2008)
7. Alberti, M., Gavanelli, M., Lamma, E., Mello, P., Torroni, P.: Modeling Interactions Using *Social Integrity Constraints*: A Resource Sharing Case Study. In: Leite, J., Omicini, A., Sterling, L., Torroni, P. (eds.) DALT 2003. LNCS (LNAI), vol. 2990, pp. 243–262. Springer, Heidelberg (2004)
8. Alechina, N., Bordini, R.H., Hübner, J.F., Jago, M., Logan, B.: Belief revision for AgentSpeak agents. In: Proc. of AAMAS 2006, pp. 1288–1290. ACM, New York (2006)
9. Alonso, G., Casati, F., Kuno, H., Machiraju, V.: Web Services. Springer, Heidelberg (2004)
10. Ancona, D., Mascardi, V., Hübner, J.F., Bordini, R.H.: Coo-AgentSpeak: Cooperation in AgentSpeak through Plan Exchange. In: Proc. of AAMAS 2004, pp. 698–705 (2004)
11. Appiani, E., Martelli, M., Mascardi, V.: A multi-agent approach to vehicle monitoring in motorway. Technical report, DISI – Università di Genova. DISI TR-00-13. Presented at the poster session of the 2nd European Workshop on Advanced Video-based Surveillance Systems, AVBS 2001 (2000)
12. Arisha, K., Eiter, T., Kraus, S., Ozcan, F., Ross, R., Subrahmanian, V.S.: IMPACT: A platform for collaborating agents. IEEE Intelligent Systems 14(2), 64–72 (1999)

13. Baldoni, M., Baroglio, C., Martelli, A., Patti, V.: Reasoning about interaction protocols for customizing web service selection and composition. JLAP, special issue on Web Services and Formal Methods 70(1), 53–73 (2007)

14. Baldoni, M., Baroglio, C., Martelli, A., Patti, V., Schifanella, C.: Reasoning on choreographies and capability requirements. International Journal of BPIM 2(4), 247–261 (2007)

15. Baldoni, M., Baroglio, C., Patti, V.: Web-based adaptive tutoring: an approach based on logic agents and reasoning about actions. Artificial Intelligence Review 22, 3–39 (2004)

16. Baldoni, M., Baroglio, C., Patti, V., Schifanella, C.: Conservative re-use ensuring matches for service selection. In: Proc. of Sixth European Workshop on Multi-Agent Systems, EU-MAS 2008, Bath, UK (December 2008)

17. Baldoni, M., Giordano, L., Martelli, A., Patti, V.: Programming Rational Agents in a Modal Action Logic. AMAI, Special issue on Logic-Based Agent Implementation 41(2-4), 207–257 (2004)

18. Baselice, S., Bonatti, P.A., Faella, M.: Policy language specification. Technical Report I2-D2, REWERSE network of excellence (2007)

19. Bazzocchi, L.: Lo SMAU scopre l'intelligenza artificiale. Office Automation, 86–90 (November 1988), http://www.bazzocchi.com/

20. Bellifemine, F.L., Caire, G., Greenwood, D.: Developing Multi-Agent Systems with JADE. Wiley, Chichester (2007)

21. Benfield, S.S., Hendrickson, J., Galanti, D.: Making a strong business case for multiagent technology. In: AAMAS 2006, pp. 10–15. ACM, New York (2006)

22. Bonatti, P.A., De Coi, J.L., Olmedilla, D., Sauro, L.: Policy-driven negotiations and explanations: Exploiting logic-programming for trust management, privacy & security. In: Garcia de la Banda, M., Pontelli, E. (eds.) ICLP 2008. LNCS, vol. 5366, pp. 779–784. Springer, Heidelberg (2008)

23. Bonatti, P.A., Duma, C., Fuchs, N.E., Nejdl, W., Olmedilla, D., Peer, J., Shahmehri, N.: Semantic web policies – A discussion of requirements and research issues. In: Sure, Y., Domingue, J. (eds.) ESWC 2006. LNCS, vol. 4011, pp. 712–724. Springer, Heidelberg (2006)

24. Bonatti1, P.A., Coi, J.L.D., Olmedilla, D.: Protunes technical specifications. Technical Report I2-D12, REWERSE (2007)

25. Bordini, R.H., Fisher, M., Visser, W., Wooldridge, M.: Verifying multi-agent programs by model checking. JAAMAS 12(2), 239–256 (2006)

26. Bordini, R.H., Hübner, J.F.: BDI agent programming in AgentSpeak using *Jason* (tutorial paper). In: Toni, F., Torroni, P. (eds.) CLIMA 2005. LNCS (LNAI), vol. 3900, pp. 143–164. Springer, Heidelberg (2006)

27. Bordini, R.H., Hübner, J.F., Wooldridge, M. (eds.): Programming Multi-Agent Systems in AgentSpeak using Jason. Wiley, Chichester (2007)

28. Bozzo, L., Mascardi, V., Ancona, D., Busetta, P.: CooWS: Adaptive BDI agents meet service-oriented computing. In: Proc. of WWW/Internet, pp. 205–209 (2005)

29. Braubach, L., Pokahr, A., Lamersdorf, W.: Jadex: A short overview. In: Main Conference Net.ObjectDays 2004, pp. 195–207 (2004)

30. Briola, D., Mascardi, V., Martelli, M.: Intelligent agents that monitor, diagnose and solve problems: Two success stories of industry-university collaboration. Journal of Information Assurance and Security 4(2), 106–116 (2009)

31. Brogi, A., Ciancarini, P.: The concurrent language, Shared Prolog. ACM Transactions on Programming Languages and Systems (TOPLAS) 13(1), 99–123 (1991)

32. Bryl, V., Mello, P., Montali, M., Torroni, P., Zannone, N.: B-tropos: Agent-oriented requirements engineering meets computational logic for declarative business process modelling and verification. In: Sadri, F., Satoh, K. (eds.) CLIMA VIII 2007. LNCS (LNAI), vol. 5056, pp. 157–176. Springer, Heidelberg (2008)

33. Burmeister, B., Arnold, M., Copaciu, F., Rimassa, G.: BDI-agents for agile goal-oriented business processes. In: Proc. of AAMAS 2008, pp. 37–44. IFAAMAS (2008)
34. Burmeister, B., Steiert, H.-P., Bauer, T., Baumgärtel, H.: Agile processes through goal- and context-oriented business process modeling. In: Eder, J., Dustdar, S. (eds.) BPM Workshops 2006. LNCS, vol. 4103, pp. 217–228. Springer, Heidelberg (2006)
35. Busetta, P., Ronnquist, R., Hodgson, A., Lucas, A.: JACK intelligent agents – components for intelligent agents in Java. AgentLink News Letter 2 (1999)
36. Casella, G., Mascardi, V.: West2East: exploiting WEb Service Technologies to Engineer Agent-based SofTware. IJAOSE 1(3/4), 396–434 (2007)
37. Chesani, F., Mello, P., Montali, M., Torroni, P.: Commitment tracking via the reactive event calculus. In: Proc. of IJCAI, pp. 91–96 (2009)
38. Cohen, P.R., Levesque, H.J.: Intention is choice with commitment. Artificial Intelligence 42 (1990)
39. Colombetti, M., Fornara, N., Verdicchio, M.: A social approach to communication in multi-agent systems. In: Leite, J., Omicini, A., Sterling, L., Torroni, P. (eds.) DALT 2003. LNCS (LNAI), vol. 2990, pp. 191–220. Springer, Heidelberg (2004)
40. Costantini, S., Mostarda, L., Tocchio, A., Tsintza, P.: Dalica: Agent-based ambient intelligence for cultural-heritage scenarios. IEEE Intelligent Systems 23(2), 34–41 (2008)
41. Costantini, S., Tocchio, A.: The DALI logic programming agent-oriented language. In: Alferes, J.J., Leite, J. (eds.) JELIA 2004. LNCS (LNAI), vol. 3229, pp. 685–688. Springer, Heidelberg (2004)
42. Cuppari, A., Guida, P.L., Martelli, M., Mascardi, V., Zini, F.: An agent-based prototype for freight trains traffic management. In: Proceedings of the FMERail Workshop 5. Springer, Heidelberg (1999)
43. Dal Palù, A., Torroni, P.: 25 Years of Applications of Logic Programming. In: Dovier, A., Pontelli, E. (eds.) 25 Years of Logic Programming in Italy, ch. 14. LNCS, vol. 6125, pp. 300–328. Springer, Heidelberg (2010)
44. Dastani, M.: 2APL: a practical agent programming language. Autonomous Agents and Multi-Agent Systems 16(3), 214–248 (2008)
45. Dastani, M., de Boer, F.S., Dignum, F., Meyer, J.-J.C.: Programming agent deliberation – an approach illustrated using the 3APL language. In: Proc. of AAMAS 2003 (2003)
46. Davies, W.H., Edwards, P.: Agent-K: An integration of AOP & KQML. In: Proceedings of the Workshop on Intelligent Information Agents (1994)
47. De Giacomo, G., Lespérance, Y., Levesque, H., Sardiña, S.: On the semantics of deliberation in IndiGolog – from theory to implementation. In: Proceedings of KR 2002, pp. 603–614. Morgan Kaufmann, San Francisco (2002)
48. De Giacomo, G., Lespérance, Y., Levesque, H.J.: Congolog, a concurrent programming language based on the situation calculus. Artificial Intelligence 121, 109–169 (2000)
49. Delato, M., Martelli, A., Martelli, M., Mascardi, V., Verri, A.: A multimedia, multichannel, and personalized news provider. In: Ventre, G., Canonico, R. (eds.) MIPS 2003. LNCS, vol. 2899, pp. 388–399. Springer, Heidelberg (2003)
50. Dennett, D.C.: The Intentional Stance. MIT Press, Cambridge (1987)
51. Denti, E., Omicini, A., Ricci, A.: Multi-paradigm Java-Prolog integration in tuProlog. Sci. Comput. Program. 57(2), 217–250 (2005)
52. Dickinson, I., Wooldridge, M.: Towards practical reasoning agents for the semantic web. In: Proc. of AAMAS 2003, pp. 827–834 (2003)
53. d'Inverno, M., Hindriks, K.V., Luck, M.: A formal architecture for the 3APL agent programming language. In: Bowen, J.P., Dunne, S., Galloway, A., King, S. (eds.) B 2000, ZUM 2000, and ZB 2000. LNCS, vol. 1878, pp. 168–187. Springer, Heidelberg (2000)

54. d'Inverno, M., Kinny, D., Luck, M., Wooldridge, M.: A formal specification of dMARS. In: Rao, A., Singh, M.P., Wooldridge, M.J. (eds.) ATAL 1997. LNCS, vol. 1365, pp. 155–176. Springer, Heidelberg (1998)

55. d'Inverno, M., Luck, M.: Engineering AgentSpeak(L): A formal computational model. Logic and Computation Journal 8(3), 1–27 (1998)

56. Dix, J., Munoz-Avila, H., Nau, D.: IMPACTing SHOP: Putting an AI planner into a Multi-Agent Environment. Annals of Mathematics and AI 4(37), 381–407 (2003)

57. Dovier, A., Pontelli, E. (eds.): 25 Years of Logic Programming in Italy. LNCS, vol. 6125. Springer, Heidelberg (2010)

58. Dung, P.M.: On the acceptability of arguments and its fundamental role in nonmonotonic reasoning, logic programming and n-person games. Artif. Intell. 77(2), 321–358 (1995)

59. Eiter, T., Mascardi, V., Subrahmanian, V.S.: Error-Tolerant Agents. In: Kakas, A.C., Sadri, F. (eds.) Computational Logic: Logic Programming and Beyond, part I. LNCS (LNAI), vol. 2407, pp. 586–625. Springer, Heidelberg (2002)

60. Eiter, T., Subrahmanian, V.S., Pick, G.: Heterogeneous active agents, I: Semantics. Artificial Intelligence 108(1-2), 179–255 (1999)

61. Evertsz, R., Fletcher, M., Jones, R., Jarvis, J., Brusey, J., Dance, S.: Implementing industrial multi-agent systems using JACK. In: Dastani, M.M., Dix, J., El Fallah-Seghrouchni, A. (eds.) PROMAS 2003. LNCS (LNAI), vol. 3067, pp. 18–48. Springer, Heidelberg (2004)

62. Fensel, D., Lausen, H., de Bruijn, J., Stollberg, M., Roman, D., Polleres, A.: Enabling Semantic Web Services: The Web Service Modeling Ontology. Springer, Heidelberg

63. Finger, M., McBrien, P., Owens, R.: Databases and executable temporal logic. In: Comission of the European Communities (ed.) Proceedings of the Annual ESPRIT Conference 1991, pp. 288–302 (1991)

64. FIPA Home Page, http://www.fipa.org/

65. Fisher, M.: A normal form for first-order temporal formulae. In: Kapur, D. (ed.) CADE 1992. LNCS, vol. 607, pp. 370–384. Springer, Heidelberg (1992)

66. Fisher, M.: Concurrent METATEM – A language for modeling reactive systems. In: Reeve, M., Bode, A., Wolf, G. (eds.) PARLE 1993. LNCS, vol. 694, pp. 185–196. Springer, Heidelberg (1993)

67. Fisher, M.: A survey of Concurrent METATEM – the language and its applications. In: Gabbay, D.M., Ohlbach, H.J. (eds.) ICTL 1994. LNCS, vol. 827, pp. 480–505. Springer, Heidelberg (1994)

68. Fisher, M.: Implementing BDI-like systems by direct execution. In: Proc. of IJCAI 1997, pp. 316–321. Morgan Kaufmann, San Francisco (1997)

69. Fisher, M.: Representing abstract agent architectures. In: Rao, A.S., Singh, M.P., Müller, J.P. (eds.) ATAL 1998. LNCS (LNAI), vol. 1555, pp. 227–241. Springer, Heidelberg (1999)

70. Fisher, M., Bordini, R., Hirsch, B., Torroni, P.: Computational logics and agents: A road map of current technologies and future trends. Computational Intelligence 23(1), 61–91 (2007)

71. Fisher, M., Ghidini, C.: Programming resource-bounded deliberative agents. In: Proc. of IJCAI 1999, pp. 200–205. Morgan Kaufmann, San Francisco (1999)

72. Fisher, M., Ghidini, C.: The ABC of rational agent programming. In: Proc. of AAMAS 2002, pp. 849–856. ACM Press, New York (2002)

73. Fisher, M., Kakoudakis, T.: Flexible agent grouping in executable temporal logic. In: Intensional Programming II (ISPLIP 1999). World Scientific Publishers, Singapore (2000)

74. Fisher, M., Wooldridge, M.: On the formal specification and verification of multi-agent systems. International Journal of Cooperative Information Systems 6(1), 37–65 (1997)

75. Forgy, C.: Ops5 user's manual. Technical Report CMU-CS-81-135, Carnegie-Mellon University (1981)

76. Fornara, N., Colombetti, M.: A commitment-based approach to agent communication. Applied Artificial Intelligence 18(9-10), 853–866 (2004)
77. Foundation for Intelligent Physical Agents (FIPA). Agent Communication Language Specifications (2002)
78. Friedman-Hill, E.: Jess in Action: Java Rule-Based Systems (In Action series). Manning Publications (2002)
79. Fritz, C., Baier, J.A., McIlraith, S.A.: ConGolog, sin trans: Compiling ConGolog into basic action theories for planning and beyond. In: Proc. of 11th Int. Conf. on PKRR, pp. 600–610 (2008)
80. Frühwirth, T.: Theory and practice of constraint handling rules. Journal of Logic Programming 37(1-3), 95–138 (1998)
81. Gavanelli, M., Alberti, M., Lamma, E.: Integration of abductive reasoning and constraint optimization in SCIFF. In: Hill, P.M., Warren, D.S. (eds.) ICLP 2009. LNCS, vol. 5649, pp. 387–401. Springer, Heidelberg (2009)
82. Georgeff, M.P., Lansky, A.L.: Reactive reasoning and planning. In: Proc. of AAAI 1987, pp. 677–682 (1987)
83. Georgeff, M.P., Rao, A.S.: A profile of the Australian AI institute. IEEE Expert 11(6), 89–92 (1996)
84. Giordano, L., Martelli, A., Schwind, C.: Ramification and causality in a modal action logic. Journal of Logic and Computation 10(5), 626–662 (2000)
85. Giordano, L., Martelli, A., Schwind, C.: Reasoning About Actions in Dynamic Linear Time Temporal Logic. Journal of the IGPL 9(2), 298–303 (2001)
86. Giordano, L., Martelli, A., Schwind, C.: Specifying and Verifying Interaction Protocols in a Temporal Action Logic. Journal of Applied Logic 5(2), 214–234 (2007)
87. Gray, R.S., Kotz, D., Cybenko, G., Rus, D.: Agent Tcl. In: Mobile Agents: Explanations and Examples. Manning Publishing (1997)
88. Harbers, M., van den Bosch, K., Meyer, J.: Enhancing training by using agents with a theory of mind. In: EduMAS 2009, Proceedings, pp. 23–30 (2009)
89. Harbers, M., van den Bosch, K., Meyer, J.-J.C.: A study into preferred explanations of virtual agent behavior. In: Ruttkay, Z., Kipp, M., Nijholt, A., Vilhjálmsson, H.H. (eds.) IVA 2009. LNCS, vol. 5773, pp. 132–145. Springer, Heidelberg (2009)
90. Hindriks, K.V., Boer, F.S.D., der Hoek, W.V., Meyer, J.-J.C.: Agent programming in 3APL. AAMAS Journal 2(4), 357–401 (1999)
91. Hindriks, K.V., de Boer, F.S., van der Hoek, W., Meyer, J.-J.C.: Formal semantics for an abstract agent programming language. In: Rao, A., Singh, M.P., Wooldridge, M.J. (eds.) ATAL 1997. LNCS, vol. 1365, pp. 215–229. Springer, Heidelberg (1998)
92. Hindriks, K.V., de Boer, F.S., van der Hoek, W., Meyer, J.-J.C.: Agent programming with declarative goals. In: Castelfranchi, C., Lespérance, Y. (eds.) ATAL 2000. LNCS (LNAI), vol. 1986, pp. 228–243. Springer, Heidelberg (2001)
93. Hinrichs, T.L., Gude, N.S., Casado, M., Mitchell, J.C., Shenker, S.: Practical declarative network management. In: WREN 2009: Proceedings of the 1st ACM workshop on Research on enterprise networking, pp. 1–10. ACM, New York (2009)
94. Hirsch, B., Fisher, M., Ghidini, C.: Organising logic-based agents. In: Hinchey, M.G., Rash, J.L., Truszkowski, W.F., Rouff, C.A., Gordon-Spears, D.F. (eds.) FAABS 2002. LNCS (LNAI), vol. 2699, pp. 15–27. Springer, Heidelberg (2003)
95. Huber, M.J.: JAM: A BDI-theoretic mobile agent architecture. In: Agents 1999, Third International Conference on Autonomous Agents, Proceedings, pp. 236–243 (1999)
96. Hübner, J.F., Bordini, R.H.: Developing a team of gold miners using Jason. In: Dastani, M.M., El Fallah Seghrouchni, A., Ricci, A., Winikoff, M. (eds.) ProMAS 2007. LNCS (LNAI), vol. 4908, pp. 241–245. Springer, Heidelberg (2008)

97. Ingrand, F.F., Georgeff, M.P., Rao, A.S.: An architecture for real-time reasoning and system control. IEEE Expert Magazine 7(6), 33–44 (1992)
98. Issarny, V., Caporuscio, M., Georgantas, N.: A perspective on the future of middleware-based software engineering. In: FOSE 2007: 2007 Future of Software Engineering, Washington, DC, USA, pp. 244–258. IEEE Computer Society, Los Alamitos (2007)
99. Jennings, N.R., Mamdani, E.H., Corera, J.M., Laresgoiti, I., Perriollat, F., Skarek, P., Zsolt Varga, L.: Using Archon to develop real-world DAI applications, part 1. IEEE Expert 11(6), 64–70 (1996)
100. Jhingran, A.: Enterprise information mashups: integrating information, simply. In: Proc. of VLDB 2006, pp. 3–4. VLDB Endowment (2006)
101. Kakas, A.C., Kowalski, R., Toni, F.: The role of abduction in logic programming. In: Gabbay, C.H.D.M., Robinson, J. (eds.) Handbook of Logic in Artificial Intelligence and Logic Programming 5, pp. 235–324. Oxford University Press, Oxford
102. Kavantzas, N., Burdett, D., Ritzinger, G., Fletcher, T., Lafon, Y.: Web services choreography description language version 1.0 (2004), http://www.w3.org/TR/ws-cdl-10/
103. Kellett, A., Fisher, M.: Automata representations for concurrent METATEM. In: Proc. of TIME 1997, pp. 12–19. IEEE Press, Los Alamitos (1997)
104. Kellett, A., Fisher, M.: Concurrent METATEM as a coordination language. In: Garlan, D., Le Métayer, D. (eds.) COORDINATION 1997. LNCS, vol. 1282, pp. 418–421. Springer, Heidelberg (1997)
105. Kinny, D.: ViP: a visual programming language for plan execution systems. In: Proc. of AAMAS 2002, pp. 721–728 (2002)
106. Klan, D., Hose, K., Sattler, K.-U.: Developing and deploying sensor network applications with anduin. In: Proc. of DMSN 2009, pp. 1–6. ACM, New York (2009)
107. Klapiscak, T., Bordini, R.H.: JASDL: A practical programming approach combining agent and semantic web technologies. In: Baldoni, M., Son, T.C., van Riemsdijk, M.B., Winikoff, M. (eds.) DALT 2008. LNCS (LNAI), vol. 5397, pp. 91–110. Springer, Heidelberg (2009)
108. Labrou, Y., Finin, T.: Semantics and conversations for an agent communication language. In: Readings in Agents, pp. 235–242. Morgan Kaufmann, San Francisco (1997)
109. Lange, D., Mitsuru, O.: Programming and Deploying Java Mobile Agents with Aglets (1998)
110. Leckie, C., Senjen, R., Ward, B., Zhao, M.: Communication and coordination for intelligent fault diagnosis agents. In: 8th IFIP/IEEE International Workshop for Distributed Systems Operations and Management, DSOM 1997, Proceedings, pp. 280–291 (1997)
111. Lee, J., Durfee, E.H.: Structured circuit semantics for reactive plan execution systems. In: Proc. of AAAI 1994, pp. 1232–1237 (1994)
112. Leite, J., Omicini, A., Sterling, L., Torroni, P. (eds.): DALT 2003. LNCS (LNAI), vol. 2990. Springer, Heidelberg (2004)
113. Levesque, H.J., Pagnucco, M.: Legolog: Inexpensive experiments in cognitive robotics. In: Proc. of CogRob 2000 (2000)
114. Levesque, H.J., Reiter, R., Lespérance, Y., Lin, F., Scherl, R.B.: Golog: A logic programming language for dynamic domains. Journal of Logic Programming 31, 59–84 (1997)
115. Machado, R., Bordini, R.H.: Running agentSpeak(L) agents on SIM_AGENT. In: Meyer, J.-J.C., Tambe, M. (eds.) ATAL 2001. LNCS (LNAI), vol. 2333, pp. 158–174. Springer, Heidelberg (2002)
116. Mascardi, V., Demergasso, D., Ancona, D.: Languages for programming BDI-style agents: an overview. In: Corradini, F., Paoli, F.D., Merelli, E., Omicini, A. (eds.) WOA 2005: Dagli Oggetti agli Agenti, Proceedings, pp. 9–15. Pitagora Editrice Bologna (2005)
117. Mascardi, V., Martelli, M., Gungui, I.: DCaseLP: a prototyping environment for multi-language agent systems. In: Dastani, M.M., El Fallah Seghrouchni, A., Leite, J., Torroni, P. (eds.) LADS 2007. LNCS (LNAI), vol. 5118, pp. 139–155. Springer, Heidelberg (2008)

118. Mascardi, V., Martelli, M., Sterling, L.: Logic-based specification languages for intelligent software agents. J. of TPLP 4(4), 429–494 (2004)
119. Banzi, M., Caire, G., Gotta, D.: Wade: A software platform to develop mission critical applications exploiting agents and workflows. In: Proc. of AAMAS 2008 (2008)
120. Mellis, W.: TWAICE: A knowledge engineering tool. Inf. Syst. 15(1), 137–150 (1990)
121. Milner, R.: Communicating and Mobile Systems: the Pi-Calculus, June 1999. Cambridge University Press, Cambridge (1999)
122. Minton, S., Knoblock, C.A., Kuokka, D.R., Gil, Y., Joseph, R.L., Carbonell, J.G.: Prodigy 2.0: The manual and tutorial. Technical Report CMU-CS-89-146, Carnegie-Mellon University (1989)
123. Montali, M.: Specification and Verification of Open Declarative Interaction Models: a Logic-Based Framework. PhD thesis, DEIS, University of Bologna, Italy (2009)
124. Montali, M., Pesic, M., van der Aalst, W.M.P., Chesani, F., Mello, P., Storari, S.: Declarative specification and verification of service choreographies. ACM Transactions on the Web (2010)
125. Montali, M., Torroni, P., Alberti, M., Chesani, F., Gavanelli, M., Lamma, E., Mello, P.: Verification from declarative specifications using logic programming. In: Garcia de la Banda, M., Pontelli, E. (eds.) ICLP 2008. LNCS, vol. 5366, pp. 440–454. Springer, Heidelberg (2008)
126. Moreira, Á.F., Vieira, R., Bordini, R.H., Hübner, J.F.: Agent-oriented programming with underlying ontological reasoning. In: Baldoni, M., Endriss, U., Omicini, A., Torroni, P. (eds.) DALT 2005. LNCS (LNAI), vol. 3904, pp. 155–170. Springer, Heidelberg (2006)
127. Morley, D., Myers, K.: The SPARK agent framework. In: Proc. of AAMAS 2004, pp. 714–721 (2004)
128. Myers, K.L., Wilkins, D.E.: The Act Formalism, Version 2.2. Technical report, SRI International AI Center Technical Report, SRI International, Menlo Park, CA (1997)
129. Nwana, H.S., Ndumu, D.T.: An introduction to agent technology. In: Nwana, H.S., Azarmi, N. (eds.) Software Agents and Soft Computing: Towards Enhancing Machine Intelligence. LNCS, vol. 1198, pp. 3–26. Springer, Heidelberg (1997)
130. OASIS. Business process execution language for web services v.1.1 (2003)
131. OASIS, eXtensible Access Control Markup Language (XACML) Version 2.0 (2005), http://docs.oasis-open.org/xacml/2.0/access_control-xacml-2.0-core-spec-os.pdf
132. Omicini, A., Denti, E.: From tuple spaces to tuple centres. Science of Computer Programming 41(3), 277–294 (2001)
133. Omicini, A., Ricci, A., Viroli, M.: Artifacts in the A&A meta-model for multi-agent systems. Autonomous Agents and Multi-Agent Systems 17(3), 432–456 (2008); Special Issue on Foundations, Advanced Topics and Industrial Perspectives of Multi-Agent Systems
134. Omicini, A., Zambonelli, F.: MAS as complex systems: A view on the role of declarative approaches. In: Leite, J., Omicini, A., Sterling, L., Torroni, P. (eds.) DALT 2003. LNCS (LNAI), vol. 2990, pp. 1–16. Springer, Heidelberg (2004)
135. Paolucci, M., Kawamura, T., Payne, T., Sycara, K.: Semantic matching of web services capabilities. In: Horrocks, I., Hendler, J. (eds.) ISWC 2002. LNCS, vol. 2342, pp. 333–347. Springer, Heidelberg (2002)
136. Peine, H.: ARA - Agents for Remote Action. In: Mobile Agents. Manning Publishing (1997)
137. Pesic, M., van der Aalst, W.M.P.: A declarative approach for flexible business processes management. In: Eder, J., Dustdar, S. (eds.) BPM Workshops 2006. LNCS, vol. 4103, pp. 169–180. Springer, Heidelberg (2006)

138. Phung-Khac, A., Beugnard, A., Gilliot, J.-M., Segarra, M.-T.: Model-driven development of component-based adaptive distributed applications. In: Proc. of SAC 2008, pp. 2186–2191. ACM, New York (2008)

139. Pistore, M., Spalazzi, L., Traverso, P.: A minimalist approach to semantic annotations for web processes compositions. In: Sure, Y., Domingue, J. (eds.) ESWC 2006. LNCS, vol. 4011, pp. 620–634. Springer, Heidelberg (2006)

140. Piunti, M., Santi, A., Ricci, A.: Programming SOA/WS systems with cognitive agents and artifact-based environments. In: Proc. of MALLOW 2009 Multi-Agent Logics, Languages, and Organisations Federated Workshops, CEUR Workshop Proceedings (2009) ISSN 1613-0073

141. Rao, A.S.: AgentSpeak(L): BDI agents speak out in a logical computable language. In: Perram, J., Van de Velde, W. (eds.) MAAMAW 1996. LNCS, vol. 1038, pp. 42–55. Springer, Heidelberg (1996)

142. Rao, A.S., Georgeff, M.P.: Asymmetry thesis and side-effect problems in linear-time and branching-time intention logics. In: Proc. of IJCAI 1991, pp. 498–504 (1991)

143. Rao, A.S., Georgeff, M.P.: Decision procedures for BDI logics. J. Log. Comput. 8(3), 293–342 (1998)

144. Reisig, W., Rozenberg, G. (eds.): APN 1998. LNCS, vol. 1491. Springer, Heidelberg (1998); the volumes are based on the Advanced Course on Petri Nets, held in Dagstuhl (September 1996)

145. Reiter, R.: On knowledge-based programming with sensing in the situation calculus. ACM Transactions on Computational Logic (TOCL) 2(4), 433–457 (2001)

146. Renz, W.: Models and multi-agent simulations of logistics networks - a case-study in self-organization by microeconomics. In: MKWI 2008. GITO-Verlag, Berlin (2008)

147. Rimassa, G., Burmeister, B.: Achieving business process agility in engineering change management with agent technology. In: WOA 2007, pp. 1–7. Seneca Edizioni Torino (2007)

148. Roland, J., Vesonder, G., Wilson, J.: C5 user manual, release 2.1. Technical report, AT&T Bell Laboratories (1990)

149. Rossi, G.: Logic Programming in Italy: A Historical Perspective. In: Dovier, A., Pontelli, E. (eds.) 25 Years of Logic Programming in Italy. LNCS, vol. 6125, pp. 1–14. Springer, Heidelberg (2010)

150. Sadri, F., Toni, F.: Computational Logic and Multi-Agent Systems: a Roadmap. Technical report, Department of Computing, Imperial College, London (1999)

151. Salasin, J., Madni, A.M.: Metrics for service-oriented architecture (soa) systems: What developers should know. J. Integr. Des. Process Sci. 11(2), 55–71 (2007)

152. Schroeder, M., de Almeida Móra, I., Pereira, L.M.: A deliberative and reactive diagnosis agent based on logic programming. In: Rao, A., Singh, M.P., Wooldridge, M.J. (eds.) ATAL 1997. LNCS, vol. 1365, pp. 293–307. Springer, Heidelberg (1998)

153. Semantic Annotations for WSDL Working Group. Semantic annotations for wsdl and xml schema. Technical report, W3C (2007)

154. Semmel, G.S., Davis, S.R., Leucht, K.W., Rowe, D.A., Smith, K.E., Boloni, L.: Space shuttle ground processing with monitoring agents. IEEE Intelligent Systems 21(1), 68–73 (2006)

155. Shapiro, S., Lespérance, Y., Levesque, H.J.: The cognitive agent specification language and verification environment for multiagent systems. In: Proc. of AAMAS 2002, pp. 19–26. ACM Press, New York (2002)

156. Shoham, Y.: Agent-oriented programming. Artificial Intelligence 60(1), 51–92 (1993)

157. Singh, M.P.: Agent communication languages: Rethinking the principles. IEEE Computer 31(12), 40–47 (1998)

158. Sloman, A., Poli, R.: SIM_AGENT: A toolkit for exploring agent design. In: Tambe, M., Müller, J., Wooldridge, M.J. (eds.) IJCAI-WS 1995 and ATAL 1995. LNCS, vol. 1037, pp. 392–407. Springer, Heidelberg (1996)

159. Sohrabi, S., Prokoshyna, N., McIlraith, S.A.: Web service composition via the customization of Golog programs with user preferences. In: Borgida, A.T., Chaudhri, V.K., Giorgini, P., Yu, E.S. (eds.) Conceptual Modeling: Foundations and Applications: Essays in Honor of John Mylopoulos, pp. 319–334. Springer, Heidelberg (2009)

160. Sterling, L., Shapiro, E.: The art of Prolog: advanced programming techniques (1986)

161. Subrahmanian, V., Bonatti, P., Dix, J., Eiter, T., Kraus, S., Özcan, F., Ross, R.: Heterogenous Active Agents, 580 pages. MIT Press, Cambridge (2000)

162. Thomas, S.R.: The PLACA agent programming language. In: Wooldridge, M.J., Jennings, N.R. (eds.) ECAI 1994 and ATAL 1994. LNCS, vol. 890, pp. 355–370. Springer, Heidelberg (1995)

163. Torroni, P.: Computational logic in multi-agent systems: Recent advances and future directions. Ann. Math. Artif. Intell. 42(1-3), 293–305 (2004)

164. Torroni, P., Chesani, F., Mello, P., Montali, M.: Social commitments in time: Satisfied or compensated. In: Baldoni, M., van Riemsdijk, M.B. (eds.) DALT 2009. LNCS, vol. 5948, pp. 228–243. Springer, Heidelberg (2010)

165. Torroni, P., Chesani, F., Mello, P., Montali, M.: Social commitments in time: satisfied or compensated. In: Baldoni, M., Bentahar, J., Lloyd, J., van Riemsdijk, M.B. (eds.) DALT 2009. LNCS, Springer, Heidelberg (2010)

166. Torroni, P., Yolum, P., Singh, M.P., Alberti, M., Chesani, F., Gavanelli, M., Lamma, E., Mello, P.: Modelling interactions via commitments and expectations. In: Handbook of Research on Multi-Agent Systems: Semantics and Dynamics of Organizational Models, Hershey, Pennsylvania, March 2009, pp. 263–284. IGI Global (2009)

167. van der Aalst, W.M.P., Dumas, M., ter Hofstede, A.H.M., Russell, N., Verbeek, H.M.W., Wohed, P.: Life after BPEL? In: Bravetti, M., Kloul, L., Zavattaro, G. (eds.) EPEW/WS-EM 2005. LNCS, vol. 3670, pp. 35–50. Springer, Heidelberg (2005)

168. Van Linder, B.: Modal Logics for Rational Agents. PhD thesis, Universiteit Utrecht, Utrecht, The Netherlands (1987)

169. van Riemsdijk, B., van der Hoek, W., Meyer, J.-J.C.: Agent programming in Dribble: from beliefs to goals using plans. In: Proc. of AAMAS 2003, pp. 393–400 (2003)

170. van Riemsdijk, M.B., de Boer, F.S., Dastani, M.M., Meyer, J.-J.C.: Prototyping 3APL in the Maude term rewriting language. In: Inoue, K., Satoh, K., Toni, F. (eds.) CLIMA 2006. LNCS (LNAI), vol. 4371, pp. 95–114. Springer, Heidelberg (2007)

171. van Riemsdijk, M.B., Wirsing, M.: Goal-Oriented and Procedural Service Orchestration. A Formal Comparison. In: AWESOME 2007, Durham, UK (September 2007)

172. Vasconcelos, W.W.: Logic-based electronic institutions. In: Leite, J., Omicini, A., Sterling, L., Torroni, P. (eds.) DALT 2003. LNCS (LNAI), vol. 2990, pp. 221–242. Springer, Heidelberg (2004)

173. W3C The Platform for Privacy Preferences 1.0 (P3P1.0) Specification (2002), http://www.w3.org/TR/P3P/

174. Weber, N., Braubach, L., Pokahr, A., Lamersdorf, W.: Agent-based semantic search at motoso.de. In: Braubach, L., van der Hoek, W., Petta, P., Pokahr, A. (eds.) MATES 2009. LNCS, vol. 5774, pp. 278–287. Springer, Heidelberg (2009)

175. Weihmayer, T., Tan, M.: Modeling cooperative agents for customer network control using planning and agent-oriented programming. In: IEEE Global Telecommunications Conference, Globecom 1992, Proceedings, pp. 537–543. IEEE, Los Alamitos (1992)

176. White, S.: Business Process Modeling Notation Specification 1.0. Technical report, OMG (2006)

177. Winikoff, M., Padgham, L., Harland, J., Thangarajah, J.: Declarative & procedural goals in intelligent agent systems. In: Proc. of KR 2002, pp. 470–481 (2002)
178. Wooldridge, M.: The Logical Model of Computational Multi–Agent Systems. PhD thesis, Department of Computation, UMIST, Manchester, UK (1992)
179. Wooldridge, M.: This is MYWORLD: The logic of an agent-oriented testbed for DAI. In: Wooldridge, M.J., Jennings, N.R. (eds.) ECAI 1994 and ATAL 1994. LNCS, vol. 890, pp. 160–178. Springer, Heidelberg (1995)
180. Wooldridge, M.: An Introduction to MultiAgent Systems. Wiley, Chichester (2002)
181. Wooldridge, M., Jennings, N.R.: Intelligent agents: Theory and practice. Knowledge Engineering Review 10(2), 115–152 (1995)
182. Wooldridge, M.J.: In: Kandzia, P., Klusch, M. (eds.) CIA 1997. LNCS, vol. 1202, pp. 1–18. Springer, Heidelberg (1997)
183. Yolum, P., Singh, M.P.: Flexible protocol specification and execution: applying event calculus planning using commitments. In: AAMAS, pp. 527–534. ACM, New York (2002)
184. Yu, Q., Liu, X., Bouguettaya, A., Medjahed, B.: Deploying and managing web services: issues, solutions, and directions. The VLDB Journal 17(3), 537–572 (2008)
185. Zaremski, A.M., Wing, J.M.: Specification matching of software components. ACM Transactions on SEM 6(4), 333–369 (1997)
186. Zöller, A., Braubach, L., Pokahr, A., Rothlauf, F., Paulussen, T.O., Lamersdorf, W., Heinzl, A.: Evaluation of a multi-agent system for hospital patient scheduling. International Transactions on Systems Science and Applications 1(4), 375–380 (2006)

Concurrent and Reactive Constraint Programming

Maurizio Gabbrielli[1], Catuscia Palamidessi[2], and Frank D. Valencia[3]

[1] Lab. Focus, INRIA and University of Bologna
[2] INRIA and LIX, Ecole Polytechnique
[3] CNRS, LIX, Ecole Polytechnique

Abstract. The Italian Logic Programming community has given several contributions to the theory of Concurrent Constraint Programming. In particular, in the topics of semantics, verification, and timed extensions. In this paper we review the main lines of research and contributions of the community in this field.

1 The Origins: From Concurrect Logic Programming to Concurrent Constraint Programming

In the 80's there had been several proposals to extend logic programming with constructs for concurrency, aiming at the development of a concurrent language which would maintain the typical advantages of logic programming: declarative reading, computations as proofs, amenability to meta-programming etc. Examples of concurrent logic languages include PARLOG [14], Concurrent Prolog [59,60], Guarded Horn Clauses (GHC) [62,63] and their so-called *flat* versions. Towards the end of the decade, Concurrent constraint programming ([53,57,58]) emerged as one of the most successful proposals in this area.

Concurrent constraint programming (ccp) presented two new perspectives on the underlying philosophy of logic programming. One is the replacement of the concept of unification over the Herbrand universe by the more general notion of *constraint* over an arbitrary domain. This is in a sense a 'natural' development, and the idea was already introduced in 'sequential' logic programming by Jaffar and Lassez ([46]). The other is the introduction of extra-logical *operators* typical of the imperative concurrent paradigms, like CCS ([48]), TCSP ([8]) and ACP ([1]); in particular, the *choice* $(+)$, the *action prefixing* $(\rightarrow)$, and the *hiding* operator $(\exists)$. Additionally, concurrent constraint programming embodies an explicit characterization of the *control* mechanisms for communication and synchronization by means of the introduction of two kinds of actions (*ask* and *tell*). Also in concurrent logic languages these control features were present, but they were hidden in various ways: the choice was represented by alternative clauses, hiding by local (existentially quantified) variables, prefixing by commitment, communication by sharing of variables, and synchronization by restrictions on the unification algorithm.

There are many advantages in an explicit representation of these concurrency control mechanisms by means of operators. First of all, they are 'isolated' and

A. Dovier, E. Pontelli (Eds.): 25 Years of Logic Programming, LNCS 6125, pp. 231–253, 2010.

therefore the laws of their behaviour can be understood better. For instance, one of the problems in studying the semantics of concurrent logic programming is that the choice mechanism is 'mixed up' with recursion, since a clause is in general a recursive definition. Second, the standard tools developed in the theory of concurrency can be applied more easily. Third, a 'reconciliation' with the declarative principles of logic programming is more feasible, once the basic limitations are well understood. For instance, the conditions which rule the behaviour of *ask* and *tell* can be described in a logical way, thus providing the synchronization mechanism with a 'declarative flavour' ([47,52]) that was missing in the 'restricted-unification' approach.

2   The ccp Paradigm

Ccp is based on the concept of *store-as-constraint*, in contrast to von Neumann's concept of *store-as-valuation*. The computation proceeds through the concurrent execution of different processes, which interact and communicate through the common store. They refine the partial information about the values of the variables by adding (*telling*) constraints to the store, and they test (*ask*) whether the store entails a constraint before proceeding in the computation.

One of the most characteristic features of the ccp paradigm is a formalization of these basic operations which allow to update and to query the common store, in terms of the logical notions of consistency, conjunction and entailment supported by a given underlying constraint system.

Here we recall briefly the syntax and semantics of ccp. Among the several variants which have been proposed in literature, we choose the simplest and most basic one, called *eventual tell* ccp. Most of the other ccp dialects can be obtained by enriching this one.

The ccp languages are defined parametrically w.r.t. to a given *cylindric constraint system*.

Definition 1

- *A constraint system is a complete algebraic lattice $\langle \mathcal{C}, \vdash, \sqcup, true, false \rangle$ where $\sqcup$ is the lub operation, and true, false are the least and the greatest elements of $\mathcal{C}$, respectively. The entailment relation $\vdash$ is the inverse ordering.*
- *Consider a (denumerable) set of variables $x, y, z, \ldots$. Assume that for each $x \in Var$ a function $\exists_x : \mathcal{C} \to \mathcal{C}$ is defined such that for any $c, d \in \mathcal{C}$:*
  (i) $c \vdash \exists_x(c)$,
  (ii) if $c \vdash d$ then $\exists_x(c) \vdash \exists_x(d)$,
  (iii) $\exists_x(c \sqcup \exists_x(d)) = \exists_x(c) \sqcup \exists_x(d)$,
  (iv) $\exists_x(\exists_y(c)) = \exists_y(\exists_x(c))$.
  Then $\mathbf{C} = \langle \mathcal{C}, \leq, \sqcup, true, false, Var, \exists \rangle$ is a cylindric constraint system.

In order to model parameter passing, *diagonal elements* ([45]) are added to the primitive constraints: We assume that, for x, y ranging in *Var*, D contains the constraints d_{xy} which satisfy the following axioms.

(i) $true \vdash d_{xx}$,

(ii) if $z \neq x, y$ then $d_{xy} = \exists_z (d_{xz} \sqcup d_{zy})$,

(iii) if $x \neq y$ then $d_{xy} \sqcup \exists_x (c \sqcup d_{xy}) \vdash c$.

Note that if $\mathbf{C}$ models the equality theory, then the elements d_{xy} can be thought of as the formulas $x = y$. In the following $\exists_x(c)$ is denoted by $\exists_x c$ with the convention that, in case of ambiguity, the scope of $\exists_x$ is limited to the first constraint subexpression. (So, for instance, $\exists_x c \sqcup d$ stands for $\exists_x(c) \sqcup d$.)

Definition 2. *Assuming a given cylindric constraint system $\mathbf{C}$ the syntax of agents is given by the following grammar:*

$$A ::= stop \mid tell(c) \mid \textstyle\sum_{i=1}^{n} ask(c_i) \to A_i \mid A \parallel A \mid \exists x A \mid p(x)$$

where the c, c_i are supposed to be finite *constraints (i.e. algebraic elements) in $\mathcal{C}$. A ccp process P is then an object of the form $D.A$, where D is a set of procedure declarations of the form $p(x) :: A$ and A is an agent.*

The *deterministic* agents are obtained by imposing the restriction $n = 1$ in the previous grammar. The standard operational model of ccp can be described by a transition system $T = (Conf, \longrightarrow)$. The configurations (in) *Conf* are pairs consisting of a process, and a constraint in $\mathcal{C}$ representing the common *store*. The transition relation $\longrightarrow \subseteq Conf \times Conf$ is described by the (least relation satisfying the) rules **R1-R6** of table 1.

The agent *stop* represents successful termination. The basic actions are given by *tell(c)* and *ask(c)* constructs which act on the common store. Given a store d, as shown by rule **R1**, the execution of *tell(c)* updates the store to $c \sqcup d$. The action *ask(c)* represents a guard, i.e. a test on the current store d, whose execution does not modify d. We say that *ask(c)* is *enabled* in d iff $d \vdash c$. According to rule **R2** the *guarded choice* operator gives rise to global non-determinism: the agent $\sum_{i=1}^{n} ask(c_i) \to A_i$ nondeterministically selects one *ask(c_i)* which is enabled in the current store, and then behaves like A_i. The external environment can then affect the choice since *ask(c)* is enabled iff the current store d entails c, and d can be modified by other agents (rule **R1**). If no guard is enabled, then the guarded choice agent *suspends*, waiting for other (parallel) agents to add information to the store. The situation in which all the components of a system of parallel agents suspend is called *global suspension* or *deadlock*. The operator $\parallel$ represents parallel composition which is described by rule **R3** as interleaving. The agent $\exists x A$ behaves like A, with x considered *local* to A. To describe locality in rule **R4** the syntax has been extended by an agent $\exists^d x A$ where d is a local store of A containing information on x which is hidden in the external store. Initially the local store is empty, i.e. $\exists x A = \exists^{true} x A$.

Rule **R5** treats the case of a procedure call when the actual parameter equals the formal parameter: in this case a simple body replacement suffices. We do not need more rules since, for the sake of simplicity, we assume that the set D of procedure declarations is closed w.r.t. parameter names.

Table 1. The transition system of ccp

R1 $\langle D.tell(c), d\rangle \longrightarrow \langle D.Stop, c \sqcup d\rangle$

R2 $\langle D. \sum_{i=1}^{n} ask(c_i) \to A_i, d\rangle \longrightarrow \langle D.A_j, d\rangle \; j \in [1, n] \; and \; d \vdash c_i$

R3 $\dfrac{\langle D.A, c\rangle \longrightarrow \langle D.A', c'\rangle}{\begin{array}{l}\langle D.A \parallel B, c\rangle \longrightarrow \langle D.A' \parallel B, c'\rangle \\ \langle D.B \parallel A, c\rangle \longrightarrow \langle D.B \parallel A', c'\rangle\end{array}}$

R4 $\dfrac{\langle D.A, d \sqcup \exists_x c\rangle \longrightarrow \langle D.B, d'\rangle}{\langle D.\exists^d x A, c\rangle \longrightarrow \langle D.\exists^{d'} x B, c \sqcup \exists_x d'\rangle}$

R5 $\langle D.p(x), c\rangle \longrightarrow \langle D.A, c\rangle$ $p(x) : -A \in D$

---

3   Semantic Aspects of ccp

In the first few years after its design, ccp had been understood just as a particular case of process algebra. Therefore, the definition of its compositional semantics had been approached by the standard methods, like failure sets and bisimulation. For instance, De Boer et al. [15,16] used tree-like structures labeled with functions on substitutions. More simple tree-like structures, labeled by constraints, were used by Gabbrielli and Levi [40]. Saraswat and Rinard [57] used similar structures modulo equivalence relations based on bisimulation.

De Boer and Palamidessi [18] realized that, due to the fact that the communication mechanism of ccp is asynchronous, the branching structures used for process algebra are not needed. In fact, which actions are enabled does not depend upon the current state of the environment, but only upon the store. In a transition system this can be made explicit by adding a passive rule that does not exist in the classical concurrent paradigms: an arbitrary assumption of a step made by the environment. This amounts to considering all the possible interactions between the given process and arbitrary environments, and it leads to a simple compositional semantics, consisting of sequences of constraints labeled by assume/tell modes. In this framework the parallel composition corresponds to zip sequences, so that the assumptions of a process match with the actions of the other, and vice-versa.

Independently, a different approach was developed in [58]. The basic idea consists in denoting processes as Scott's closure operators, which have the nice property of being representable by the set of their fixpoints. The operators of the language can then be described as operations on those sets. In particular, parallelism can be modeled simply by intersection.

The semantics developed in [18] and in [58] are based on very different points of view. The one in [18] is more general, in the sense that it applies, without essential modifications, to many variants of ccp, including the atomic and non-deterministic versions. The one in [58] is very ingenious and elegant, and can be considered one of the principal reasons of the success of ccp. However, it works well only in the basic fragment, the *deterministic eventual tell ccp*, which is obtained from Definition 6 by imposing $n = 1$ in the summation. Both semantics are fully abstract, and therefore in the basic fragment they are equivalent. The precise correspondence was delineated in [19].

One question that had remained open in [18] was how to model infinite computations in an abstract way, i.e. by considering only the limit of the answer substitution. When nondeterminism is present, the denotational characterization of infinite computation is actually a non trivial problem: The semantics based on Smith, Hoare and Plotkin's powerdomains constitute only a partial solution to this problem (in the sense that they identify too much), and the semantics based on metric domains are far from being abstract. This problem was solved in [25] by considering a categorical construction called *Lehmann's powerdomain*, which can be regarded as an extension of Smith's powerdomain. This structure contains more information than the powerdomains, enough to achieve compositionality.

3.1 Analysis and Verification

De Boer et al. developed in [20] a system based on the closure operators semantics to prove correctness assertions about concurrent constraint programs. Thanks to the strong properties of ccp, this system is much simpler than the ones developed for other parallel languages. In particular, only the strongest post-condition w.r.t. *True* needs to be considered, and parallel composition is modeled simply by logical conjunction.

Falaschi et al. investigated in [34] various fragments of ccp. Some of them have a very simple semantics based on closure operators. Such semantics can be considered as approximated semantics of ccp, and they were used as a basis for static analysis [33,35], by means of *abstract interpretation* techniques. These techniques allow to statically optimize programs and to approximate several important semantic properties, such as deadlock detection, groundness propagation etc.

One interesting fragment is *ccp with local choice*: This corresponds in fact to *CLP with delay*, an extension of Constraint Logic Programming which allows efficient implementations. Falaschi et al. [36] and De Boer et al. [23] used this observation for developing the semantics foundations and a verification system of CLP with delay, by means of techniques based on closure operators.

Another approach to the analysis of *ccp* was pursed in [66,67] where it was extended to *ccp* languages the *generalized semantics* approach to static analysis, initially proposed in [42] for sequential CLP languages. [66] shows that such an extension can be easily achieved for approximations that are closed under anti-entailment: applications include analyses that can identify definite suspensions, e.g., to compute upper bounds to the degree of concurrency in a *ccp* program. For the more common case of entailment closed properties (that are of interest for, e.g., proving suspension freeness), it is shown in [67] that correctness can only be achieved by modifying the generalized semantics approach so as to introduce a domain-dependent approximation of the synchronization primitive, which cannot be modeled as an entailment test on the abstract domain.

3.2   Fold/Unfold Transformations of *ccp*

Unfold/fold are source-to source transformation techniques which were first introduced in functional programming by Burstall and Darlington [10], and then adapted to logic programming both for program synthesis and for program specialization and optimization. As shown by a number of applications, these techniques provide a powerful methodology for the development and optimization of large programs, and can be regarded as the basis to be used for partial evaluation.

Despite a large amount of literature in the field of declarative sequential languages, the applications of unfold/fold transformations to concurrent languages are relatively rare. This is partially due to the fact that the nondeterminism and the synchronization mechanisms present in concurrent languages substantially complicate their semantics, thus complicating also the definition of *correct* transformation systems. Nevertheless, these transformation techniques can be very useful also for concurrent languages, since they allow further optimizations related to the simplification of synchronization and communication mechanisms.

One of the few papers addressing this issue is [32], where a transformation system for concurrent constraint programming (*ccp*) was introduced. This systems was inspired by that one of Tamaki and Sato [61], a general framework for the unfold/fold transformation of logic programs, which has remained over the years the main historical reference of the field.

Compared to its predecessors, the system in [32] improves by eliminating the limitation that in a folding operation the *folding rule* has to be non-recursive. Moreover, following de Francesco and Santone [39], the applicability conditions for this operation are based on the notion of "guardedness" and can be checked locally on the program to be folded (rather than on the transformation history). This makes the operation much easier to understand and to implement. Besides folding and unfolding, the transformation system for *ccp* includes several other new operations, namely backward instantiation, ask and tell simplification, branch elimination, conservative ask elimination and distribution. The declarative nature of *ccp* allows one to define reasonably simple applicability conditions for these operations which ensure the total correctness of the system: the original and the transformed program have the same semantics when considering both input/output pairs and (under different applicability conditions) traces, and distinguishing successful, deadlocked, and failed derivations.

From the correctness result follows that the original program is deadlock-free iff the transformed one is, and this allows us to employ the transformation system as an effective tool for proving deadlock-freeness of *ccp* programs. Moreover, the systems allows to optimize programs by eliminating communication channels and synchronization points, by transforming nondeterministic computations into deterministic ones, and by saving of computational *space*. Some of these improvements were possible already in the context of GHC programs by using the system defined in Ueda and Furukawa [64].

Following the above line of research, [3] investigated transformation techniques based on the *replacement*. This is a powerful operation which can mimic the most common transformation operations such as unfold, fold, switching, distribution. Because of this flexibility, it can be incorrect if used without specific applicability conditions. The above paper presented applicability conditions for ccp and it showed that, under these conditions, the replacement generalizes both the unfolding operation as well as a restricted form of folding operation.

4   Timed Reactive CCP

The *tcc* model is a timed reactive ccp framework introduced by Saraswat et al [54] as an extension of *deterministic* ccp. This model is aimed at programming and modeling timed reactive systems and it elegantly combines deterministic ccp with ideas from the paradigms of Synchronous Languages [2].

In order to increase the specification expressiveness of *tcc*, Nielsen et al [50] introduced a non-deterministic extension of *tcc*, called the *ntcc* calculus. As its predecessor, the *ntcc* calculus takes the view of reactive computation as proceeding in discrete time units (or *time intervals*). Time is conceptually divided into discrete intervals. In each time interval a ccp process receives a *stimulus*, represented as a constraint, from the environment, it executes with this stimulus as the initial store, and when it reaches its resting point, it *responds* to the environment with the final store. Furthermore, the resting point determines a residual process, which is then executed in the next time interval.

As illustrated in [50], this view of reactive computation is particularly appropriate for modeling and programming reactive systems such as robotic devices and micro-controllers. These systems typically operate in a cyclic fashion; in each cycle they receive and input from the environment, compute on this input, and then return the corresponding output to the environment.

4.1   Syntax and Operational Semantics of *ntcc*

The *ntcc* calculus introduces operators to specify temporal executions. The *unit-delay* operation *next A*, also present in *tcc*, specifies that *A* should be executed in the next time interval, and the *unbounded* delay operation $\star A$ specifies that *A* will be *eventually* executed. The *time-out* operation *unless c next A*, also present in *tcc*, specifies that unless *c* can be inferred from the final store in the current time unit, *A* should be executed in the next time unit.

Furthermore, to ensure that only terminating processes can be executed within time intervals, procedures are replaced with the simpler replicated form !*A*. The replication operation !*A* specifies that *A* will be executed now and in each future time interval. Thus, !*A* can be viewed as *A* ∥ *next A* ∥ *next (next A)* ∥ . . .

All in all, the agents of *ntcc* include those of ccp in Definition 2 except for procedures, plus the above-mentioned temporal operators. More precisely,

Definition 3. *Assuming a given cylindric constraint system* **C** *the syntax of ntcc agents is given by the following grammar:*

$$A ::= stop \mid tell(c) \mid \sum_{i=1}^{n} ask(c_i) \rightarrow A_i \mid A \parallel A \mid \exists x A$$
$$\mid next\ A \mid \star A \mid unless\ c\ next\ A \mid !A$$

where the c, c_i *are supposed to be* finite *constraints (i.e. algebraic elements) in* $\mathcal{C}$. *For the sake of consistency with Definition 2, an ntcc process* P *can be interpreted as an object of the form* $D.A$ *by decreeing that* $D = \emptyset$; *i.e., the empty set of procedure declarations.*

4.2   Reduction Relations

The operational semantics of *ntcc* is given in terms of an internal reduction relation $\longrightarrow$ given by the rules in Table 1 plus the rules in Table 2 and the observable reduction relation $\Longrightarrow$ given in Table 2.

The *internal* transition $\gamma \longrightarrow \gamma'$ specifies the internal steps much like the ccp transitions $\longrightarrow$ in the previous section. The additional rules **R6-R8** in Table 2 realize the above intuitions about the temporal operators.

The *observable transition* $P \overset{(c,d)}{\Longrightarrow} R$ should be read as "P on input c from the environment, reduces in one *time unit* to R and outputs d to the environment". The rule **ROBS** realizes the above intuition by stating that an observable transition from $P = D.A$ labeled by (c, d) is obtained by performing a sequence of internal transitions from the initial configuration $\langle P, c \rangle$ to a final configuration $\langle Q, d \rangle$ with $Q = D.A'$ in which no further internal evolution is possible. The residual process R to be executed in the next time interval is equivalent to $D.F(A')$, where $F(A')$ represents the "future" of A'. The process $F(A')$, given in Definition 4, is obtained by removing from A' summations that did not trigger activity within the current time interval and any local information which has been stored in A', and by "unfolding" the sub-terms within "next" and "unless" expressions. This "unfolding" specifies the evolution across time intervals of processes of the form *next* B and *unless* c *next* B.

Definition 4 (Future Function). *Let* F *be the partial function defined by*

$$F(A) = \begin{cases} stop & if\ A = \sum_{i \in I} c_i \rightarrow A_i \\ F(A_1) \parallel F(A_2) & if\ A = A_1 \parallel A_2 \\ \exists x F(B) & if\ A = \exists^d x B \\ B & if\ A = next\ B\ or\ A = unless\ c\ next\ B \end{cases}$$

4.3   A Simple Example of Weak Pre-emption

In spite of its simplicity, the *tcc* and *ntcc* extensions to ccp are far-reaching. Many interesting temporal constructs can be expressed (see e.g. [54]). For example, *tcc* allows processes to be "clocked" by other processes. This provides meaningful pre-emption constructs and the ability to define *multiple forms of time* instead of only having a unique global clock.

Table 2. Additional rules for the transitions of *ntcc* processes. The internal reduction $\longrightarrow$ is given by the rules in Table 1 and Rules **R6-R8**. The observable reduction $\Longrightarrow$ is given by Rule **ROBS**. The relation $\longrightarrow^*$ denotes transitive and reflexive closure of $\longrightarrow$. $\gamma \nrightarrow$ holds iff that is no γ' such that $\gamma \longrightarrow \gamma'$. The function F is given in Definitions 4.

R6	$\langle D.\star A, d \rangle \longrightarrow \langle D.A^n, d \rangle$	$n \geq 0$
R7	$\langle D.unless\ c\ next\ A, d \rangle \longrightarrow \langle D.stop, d \rangle,\ d \vdash c$	
R8	$\langle D.!A, d \rangle \longrightarrow \langle D.A \parallel next\ !A, c \sqcup d \rangle$	

$$\textbf{ROBS} \quad \frac{\langle D.A, c \rangle \longrightarrow^* \langle D.A', c' \rangle \nrightarrow}{D.A, c \stackrel{(c,c')}{\Longrightarrow} D.F(A')}$$

A rather simple example is the specification of a power-saver:

$$A = !\ unless\ (LightsOff)\ next\ \star\ tell(LightsOff)$$

The power-saver agent A runs forever, hence it is replicated. Furthermore, unless A can infer that the lights are already off in the current time interval, A should turn them off either in the next time unit or sometime later.

Notice that because of the weak pre-emption nature of the time-out operation in *ntcc*, it is not possible to specify that the lights should be turned off within the current time interval unless they are already off.

The work in [55] introduces *Default tcc* as an extension of *tcc* with the ability to define strong pre-emption. In this model, the time-out operation can trigger activity in the current time interval. Strong pre-emption is useful when an action must be triggered immediately on the absence of a constraint c rather than delayed to the next interaction.

4.4   Observables and Their Characterizations

Let us consider an infinite sequence of observable transitions:

$$P = P_1 \stackrel{(c_1, c_1')}{\Longrightarrow} P_2 \stackrel{(c_2, c_2')}{\Longrightarrow} P_3 \stackrel{(c_3, c_3')}{\Longrightarrow} \ldots$$

Intuitively, at time interval i, with $i \geq 0$, the process P_i gets a stimulus c_i and then it provides a response c_i' and evolves into P_{i+1}. We shall also represent this run as $P \stackrel{(\alpha, \alpha')}{\Longrightarrow}$ where $\alpha = c_1.c_2.c_3.\ldots$ and $\alpha' = c_1'.c_2'.c_3'.\ldots$.

The observable *input-output* behaviour of an *ntcc* process is its set of stimulus-response sequences. The *strongest-postcondition*, or *quiescent behaviour*, of a process P is the set of sequences on input of which P can run without adding any information whatsoever. More precisely,

Definition 5 (Observables of $ntcc$**).** *Let* P *be a process. The input-output behaviour of* P *is given by* $\mathcal{O}_{io}(P) = \{(\alpha, \alpha') \mid P \xrightarrow{(\alpha,\alpha')}\}$. *The strongest postcondition of* P *is given by* $\mathcal{O}_{sp}(P) = \{\alpha \mid P \xrightarrow{(\alpha,\alpha)}\}$.

As shown in [50] the observable input-output behaviour of deterministic $ntcc$ processes (i.e., tcc processes) can be compositionally specified as closure operators over sequences of constraints much like for the deterministic ccp case. Also, by building on the strongest-postcondition semantics for ccp in [20], the work in [50] includes a compositional characterization of the quiescent behaviour of $ntcc$ processes as well as a proof system for their temporal properties. The $ntcc$ proof system is similar to Dijkstra's proof system for the strongest postcondition of imperative programs.

In [49] the authors provided a hierarchy of $ntcc$ variants based on the input-output behaviour. A variant C is said to be *as expressive as* a variant C' if for every process P in C', one can compute a process $E(P)$ in C such that $\mathcal{O}_{io}(P) = \mathcal{O}_{io}(E(P))$. The variants were obtained by replacing replication with alternative mechanisms to specify infinite behaviour: Namely, procedure definitions, static-scoping parameterless recursion, and dynamic-scoping parameterless recursion. It was shown that $ntcc$ is equally expressive to the variant with static-scoping parameterless recursion. These variants were also shown to be strictly less expressive than the variant with parametric procedures which in turn was shown to be equally expressive to the variant with dynamic-scoping parameterless recursion. The authors also showed that the input-output behavior of every $ntcc$ processes is *omega-regular*; i.e. it can be specified by a finite-state Büchi automaton [9].

In [37] it is defined a framework for the declarative debugging of $ntcc$ programs, which is based on a fixpoint semantics for this language. A general framework, parametric w.r.t an abstract domain, for the static analysis of tcc programs is provided in [38].

5   Another Timed *ccp* Language

A different timed extension of *ccp*, called *tccp*, was proposed in [21]. Similarly to the previously mentioned timed languages (*tcc*) [54] and *default tcc* [55], *tccp* is a language for reactive programming where computation takes a bounded period of time rather than being instantaneous (as it is in ESTEREL [2]). However, differently from *tcc* and *default tcc*, which are inspired by the deterministic synchronous languages, *tccp* follows the guidelines of the timed process algebras approach and allows for non-determinism. This corresponds to a different view and use of a timed language: deterministic languages can be used for programming "kernels" of real-time systems, since deterministic systems are simpler to specify, debug and analyze. However, non-determinism arises when considering larger reactive systems involving several processes running on different processors and communicating via asynchronous links. These (timed) systems can be naturally specified and programmed by using a non-deterministic language.

Indeed all the existing timed process algebras and almost all the variants of Statecharts admit non-determinism.

Notice that the *ntcc* calculus discussed in the previous section, is also a non-deterministic timed ccp language. However, *ntcc* is an orthogonal non-deterministic extension of *tcc*, while *tccp* is an orthogonal timed nondeterministic extension of *ccp*. That means that, unlike in *tccp*, in *ntcc* computation proceeds as in the synchronous languages.

Below we first describe the *tccp* language and its operational semantics. Then we define a fix-point semantics for it which is based on reactive sequences and which is fully abstract w.r.t. the input/output notion of observables. All the technical definitions and results in this section are from [21].

5.1   Syntax and Operational Semantics of *tccp*

When querying the store for some information which is not present (yet) a *ccp* agent will simply suspend until the required information has arrived. In many applications involving time, however, often one cannot wait indefinitely for an event. Consider for example the case of a bank teller machine: if there is a problem with the authorization of the bank, after a reasonable amount of time the card should be given back to the customer. In order to model such a situation then the language should allow us to specify that, in case a given time bound is exceeded (i.e. a time-out occurs), the wait is interrupted and an alternative action is taken. Moreover, in some cases it is also necessary to abort an active process A and to start a process B when a specific event occurs (this is usually called *preemption* of A). For example, according to a typical pattern, A is the process controlling the normal activity of some physical device, the event indicates some abnormal situation and B is the exception handler.

In order to enrich *ccp* agents with such timing mechanisms, we introduce a *discrete global clock* and assume that *ask and tell actions take one time-unit*. Computation evolves in steps of one time-unit, so called clock-cycles, and action prefixing is the syntactic marker which distinguishes a time instant from the next one.

Furthermore, we make the assumption that parallel processes are executed on different processors, which implies that at each moment every enabled agent of the system is activated. This assumption, which is common to many timed process algebras, gives rise to what is called *maximal parallelism*.

Since the store is monotonically increasing and one can have dynamic process creation, clearly the previous assumptions in principle imply that the constraint solver takes a constant time (no matter how big the store is) and that there is an unbound number of processors. In practice, however, one can impose suitable restrictions on programs, thus ensuring that the (significant part of the) store and the number of processes do not exceed a fixed bound.

In order to express *time-out* and *preemption* which, as previously mentioned, are essential to many applications, the language is enriched by introducing a more basic timing construct of the form

$$now \ c \ then \ A \ else \ B.$$

This construct is similar to the analogous one used in [54], even though here it has a different interpretation: If c is entailed by the store then the above agent behaves as A at the *current* time instant, otherwise it behaves as B (at the current time instant). Note that the ability to detect the *absence* of an event is essential here.

Thus, we end up with the following syntax.

Definition 6 (*tccp* Language). *Assuming a given cylindric constraint system* $\mathbf{C}$ *the syntax of agents is given by the following grammar:*

$$A ::= stop \mid tell(c) \mid \sum_{i=1}^{n} ask(c_i) \rightarrow A_i \mid now \ c \ then \ A \ else \ B \mid A \parallel B \mid \exists x \ A \mid p(x)$$

where the c, c_i *are supposed to be finite constraints (i.e. algebraic elements) in* $\mathcal{C}$. *A tccp process P is then an object of the form $D.A$, where D is a set of procedure declarations of the form $p(x) : -A$ and A is an agent.*

In order to simplify the notation, in the following we will omit the $\sum_{i=1}^{n}$ whenever $n = 1$ and we will use $tell(c) \rightarrow A$ as a shorthand for $tell(c) \parallel (ask(true) \rightarrow A)$.

The operational model of *tccp* can be formally described by a transition system $T = (Conf, \longrightarrow)$ where we assume that each transition step takes exactly one time-unit. Configurations (in) *Conf* are pairs consisting of an agent and a constraint in $\mathcal{C}$ representing the common *store*. The transition relation $\longrightarrow \subseteq Conf \times Conf$ is the least relation satisfying the rules **R1, R2, R4 and R5** in Table 1 plus the rules in Table 3.

Notice that the rules now characterizes also the temporal evolution of the system, so $\langle A, c \rangle \longrightarrow \langle B, d \rangle$ means that if at time t we have the agent A and the store c then at time $t + 1$ we have the agent B and the store d.

In particular, Rule **R1** (in Table 1) shows that the evaluation of a tell action takes one time-unit, thus the updated store $c \sqcup d$ will be visible only starting from the next time instant. Analogously, also the evaluation of an ask action takes one time-unit (rule **R2**).

Let us now briefly discuss the new rules in Table 3.

Rules **R3bis** and **R3ter**, which replace rule **R3** of Table 1, model the parallel composition operator in terms of *maximal parallelism*: The agent $A \parallel B$ executes in one time-unit all the initial enabled actions of A and B.

The rules **R9-R12** show that the agent *now c then A else B* behaves as A or B depending on the fact that c is or is not entailed by the store. Note that here, differently from the case of the ask, the evaluation of the guard is instantaneous. Since A and B could contain nested *now then else* agents, a limit for the number of these nested agents should be fixed. However, for recursive programs such a limit is ensured by the presence of the procedure call, since we assume that the evaluation of such a call takes one time unit.

Using the transition system described by (the rules in) Table 1 we can define the following notion of observables which considers the input/output of terminating computations, including the deadlocked ones. Here and in the sequel $\longrightarrow^*$ denotes the reflexive and transitive closure of the relation $\longrightarrow$.

Definition 7 (Observables). *Let A be an agent. We define $\mathcal{O}_{io}(A) = \{\langle c, d \rangle \mid \langle A, c \rangle \longrightarrow^* \langle B, d \rangle \not\rightarrow \}$.*

Table 3. The additional rules for *tccp*

R3bis	$\dfrac{\langle A, c \rangle \longrightarrow \langle A', c' \rangle \quad \langle B, c \rangle \longrightarrow \langle B', d' \rangle}{\langle A \parallel B, c \rangle \longrightarrow \langle A' \parallel B', c' \sqcup d' \rangle}$
R3ter	$\dfrac{\langle A, c \rangle \longrightarrow \langle A', c' \rangle \quad \langle B, c \rangle \longarrownot\longrightarrow}{\begin{array}{c}\langle A \parallel B, c \rangle \longrightarrow \langle A' \parallel B, c' \rangle \\ \langle B \parallel A, c \rangle \longrightarrow \langle B \parallel A', c' \rangle\end{array}}$
R9	$\dfrac{\langle A, d \rangle \longrightarrow \langle A', d' \rangle}{\langle now\ c\ then\ A\ else\ B, d \rangle \longrightarrow \langle A', d' \rangle} \quad d \vdash c$
R10	$\dfrac{\langle A, d \rangle \longarrownot\longrightarrow}{\langle now\ c\ then\ A\ else\ B, d \rangle \longrightarrow \langle A, d \rangle} \quad d \vdash c$
R11	$\dfrac{\langle B, d \rangle \longrightarrow \langle B', d' \rangle}{\langle now\ c\ then\ A\ else\ B, d \rangle \longrightarrow \langle B', d' \rangle} \quad d \not\vdash c$
R12	$\dfrac{\langle B, d \rangle \longarrownot\longrightarrow}{\langle now\ c\ then\ A\ else\ B, d \rangle \longrightarrow \langle B, d \rangle} \quad d \not\vdash c$

5.2 Programming Example

We show now how some typical reactive programming idioms can be derived from the basic combinators of *tccp*. Then we use these in a programming example.

Time-out. The timed guarded choice agent

$$\sum_{i=1}^{n} ask(c_i) \rightarrow A_i\ time\text{-}out(m)\ B$$

waits at most m time-units ($m \geq 0$) for the satisfaction of one of the guards. Before this time-out the process behaves just like the guarded choice: As soon as there exist enabled guards, one of them and the corresponding branch is nondeterministically selected. After waiting for m time-units, if no guard is enabled, the timed choice agent behaves as B. This agent can be defined inductively as follows. Let us denote by A the agent $\sum_{i=1}^{n} ask(c_i) \rightarrow A_i$. In the base case, $m = 0$, we define $\sum_{i=1}^{n} ask(c_i) \rightarrow A_i\ time\text{-}out(0)\ B$ as the agent

$$now\ c_1\ then\ A\ else$$
$$(\ now\ c_2\ then\ A\ else$$
$$\vdots$$
$$(\ now\ c_n\ then\ A\ else\ ask(true) \rightarrow B)\ldots)$$

For the inductive step we define $\sum_{i=1}^{n} ask(c_i) \rightarrow A_i\ time\text{-}out(m)\ B$ as

$$\sum_{i=1}^{n} ask(c_i) \rightarrow A_i\ time\text{-}out(0) \left(\sum_{i=1}^{n} ask(c_i) \rightarrow A_i\ time\text{-}out(m\text{-}1)\ B \right).$$

Watchdogs. These are typical preemption primitives of such languages as ESTEREL and are used to interrupt the activity of a process on signal from a specific event. Since events are expressed by constraints, a watchdog can be defined as the process

$$do\ A\ watching\ c\ else\ B$$

which behaves as A, as long as c is not entailed by the store; when c is entailed, the process A is immediately aborted and process B is started. We have here a form of weak preemption in which the abortion of A is performed in the next time interval. In fact, even though A is aborted at the *same* time instant of the detection of the entailment of c, if c is detected at time t then c has to be produced at time t' with $t' < t$.

Previous watchdog agent can be defined (by induction on the structure of process A) in terms of the other constructs of the language (see [21]). For example in case of the *tell* process one has the following translation

$$do\ tell(d)\ watching\ c\ else\ B\ \Rightarrow\ now\ c\ then\ B\ else\ tell(d),$$

As a simple example of a *tccp* program let us now consider a system $s(Ex)$ consisting of two processes $p1$ and $p2$ which perform some time critical activities, reacting to external inputs transmitted on the channel Ex. The system is continuously checked by a controller which receives a stream of *ok* messages by each process pi. Each *ok* message is sent at unpredictable time instants, however it is assumed that each pi is working correctly iff it sends the next ok within n time-units from the previous one. When this limit is exceeded by a process pi the controller aborts the whole system, starts a recovery routine rr for the activity of pi and then restart the system. The system $s(Ex)$ is implemented by the following program where the specific tasks of the pi's and of the recovery routines are not specificed:

$s(Ex):$- ∃ $Alarm,O1,R1,O2,R2$
       $((do\ p1(Ex,O1,R1)\ \|\ p2(Ex,O2,R2)\ watching\ Alarm = on)$
       $\|\ controller(O1,O2,R1,R2))$

$controller(O1,O2,R1,R2):$- ∃ $A1,A2$
              $(do\ c(O1,A1)\ \|\ c(O2,A2)\ watching\ Alarm = on\ else$
              $(now\ (A1 = on \sqcup A2 = on)\ then\ rr(R1)\ \|\ rr(R2)\ else$
              $now\ A1 = on\ then\ rr(R1)\ \ else$
              $now\ A2 = on\ then\ rr(R2))$
              $\|\ restart(Ex))$

$c(O,A):$- ask (∃ $Y.O=[ok|\,Y]) \rightarrow$ (∃ $Y\ tell(O=[ok|\,Y]) \rightarrow c(Y,A))$
       $timeout(n)\ tell(Alarm = on \sqcup A = on)$

5.3   The Denotational Model

It is easy to see that the operational semantics which associates to an agent A its observables $\mathcal{O}_{io}(A)$ is not compositional. A compositional characterization of the operational semantics can be obtained by using sequences of pairs of finite constraints, so called *timed reactive sequences*, analogous to those that we have seen in the semantics of *ccp*.

However, a reactive sequence is now provided with a different interpretation which accounts for the timing aspects. In fact such a sequence has the form

$$\langle c_1, d_1 \rangle \cdots \langle c_n, d_n \rangle \langle d, d \rangle$$

and each pair of constraints $\langle c_i, d_i \rangle$ now represents a computation step performed by the agent A which, at time i, assuming c_i as input constraint produces the constraint d_i. The last pair denotes a "stuttering step" in which no further information can be produced by the agent, thus indicating that a "resting point" has been reached.

Since in *tccp* computations the store evolves monotonically and the constraints arising from computation steps are finite, it is natural to assume that reactive sequences are monotonically increasing and contains only finite constraints. The set of all reactive sequences is denoted by $\mathcal{S}$ and its typical elements by $s, s_1 \ldots$, while sets of reactive sequences are denoted by $S, S_1 \ldots$ and ε indicates the empty reactive sequence. The semantics R which associates to an agent the reactive sequences that it generates can be defined by a fixpoint construction as follows.

Definition 8. *The semantics $R \in Agent \to \mathcal{P}(\mathcal{S})$ is defined as the least fixed-point of the operator $\Phi \in (Agent \to \mathcal{P}(\mathcal{S})) \to Agent \to \mathcal{P}(\mathcal{S})$ defined by*

$$\Phi(I)(A) = \{\langle c, d \rangle \cdot w \in \mathcal{S} \mid c \in \mathcal{C}, \langle A, c \rangle \to \langle B, d \rangle \text{ and } w \in I(B)\}$$
$$\cup$$
$$\{\langle c, c \rangle \cdot w \in \mathcal{S} \mid \langle A, c \rangle \not\to \text{ and } w \in I(A) \cup \{\varepsilon\}\}.$$

The ordering on $Agent \to \mathcal{P}(\mathcal{S})$ is that of (point-wise extended) set-inclusion and it is straightforward to check that Φ is continuous, so standard results allows us to construct the least fixpoint in ω steps.

It is possible to show that the above semantics is correct (w.r.t. the input/ouput observables) and compositional, however is not fully abstract, since it distinguishes *tccp* agents whose observables are the same under any possible context. In order to obtain a fully abstract model one needs to introduce a suitable abstraction on traces, however, due to the presence of the *now then else* construct and of maximal parallelism, one cannot use here the abstraction which has been used in [24] for *ccp* since this would be incorrect (it would identify agents which can be distinguished by a context). This semantic difference has also an expresiveness counterpart, indeed one can show [21] that *tccp* is strictly more expressive than. *ccp*.

So, the full abstraction problem for *tccp* cannot be reduced to that one for *ccp*. Indeed, differently from the case of ccp, the definition of a fully abstract semantics for *tccp* requires the ability to specify the "difference" $c_i \setminus d_{i-1}$ between an assumption c_i (at time i) and the previous contribution d_{i-1} (at time

$i - 1$). Such a difference is formalized by using the algebraic notion of weak relative pseudo-complement [43,4]. Using this difference the abstraction α on set of sequences can be defined as follows.

Definition 9 (Abstraction). *Let s, s' be reactive sequences. Then the $\preceq$ relation is defined as follows:*

- *$s \preceq s'$ iff for some sequences s_1 and s_2 one has that $s = s_1 \cdot \langle a, b \rangle \langle c, d \rangle \cdot s_2$, $s' = s_1 \cdot \langle a, b' \rangle \langle c, d \rangle \cdot s_2$ and $(c \setminus b') \leq (c \setminus b)$.*

Moreover the (equivalence) relation $\simeq$ is defined as follows

- *$s \simeq s'$ iff the sequences s and s' differ only in the number of repetitions of the last element.*

Given a set of reactive sequences S, $\alpha(S)$ denotes the least set S' such that the following holds:

(i) $S \subseteq S'$,
(ii) *if $s' \in S'$ and either $s \preceq s'$ or $s \simeq s'$, then $s \in S'$.*

The fully abstract semantics R_α is obtained by simply applying the function α to $R(A)$. One can show that the semantics obtained in this way is compositional (w.r.t. all the operators of the language) and correct (since it allows to reconstruct the observables $O_{io}(A)$). Moreover it is also fully abstract, as shown by the following theorem.

Theorem 1 (Full abstraction). *Assume that the constraint system is weakly relative pseudo-complemented. Then, for any pair of tccp agents A and B, $\alpha(R(A)) = \alpha(R(B))$ iff $\mathcal{O}_{io}(C[A]) = \mathcal{O}_{io}(C[B])$ for each context $C[\cdot]$.*

FInally it is worth noting that a temporal logic for reasoning on *tccp* programs, inpired by this semantics, has been defined in [22].

6   Other Extensions of *ccp*

In this section we survey some more recent extensions of *ccp* which mainly deal with probabilistic and uncertainty aspects.

6.1   Probabilistic *ccp*

In [27] the concurrent constraint programming paradigm is extended with a probabilistic choice construct which replaces the nondeterministic choice of the original paradigm; this allows a program to make stochastic moves during its execution, so that it may be seen as a stochastic process. This embedding of randomness within the semantics of a well structured programming paradigm, like *ccp*, also aims at providing a sound framework for formalising and reasoning about randomised algorithms and programs. For the resulting language called probabilistic *ccp*, a fixpoint semantics is given in [26,28], which is based on vector spaces and the Brouwer's fixpoint theorem. The addition of probabilities allows for a natural formulation of the average behaviour of a program, whose specification and analysis is particularly important in the study of system performance and reliability. It also allows for an average-case analysis of programs as opposite to the worst case analysis common to the classical static analysis approaches [30].

Concurrent Constraint Programming has been used as a reference programming paradigm for the introduction of a general theory of probabilistic abstract interpretation, which re-formulates the classical theory of abstract interpretation in a setting suitable for a quantitative reasoning about programs. In this setting, linear spaces replace the classical order-theoretic domains, and the notion of the so-called *Moore-Penrose pseudo-inverse* of a linear operator replaces the classical notion of a Galois connection. The resulting abstractions turn out to be *close* approximations of the concrete semantics, so that closeness becomes a quantitative replacement for classical safety [29].

6.2   *ccp* for Service Level Agreement

Service Oriented Computing is an emerging paradigm that builds upon the notion of services as interoperable elements that can be described, published, searched and composed. Services may expose both functional properties (i.e. what they do) and non-functional properties (i.e. the way they are supplied). A Service Level Agreement (SLA) is a contract between two parties, usually a service provider and a customer, that records non-functional properties about a service like performance, availability, and cost.

Recently several extensions of the pure *ccp* language have been proposed for dealing with Service Level Agreement aspects. Here we briefly describe the main proposals in this area.

The concurrent constraint pi-calculus (cc-pi calculus) [12] is a model of Service Level Agreement negotiations that is inspired by both *ccp* and name-passing calculi. Specifically, the cc-pi calculus combines basic operations of concurrent constraint programming, such as *ask* and *atomic tell*, with a symmetric, synchronous mechanism of interaction between senders and receivers, where the sent name is 'fused' (i.e. identified) to the received name and such an *explicit fusion* enables using interchangeably the two names. The cc-pi calculus is parametric with respect to the choice of an underlying constraint system that is defined using a suitable semiring structure, equipped with a notion of names. Moreover, cc-pi includes a restriction operation that allows for local stores of constraints. Synchronisations of interacting processes may have the effect of combining local into global stores.

Some semantic aspects of the cc-pi calculus are studied in [13], where its is defined a notion of open bisimilarity *à la* pi-calculus for cc-pi. Essentially, two processes are open bisimilar if they have the same stores of constraints - which can be statically checked - and if their moves can be mutually simulated. In [13] it is also shown that the polyadic Explicit Fusion calculus introduced by

Gardner and Wischik can be translated into monadic cc-pi and such a transition preserves open bisimilarity.

In [11] a further extension of the cc-pi calculus is defined by including primitives for distributed nested commits, inspired by the cjoin calculus (introduced by Bruni, Melgratti, and Montanari). The two key operations of cjoin are: the 'abort with compensation', to stop a negotiation and activate a compensating process, and the 'commit', to store a partial agreement among the parties before moving to the next negotiation phase. This extended cc-pi calculus comes equipped with both a small- and a big-step operational semantics which are proved to coincide.

A different line of research is focused on the use of, so called soft constraint, to model qualitative aspects of Service Level Agreement in the context of the *ccp* paradigm. As described in more detail in another chapter of this book [41], soft constraints extend classical constraints to represent multiple consistency levels, and thus provide a way to express preferences, fuzziness, and uncertainty. An extension of the *ccp* framework which allows soft constraints in the calculus has been proposed in [6]. In this extension it is permitted to add (tell) or check (ask) for soft constraints and the language is enriched with tell/ask thresholds which can express the level of consistency of the store, thus allowing to prune and direct the search for a solution (when some consistency levels are not satisfied). The resulting language, called soft cc (scc), can be also very useful in many web-related scenarios, since allows web agents to express their interaction and negotiation protocols, and also to post their requests in terms of preferences. Differently from the case of "hard" (or "crisp") constraints, the underlying soft constraint solver here can find an agreement among the agents even if their requests are incompatible.

A timed extension of scc has been proposed in [5] in order to be able to express also Quality of Service aspects which involve time. As in the case of scc, tell and ask agents are equipped with a preference (or consistency) threshold which is used to determine their success or suspension. The time and the semantic model of this extension follows the lines of the tccp language presented in Section 5.

Another extension of scc, which allows the nonmonotonic evolution of the constraint store, is defined in [7]. To accomplish this, some new operations are introduced: the retract(c) reduces the current store by c; the updateX(c) transactionally relaxes all the constraints of the store that deal with the variables in X set, and then adds a constraint c; the nask(c) tests if c is not entailed by the store. This language allows the management of resources that need a given Quality of Service: the requirements of all the parties should converge, through a negotiation process (which involves retract of information), on a formal agreement defined as the Service Level Agreement, which specifies the contract that must be enforced.

7   Some Working ccp Systems

In this section we shall briefly survey some existing working ccp systems.

The programming language *jcc* [56] was designed as an integration of *default tcc* into Java and is intended for embedded reactive systems. In jcc users can define their own constraint system and thus specialize the language to particular domains. The main purpose of *jcc* is to provide a model of loosely-coupled concurrent programming in Java. The language introduces the notion of a *vat*. A *vat* can be thought of as encapsulating a single synchronous, reactive *tcc* computation. A computation consists of a dynamically changing collection of interacting *vats*, communicating with each other through shared, mutable objects called *ports*. Asking and telling objects can read from and write into the port, respectively and the temporal constructs from the underlying *tcc* model allow an object to specify code whose execution should be delayed.

In the *hybrid concurrent constraint programming* language, *hcc* [44], it is possible to express discrete and continuous evolution of time. More precisely, there are points at which discontinuous change may occur (i.e. the execution proceeds as burst of activity) and open intervals in which the state of the system changes continuously (i.e. the system evolves continuously and autonomously). The notion of *continuous constraint system* (a real-time extension of constraint systems) is introduced to describe the continuous evolution. The syntax of *hcc* extends that of *tcc* with the construct *hence P*, asserting that *P* holds continuously beyond the current instant. An interpreter of *hcc* can be found at http://www-cs-students.stanford.edu/~vgupta/hcc/hcc.html

NtccSim is a simulation tool developed in Oz for *ntcc*, one of the temporal models previously described . Constraints over finite domains and real intervals have been used to implement models of biological systems. *NtccSim* can be found at http://cic.javerianacali.edu.co/wiki/doku.php?id=grupos:avispa:ntccsim. An implementation of the other temporal model previously described, *tccp*, can be found at http://users.dsic.upv.es/~villanue/tccpInterpreter

The LMNTAL model [65] provides a scalable, uniform view of concurrent programming concepts such as processes, messages, synchronous and asynchronous computation. It inherits ideas from the concurrent constraint language GHC and from Janus. Communication is based on constraints over logical variables. Processes sharing variables are thought of as been connected. Multisets of nested nodes and links are a first-class notion in LMNtal. Transformations are rules, much like in Janus. LMNtal provides both channel mobility and process mobility: it allows dynamic reconfiguration of process structures as well as the migration of nested computations. An implementation of LMNtal can be found at http://www.ueda.info.waseda.ac.jp/lmntal/

CORDIAL [51] is a visual language intended as a user transparent integration of constraints and objects. The language is based on a ccp calculus extended with the notion of objects and classes. Methods are represented as windows. Objects within methods are represented by closed contours. Object methods launch ccp processes that, in addition to the usual ask and tell operations, can send messages to other objects. Messages are objects connected by links to object mailboxes. Objects are identified by an associated constraint parametrized on a local variable (so-called *self*). Senders willing to invoke some object method post a constraint involving some variable, say X, and then send the message to X. Any object such that its associated constraint can be entailed by the store conjoined with the constraint *self* $= X$, is eligible to accept the message. Some eligible object is then non-deterministically chosen to handle the message. This scheme allows very complex patterns of communication and mobility.

References

1. Bergstra, J., Klop, J.: Process algebra: specification and verification in bisimulation semantics. In: Mathematics and Computer Science II. CWI Monographs, pp. 61–94. North-Holland, Amsterdam (1986)
2. Berry, G., Gonthier, G.: The esterel synchronous programming language: Design, semantics, implementation. Sci. Comput. Program. 19(2), 87–152 (1992)
3. Bertolino, M., Etalle, S., Palamidessi, C.: The replacement operation for CCP programs. In: Bossi, A. (ed.) LOPSTR 1999. LNCS, vol. 1817, pp. 216–233. Springer, Heidelberg (2000)
4. Birkhoff, G.: Lattice theory, XXV. AMS Colloquium Publications (1967)
5. Bistarelli, S., Gabbrielli, M., Meo, M.C., Santini, F.: Timed soft concurrent constraint programs. In: Lea, D., Zavattaro, G. (eds.) COORDINATION 2008. LNCS, vol. 5052, pp. 50–66. Springer, Heidelberg (2008)
6. Bistarelli, S., Montanari, U., Rossi, F.: Soft concurrent constraint programming. ACM Trans. Comput. Log. 7(3), 563–589 (2006)
7. Bistarelli, S., Santini, F.: A nonmonotonic soft concurrent constraint language for sla negotiation. Electr. Notes Theor. Comput. Sci. 236, 147–162 (2009)
8. Brookes, S., Hoare, C., Roscoe, W.: A theory of communicating sequential processes. Journal of ACM 31, 499–560 (1984)
9. Buchi, J.R.: On a decision method in restricted second order arithmetic. In: Proc. Int. Cong. on Logic, Methodology, and Philosophy of Science, pp. 1–11. Stanford University Press (1962)
10. Burstall, R.M., Darlington, J.: A transformation system for developing recursive programs. J. ACM 24(1), 44–67 (1977)
11. Buscemi, M., Melgratti, H.: Transactional service level agreement. In: Barthe, G., Fournet, C. (eds.) TGC 2007. LNCS, vol. 4912, pp. 124–139. Springer, Heidelberg (2008)
12. Buscemi, M., Montanari, U.: Cc-pi: A constraint-based language for specifying service level agreements. In: De Nicola, R. (ed.) ESOP 2007. LNCS, vol. 4421, pp. 18–32. Springer, Heidelberg (2007)
13. Buscemi, M., Montanari, U.: Open bisimulation for the concurrent constraint picalculus. In: Drossopoulou, S. (ed.) ESOP 2008. LNCS, vol. 4960, pp. 254–268. Springer, Heidelberg (2008)
14. Clark, K., Gregory, S.: PARLOG: parallel programming in logic. ACM Trans. on Programming Languages and Systems 8(1), 1–49 (1986)
15. de Boer, F., Kok, J., Palamidessi, C., Rutten, J.: Control flow versus logic: a denotational and a declarative model for Guarded Horn Clauses. In: Kreczmar, A., Mirkowska, G. (eds.) MFCS 1989. LNCS, vol. 379, pp. 165–176. Springer, Heidelberg (1989)

16. de Boer, F., Kok, J., Palamidessi, C., Rutten, J.: Semantic models for a version of PARLOG. In: Levi, G., Martelli, M. (eds.) Proc. of the Sixth International Conference on Logic Programming, Lisboa. Series in Logic Programming, pp. 621–636. MIT Press, Cambridge (1989); Extended version in [17]

17. de Boer, F., Kok, J., Palamidessi, C., Rutten, J.: Semantic models for Concurrent Logic Languages. Theoretical Computer Science 86(1), 3–33 (1991); A short version appeared on Proceedings of the Seventh International Conference on Logic Programming, Lisboa (1989)

18. de Boer, F., Palamidessi, C.: A Fully Abstract Model for Concurrent Constraint Programming. In: Abramsky, S., Maibaum, T. (eds.) CAAP 1991 and TAPSOFT 1991. LNCS, vol. 493, pp. 296–319. Springer, Heidelberg (1991)

19. de Boer, F., Palamidessi, C.: On the semantics of concurrent constraint programming. In: Broda, K. (ed.) Proc. of ALPUK 1992, Workshops in Computing, pp. 145–173. Springer, Heidelberg (1992)

20. de Boer, F.S., Gabbrielli, M., Marchiori, E., Palamidessi, C.: Proving concurrent constraint programs correct. ACM Transactions on Programming Languages and Systems 19(5), 685–725 (1997)

21. de Boer, F.S., Gabbrielli, M., Meo, M.C.: A timed concurrent constraint language. Inf. Comput. 161(1), 45–83 (2000)

22. de Boer, F.S., Gabbrielli, M., Meo, M.C.: A temporal logic for reasoning about timed concurrent constraint programs. In: TIME, pp. 227–233 (2001)

23. de Boer, F.S., Gabbrielli, M., Palamidessi, C.: Proving correctness of constraint logic programs with dynamic scheduling. In: Cousot, R., Schmidt, D.A. (eds.) SAS 1996. LNCS, vol. 1145, pp. 83–97. Springer, Heidelberg (1996)

24. de Boer, F.S., Palamidessi, C.: On the asynchronous nature of communication in concurrent logic languages: A fully abstract model based on sequences. In: Baeten, J.C.M., Klop, J.W. (eds.) CONCUR 1990. LNCS, vol. 458, pp. 99–114. Springer, Heidelberg (1990)

25. de Boer, F.S., Pierro, A.D., Palamidessi, C.: Nondeterminism and infinite computations in constraint programming. Theoretical Computer Science 151(1), 37–78 (1995)

26. Di Pierro, A., Wiklicky, H.: A Banach Space Based Semantics for Probabilistic Concurrent Constraint Programming. In: Lin, X. (ed.) Proc. 4th Australasian Theory Symposium, CATS 1998, Singapore. Australian Computer Science Communications, vol. 20 – 3, pp. 245–259. Springer, Heidelberg (1998)

27. Di Pierro, A., Wiklicky, H.: An Operational Semantics for Probabilistic Concurrent Constraint Programming. In: Iyer, Y.C.P., Schmidt, D. (eds.) Proc. ICCL 1998 – International Conference on Computer Languages, Chicago. IEEE Computer Society and ACM SIGPLAN, pp. 174–183. IEEE Computer Society Press, Los Alamitos (1998)

28. Di Pierro, A., Wiklicky, H.: Probabilistic Concurrent Constraint Programming: Towards a Fully Abstract Model. In: Brim, L., Gruska, J., Zlatuška, J. (eds.) MFCS 1998. LNCS, vol. 1450, p. 446. Springer, Heidelberg (1998)

29. Di Pierro, A., Wiklicky, H.: Concurrent Constraint Programming: Towards Probabilistic Abstract Interpretation. In: Gabbrielli, M., Pfenning, F. (eds.) Proceedings of PPDP 2000 – Priciples and Practice of Declarative Programming, Montréal, Canada, September 2000. ACM SIGPLAN, pp. 127–138. Association of Computing Machinery, New York (2000)

30. Di Pierro, A., Wiklicky, H.: Quantitative observables and averages in Probabilistic Concurrent Constraint Programming. In: Apt, K.R., Kakas, A.C., Monfroy, E.,

Rossi, F. (eds.) Compulog Net WS 1999. LNCS (LNAI), vol. 1865, pp. 212–236. Springer, Heidelberg (2000)

31. Dovier, A., Pontelli, E. (eds.): 25 Years of Logic Programming in Italy. LNCS, vol. 6125. Springer, Heidelberg (2010)

32. Etalle, S., Gabbrielli, M., Meo, M.C.: Transformations of ccp programs. ACM Trans. Program. Lang. Syst. 23(3), 304–395 (2001)

33. Falaschi, M., Gabbrielli, M., Marriott, K., Palamidessi, C.: Compositional analysis for concurrent constraint programming. In: Proc. of the Eight Annual IEEE Symposium on Logic in Computer Science, pp. 210–221. IEEE Computer Society Press, Los Alamitos (1993)

34. Falaschi, M., Gabbrielli, M., Marriott, K., Palamidessi, C.: Confluence in concurrent constraint programming. Theoretical Computer Science 183(2), 281–315 (1997)

35. Falaschi, M., Gabbrielli, M., Marriott, K., Palamidessi, C.: Confluence in concurrent constraint programming. Theoretical Computer Science 183(2), 281–315 (1997)

36. Falaschi, M., Gabbrielli, M., Marriott, K., Palamidessi, C.: Constraint Logic Programming with Dynamic Scheduling: A Semantics Based on Closure Operators. Information and Computation 137(1), 41–67 (1997)

37. Falaschi, M., Olarte, C., Palamidessi, C., Valencia, F.: Declarative diagnosis of temporal concurrent constraint programs. In: Dahl, V., Niemelä, I. (eds.) ICLP 2007. LNCS, vol. 4670, pp. 271–285. Springer, Heidelberg (2007)

38. Falaschi, M., Olarte, C., Valencia, F.: A framework for abstract interpretation of timed concurrent constraint programs. In: Proc. of PPDP 2009, ACM Sigplan, pp. 107–118 (2009)

39. Francesco, N.D., Santone, A.: Unfold/fold transformations of concurrent processes. In: Kuchen, H., Swierstra, S.D. (eds.) PLILP 1996. LNCS, vol. 1140, pp. 167–181. Springer, Heidelberg (1996)

40. Gabbrielli, M., Levi, G.: Unfolding and fixpoint semantics for concurrent constraint logic programs. In: Kirchner, H., Wechler, W. (eds.) ALP 1990. LNCS, vol. 463, pp. 204–216. Springer, Heidelberg (1990)

41. Gavanelli, M., Rossi, F.: Constraint Logic Programming. In: Dovier, A., Pontelli, E. (eds.) 25 Years of Logic Programming in Italy, ch. 4. LNCS, pp. 64–85. Springer, Heidelberg (2010)

42. Giacobazzi, R., Debray, S.K., Levi, G.: Generalized semantics and abstract interpretation for constraint logic programs. Journal of Logic Programming 25(3), 191–247 (1995)

43. Giacobazzi, R., Palamidessi, C., Ranzato, F.: Weak relative pseudo-complements of closure operators. Algebra Universalis 36(3), 405–412 (1996)

44. Gupta, V., Jagadeesan, R., Saraswat, V.: Computing with continuous change. Science of Computer Programming 30(1-2), 3–49 (1998)

45. Henkin, L., Monk, J., Tarski, A.: Cylindric Algebras (Part I). North-Holland, Amsterdam (1971)

46. Jaffar, J., Lassez, J.-L.: Constraint logic programming. In: Proc. of ACM Symposium on Principles of Programming Languages, pp. 111–119. ACM, New York (1987)

47. Maher, M.J.: Logic semantics for a class of committed-choice programs. In: Lassez, J.-L. (ed.) Proc. of the Fourth International Conference on Logic Programming, Melbourne. Series in Logic Programming, pp. 858–876. MIT Press, Cambridge (1987)

48. Milner, R.: A Calculus of Communication Systems. LNCS, vol. 92. Springer, Heidelberg (1980)
49. Nielsen, M., Palamidessi, C., Valencia, F.: On the expressive power of concurrent constraint programming languages. In: Proc. of PPDP 2002, pp. 156–167. ACM Press, New York (2002)
50. Nielsen, M., Palamidessi, C., Valencia, F.: Temporal concurrent constraint programming: Denotation, logic and applications. Nordic Journal of Computing 9(2), 145–188 (2002)
51. Rueda, C., Alvarez, G., Quesada, L., Tamura, G., Valencia, F., Diaz, J., Assayag, G.: Integrating constraints and concurrent objects in musical applications: A calculus and its visual language. Constraints 6(1) (2001)
52. Saraswat, V.: A somewhat logical formulation of CLP synchronization primitives. In: Kowalski, R.A., Bowen, K.A. (eds.) Proc. of the Fifth International Conference on Logic Programming, Seattle, USA. Series in Logic Programming, pp. 1298–1314. MIT Press, Cambridge (1988)
53. Saraswat, V.: Concurrent Constraint Programming. PhD thesis, Carnegie-Mellon University, January 1989. ACM distinguished dissertation series. The MIT Press, Cambridge (1993)
54. Saraswat, V., Jagadeesan, R., Gupta, V.: Foundations of timed concurrent constraint programming. In: LICS, pp. 71–80. IEEE Computer Society, Los Alamitos (1994)
55. Saraswat, V., Jagadeesan, R., Gupta, V.: Timed default concurrent constraint programming. J. Symb. Comput. 22(5/6), 475–520 (1996)
56. Saraswat, V., Jagadeesan, R., Gupta, V.: jcc: Integrating timed default concurrent constraint programming into java. In: Pires, F.M., Abreu, S.P. (eds.) EPIA 2003. LNCS (LNAI), vol. 2902, pp. 156–170. Springer, Heidelberg (2003)
57. Saraswat, V., Rinard, M.: Concurrent constraint programming. In: Proc. of the seventeenth ACM Symposium on Principles of Programming Languages, pp. 232–245. ACM, New York (1990)
58. Saraswat, V., Rinard, M., Panangaden, P.: Semantics foundations of Concurrent Constraint Programming. In: Proc. of the eighteenth ACM Symposium on Principles of Programming Languages. ACM, New York (1991)
59. Shapiro, E.: A subset of Concurrent Prolog and its interpreter. Technical Report TR-003, Institute for New Generation Computer Technology (ICOT), Tokyo (1983)
60. Shapiro, E.: Concurrent Prolog: A progress report. Computer 19(8), 44–58 (1986)
61. Tamaki, H., Sato, T.: Unfold/fold transformation of logic programs. In: ICLP, pp. 127–138 (1984)
62. Ueda, K.: Guarded Horn Clauses. In: Shapiro, E. (ed.) Concurrent Prolog: Collected Papers. Series in Logic Programming. MIT Press, Cambridge (1987)
63. Ueda, K.: Guarded Horn Clauses, a parallel logic programming language with the concept of a guard. In: Nivat, M., Fuchi, K. (eds.) Programming of Future Generation Computers, pp. 441–456. North Holland, Amsterdam (1988)
64. Ueda, K., Furukawa, K.: Transformation rules for ghc programs. In: FGCS, pp. 582–591 (1988)
65. Ueda, K., Kato, N., Hara, K., Mizuno, K.: LMNtal as a unifying declarative language: Live demonstration. In: Etalle, S., Truszczyński, M. (eds.) ICLP 2006. LNCS, vol. 4079, pp. 457–458. Springer, Heidelberg (2006)
66. Zaffanella, E.: Domain Independent Ask Approximation in CCP. In: Montanari, U., Rossi, F. (eds.) CP 1995. LNCS, vol. 976, pp. 362–379. Springer, Heidelberg (1995)
67. Zaffanella, E., Giacobazzi, R., Levi, G.: Abstracting synchronization in concurrent constraint programming. Journal of Functional and Logic Programming 1997(6) (November 1997)

Proof-Theoretic and Higher-Order Extensions of Logic Programming

Alberto Momigliano[1,2] and Mario Ornaghi[1]

[1] Dipartimento di Scienze dell'Informazione, Università degli Studi di Milano, Italy
{momiglia,ornaghi}@dsi.unimi.it
[2] Laboratory for the Foundations of Computer Science, School of Informatics,
The University of Edinburgh, Scotland

Abstract. We review the Italian contribution to proof-theoretic and higher-order extensions of logic programming; this originated from the realization that Horn clauses lacked standard abstraction mechanisms such as higher-order programming, scoping constructs and forms of information hiding. Those extensions were based on the Deduction and Computation paradigm as formulated in Miller et al's approach [51], which built logic programming around the notion of *focused uniform proofs* The Italian contribution has been both foundational and applicative, in terms of language extensions, implementation techniques and usage of the new features to capture various computation models. We argue that the emphasis has now moved to the theory and practice of logical frameworks, carrying with it a better understanding of the foundations of proof search.

1   Introduction and Motivation

We start by trying to clarify the reasons behind our choice, discussion and classification of the literature stemming from the Italian contribution to proof-theoretic and higher-order extensions of logic programming (LP). These papers belong to the multitude of proposals of extensions of the foundations of logic programming, i.e. Horn clauses (HC). We can trace that both to the purported limited expressibility of HC — see the thorny issue of a logically motivated notion of negation — and to the lack of abstraction mechanisms that are present in modern programming languages to support the modular construction of software. Here we are referring to higher-order programming, modules, abstract datatypes, scoping constructs, state encapsulation and other forms of information hiding. One can argue that from the very beginning this has led to the introduction of "impure", i.e. extra-logical, features, such as cut, negation-as-failure or assert/retract. This outcome is not specific to LP and has been named *"recreating the Turing Machine" syndrome* [48]: starting from a computationally clean and semantically motivated language, one tends to add external mechanisms in order to make it suitable for programming-in-the-large. This inevitably tends to clutter the formal definition of the language (if any), making trusting the language itself and thus reasoning about it problematic.

Hence the opposing trend in the literature to go back to the original setting and base new constructs on more solid theoretical grounds, in our case, logic. A well-known

A. Dovier, E. Pontelli (Eds.): 25 Years of Logic Programming, LNCS 6125, pp. 254–270, 2010.

(and somewhat worn out) example is again the logical foundations of negation. More in general, it is by now usual to contrast the traditional model-theoretic approach (see Chapter [11] in this volume) to the proof-theoretic one, which "happens" to be at the core of most of the work about higher-order extensions of logic programming.

Arguably, many theoretical developments in logic have had an important impact in Computer Science. Concerning proof theory, we can isolate two different research directions, broadly corresponding to two different paradigms: Proofs as Programs and Deductions as Computations (DAC). In the Proof as Programs setting, proofs can be seen as programs (a.k.a. λ-terms), while computations correspond to (β)-reductions in a λ-calculus. The proof-theoretic basis is the normalisation or cut-elimination procedure. This approach fits with the foundations of functional programming, as well as of constructive program synthesis. In DAC, proofs themselves become the computations, while programs are specifications of non-logical symbols within the logic. Here, cut-elimination is the conditio sine qua non and proof-theory offers sophisticated restrictions to proof search in a cut-free system, while preserving completeness: a computation is modeled as a search for a proof, under suitable "uniformity" assumptions [51]. LP naturally falls in the DAC approach, which has been eloquently argued as one of its possible logical foundations elsewhere, e.g. [59].

In DAC, we distinguish between a non-logical signature, related to the problem domain, and the domain independent logical language. Each extension of the logical language has a corresponding extension of the proof system, bringing at the level of logic aspects that pertain to the computational level and allowing us to reason about them logically. A paradigmatic example of DAC is Miller et al's approach, which led to λProlog in the late 80's. The paper [47] clearly illustrates the basic ideas, starting from a precise notion of *uniform* proofs (to be defined shortly) and characterizing as "abstract logic programming systems" those where each goal has a uniform proof. The paper proves that (first-order) HC is an abstract LP system and then considers various extensions. In particular, it is shown how scoping and encapsulation can be modeled at the logical level, as well as how interesting higher-order programming techniques can be supported. Essentially, the idea behind abstract LP systems is that a sequent such as $\Sigma : \Gamma \longrightarrow G$ represents the state of an idealized LP interpreter with current program Γ, goal G and signature Σ. Both Γ and Σ may dynamically change during the computation. A goal-directed or uniform proof is then a cut-free proof in which every occurrence of a sequent whose right-hand side is non-atomic is the conclusion of a right-introduction rule. It uses a suitable backchaining rule to "invoke" the definitions of the non-logical symbols provided by Γ when an atomic goal A is reached. Examples of right (introduction) rules are $\forall_R$ and $\supset_R$, while BC is the backchain rule.

$$\frac{\Sigma, c : \Gamma \longrightarrow G(c)}{\Sigma : \Gamma \longrightarrow \forall x.\, G(x)} \,\forall_R \qquad \frac{\Sigma : \Gamma, D \longrightarrow G}{\Sigma : \Gamma \longrightarrow D \supset G} \,\supset_R \qquad \frac{\Sigma : \Gamma \longrightarrow G}{\Sigma : \Gamma \longrightarrow A} \, BC, G \supset A \in \langle \Gamma \rangle$$

The $\forall_R$ rule augments the signature by a new constant c of the type of x, while $\supset_R$ augments the program by the clause D. The backchaining rule *selects* a program clause $G \supset A$ in the *closure* $\langle \Gamma \rangle$ of Γ under the $\forall_L, \wedge_L$ rules and backchains on it (see Section 2.1 when this idea is realized via *focusing*). An abstract logic programming language is then a logical system for which uniform proofs are complete. To make our discussion more concrete we consider an example taken from [44], illustrating scoping and modularity.

Example 1. Consider the well known Prolog reverse program;

```
reverse(L,R) :- r(L,R,[]).
r([],Ys,Ys).
r([X|Xs],R,Ys) :- r(Xs,R,[X|Ys]).
```

reverse/2 uses an auxiliary accumulator-based predicate r/3 to implement the following simple algorithm: *start with the pair ⟨L, []⟩ and iteratively push the elements of the first list into the second one.* This example shows two problems. Firstly, the definition of r ought to be used *locally*, inside the scope of the main predicate, but Prolog cannot (declaratively) hide it against undesired redefinitions. Secondly, the simple reverse algorithm needs only the variables L and Ys of r(L,R,Ys): R merely captures the final result and passes it to the reverse predicate. Both problems can be solved by introducing suitable scoping mechanisms. The following shows how this can be accounted for using higher-order universal quantification and embedded implication to provide scope to the definition of the auxiliary predicate and to the individual variables used in it.

```
reverse(L,R) :-
 all rev\ (
        (rev([],R),
        all X,Xs,Ys\ (rev([X|Xs],Ys) :- rev(Xs,[X|Ys])))
            => rev(L,[])
 )
```

The notation follows [44], in particular all r\ G(r) is concrete syntax for $\forall x : \tau. G(x)$. ◇

Roughly, an interpreter based on uniform proof search will proceed as follows. To prove a goal, such as reverse([1,2],V), it will replace r by a *new* binary predicate symbol, say c, and add to the current program the clauses:

```
c([],V), (all X,Xs,Ys\ c([X|Xs],Ys) :- c(Xs,[X|Ys])).
```

Then it will try to prove c([1,2],[]) backchaining on the rightmost clause. The variable V will be instantiated to [2,1] with two further backchain steps, when the computation will eventually succeed with the goal c([],[2,1]).

Logically, the module corresponds to the following formula, where *ls* is the sort of lists, *i* the sort of integers, and *o*, as usual, is the type of propositions:

$$D_{rev} : \forall_{ls} l\ r.\ (\forall_{ls \to ls \to o} rev.\ rev([], r) \wedge$$
$$\forall_i x.\ \forall_{ls} x_s\ y_s.\ rev(x_s, [x|y_s]) \supset rev([x|x_s], y_s)) \supset rev(l, []) \supset (reverse(l, r))$$

In terms of logical rules, the behaviour of the interpreter corresponds to the gradual construction of the following proof tree, where we informally label the clauses on which we backchain:

$$\cfrac{\cfrac{\cfrac{\cfrac{\Sigma, c : \Gamma, D_{c1} : c([], V), D_{c2} : \forall\, x\, y_s\, x_s.\, (c(x_s, [x|y_s]) \supset c([x|x_s], y_s)) \longrightarrow c([], [2, 1])}{\Sigma, c : \Gamma, D_{c1} : c([], V), D_{c2} : \forall\, x\, y_s\, x_s.\, (c(x_s, [x|y_s]) \supset c([x|x_s], y_s)) \longrightarrow c([2], [1])} \; BC, D_{c2}}{\Sigma, c : \Gamma, D_{c1} : c([], V), D_{c2} : \forall\, x\, y_s\, x_s.\, (c(x_s, [x|y_s]) \supset c([x|x_s], y_s)) \longrightarrow c([1, 2], [1])} \; BC, D_{c2}}{\Sigma, c : \Gamma \longrightarrow c([], V) \wedge \forall\, x\, y_s\, x_s.\, (c(x_s, [x|y_s]) \supset c([x|x_s], y_s)) \supset c([1, 2], [])} \; \supset_R, \wedge_L}{\longrightarrow \forall\, rev.\, (rev([], V) \wedge \forall x\, y_s\, x_s.\, (rev(x_s, [x|y_s]) \supset rev([x|x_s], y_s)) \supset rev([1, 2], []))} \; \forall_R}{\Sigma : \Gamma \longrightarrow reverse([1, 2], V)} \; BC, D_{rev}$$

We remark that the generation of new names required by the proof rule for $\forall$ protects the definition of r, since different uses will employ different names. Here, its definition is visible only to calls to `reverse` and will be discharged upon success. Furthermore, the possibility of using the definition of a predicate in the body of a clause and the explicit use of quantifier `all X,Xs,Ys` allows us to link the variable R in the definition of `reverse` precisely to the variable R of the predicate that will accumulate the final result, i.e., to `c([],R)`.

The previous example typifies our viewpoint: seeking extensions of LP in terms of languages endowed with a notion of uniform proofs, more precisely *focused* uniform proofs [2]. This shows a twofold *duality*:

- Between goals and clauses: a negative subformula of a goal is a program clause and a negative subformula of a program clause is a goal.
- Between goal-oriented proof search and clause selection (focusing), once backchaining is seen in a more general light.

This duality is more clearly seen in linear logic, where following Andreoli [2], each connective carries an unique intrinsic attribute called a *polarity* that determines its behaviour under search. Hence connectives can can be partitioned into *asynchronous* (those whose right rule is invertible) and *synchronous* (those whose left rule is invertible). This yields a highly normalized proof search mechanism, based on a systematic interleaving between asynchronous and synchronous reductions: one decomposes the asynchronous formulas until none remains, then picks a synchronous formula and decomposes it until new asynchronous subformulae arise, and so on. Proofs of this kind are called *focused proofs* and can be shown to be complete for entire classical linear logic. In the linear setting the polarity of a connective coincides with its being *positive/negative*: however Andreoli noted that an *arbitrary*, albeit fixed, assignment of polarity to atoms (a *bias*) will preserve completeness of focusing, with the understanding that a [negative] positive bias denote [a]syncronous behaviour. Notwithstanding its asymmetry, this observation applies to intuitionistic logic as well. In fact, it can be shown that, for the Horn fragment, a positive bias to atoms yields hyper-resolution (forward chaining), while a negative one SLD-resolution (backward chaining) [25]. More in general, uniform proofs can be seen as a special case of focusing, where atoms are given negative bias, which happens to be complete only when existentials and disjunctions are excluded from the syntax. These observation have been significantly generalized in [42].

However, there is another angle to "higher-order" extensions to which we have not done justice yet: work related to languages based on some form of λ-calculus. This is indeed the second way a language such as λProlog extends ordinary LP, an issue which was often argued for, when not distrusted since the early 80's [64]. The original rationale was adding some of the higher-order features of functional programming, namely handling functions (here predicates) as first-class citizens, without changing the computation paradigm. A classic example is the *mappred* predicate, corresponding to the `map` combinator in a language such as Standard ML:

Example 2

```
mappred(P,[],[]).
mappred(P,[X|Xs],[Y|Ys]) :- P(X,Y), mappred(P,Xs,Ys).
```

A sample goal could be

```
P = (lambda x y\ reverse(x,y)), mappred(P,[[1,2],[3,4]],Ys).
```

with answer substitution `Ys = [[2,1],[4,3]]`.   ◇

Although some nifty applications based on these features emerged early on, e.g. [32], predicate-as-values, we argue, never managed to attain the same prominence that it has in functional programming. Functional quantification instead has had a pivotal role in the theory and practice of *logical frameworks* [60], in so much as it supports higher-order abstract syntax (HOAS) [61]. This is a declarative treatment of the syntax of object logics, whose binding operators are all rendered via the λ-abstraction of the meta-logic, while bound variables of the object and meta-logic are identified. In this way seemingly banal but tediously complicated issues induced by α-equivalence and substitution principles are taken care once and for all by the meta-logic, making the specification and reasoning over object logic more concise and effective. This opened up an all new field, as we briefly touch upon in the Conclusions.

The rest of this overview is organized as follows: Section 2 succinctly presents the syntax and proof rules underlying the main LP language that we consider in separate subsections. In Section 3 we follow the same schema, highlighting the Italian contribution to the corresponding broad areas. Section 4 concludes by trying to evaluate the impact of these works on LP and computational logic more in general.

2   Calculi for Intuitionistic and Linear Logic Programming

Uniform proofs and abstract LP systems were presented in [51] as the basis for proof-theoretic extensions of LP. At about the same time, Girard's 1987 "Linear Logic" paper had a rippling effect in computer science and logic programming was quick to follow suit. In his 1990 thesis Andreoli established the foundation of focusing proofs in linear logic [2]. In 1991 the uniform proof approach was extended to linear logic programming by Miller & Hodas [41]. We start with the logic underlying λProlog.

$$\frac{}{\Sigma : \Gamma \longrightarrow \top} \vdash \top \qquad \frac{\Sigma : \Gamma \longrightarrow G_1 \qquad \Sigma : \Gamma \longrightarrow G_2}{\Sigma : \Gamma \longrightarrow G_1 \wedge G_2} \vdash \wedge$$

$$\frac{\Sigma : (\Gamma, D) \longrightarrow G}{\Sigma : \Gamma \longrightarrow D \supset G} \vdash \supset \qquad \frac{(\Sigma, c{:}A) : \Gamma \longrightarrow [c/x]G}{\Sigma : \Gamma \longrightarrow \forall x{:}\tau. G} \vdash \forall^c$$

$$\frac{\Sigma : \Gamma \overset{D}{\longrightarrow} A}{\Sigma : \Gamma \longrightarrow A} \vdash \mathrm{fcs}, D \in \Gamma$$

. . . .

$$\frac{\Sigma : \Gamma \vdash A_r \overset{.}{=} A : o}{\Sigma : \Gamma \overset{A_r}{\longrightarrow} A} \mathrm{fcsAt} \qquad \frac{\Sigma : \Gamma \overset{D_i}{\longrightarrow} A}{\Sigma : \Gamma \overset{D_1 \wedge D_2}{\longrightarrow} A} \mathrm{fcs}\wedge_i$$

$$\frac{\Sigma : \Gamma \overset{[t/x]D}{\longrightarrow} A}{\Sigma : \Gamma \overset{\forall x{:}\tau. D}{\longrightarrow} A} \mathrm{fcs}\forall, \Sigma \vdash t : \tau \qquad \frac{\Sigma : \Gamma \longrightarrow G \qquad \Sigma : \Gamma \overset{D}{\longrightarrow} A}{\Sigma : \Gamma \overset{G \supset D}{\longrightarrow} A} \mathrm{fcs} \supset$$

Fig. 1. Focused intuitionistic proofs for HOHF

2.1 λProlog

It is based on the so-called Higher-Order Hereditary Harrop Formulas, an intuitionistic
fragment of Church higher-order logic. As we have mentioned in the Introduction, it
enhances Prolog in two directions. The term language is extended to allow arbitrary λ-
terms under full higher-order unification and the formula language is extended to allow
usage of arbitrarily nested universal quantifiers and implications. It can be synthesized
by the following grammar:

$$\textit{Terms} \quad t ::= c \mid x \mid \lambda x{:}\tau. \, t \mid t_1 \, t_2$$
$$\textit{Atoms} \; A ::= A_r \mid A_f$$
$$\textit{Clauses} \; D ::= A_r \mid G \supset D \mid D_1 \wedge D_2 \mid \forall x{:}\tau. D$$
$$\textit{Goals} \; G ::= A \mid \top \mid G_1 \wedge G_2 \mid D \supset G \mid \forall x{:}\tau. G$$
$$\textit{Signatures} \; \Sigma ::= \cdot \mid \Sigma, x{:}\tau$$
$$\textit{Programs} \; \Gamma ::= \cdot \mid \Gamma, D$$

We shall be fairly loose with typing issues, noting only that we use a ML-like prenex
polymorphic system, so that for example universal quantification is given the type
$\forall \alpha. \, (\alpha \supset o) \supset o$. To preserve the operational reading of logic programs as predicate
definitions we require clause heads to be *rigid* atoms, denoted A_r, i.e. the head symbol
is not a (free) variable.[1] Otherwise, we call the atom *flexible*, denoted A_f. Note that we
could add existentials and disjunctions to the syntax of goals, but with no real expressive
enhancement—see [56] for an investigation into *maximal* abstract logic programming
languages.

---

[1] We gloss over other minor syntactic restrictions of occurrences of logical connectives in the
scope of rigid atoms required to preserve goal-orientedness during proof search.

Some terminology: the above language is named HOHF; with H^{f}OHF we denote its restriction to quantification over variable and function symbols, that is o is only allowed as a range type. Examples of H^{f}OHF are Miller's L_λ [45] and LF [39]. FOHF is the further restriction to first-order quantification.

We now introduce a focused version of the uniform proofs system of [51] (Fig. 1); it defines the following judgements, where Γ contains the program and the possible dynamic assumptions; the judgment $\Sigma : \Gamma \vdash A_r \doteq A : o$, whose definition we omit and refer to the judgmental version in [22]), denotes higher-order unification.

$$\Sigma : \Gamma \longrightarrow G \quad \text{Program } \Gamma \text{ under signature } \Sigma \text{ uniformly entails goal } G.$$

$$\Sigma : \Gamma \xrightarrow{D} A \quad \text{Focused clause } D \text{ from } \Gamma \text{ under signature } \Sigma \text{ entails atom } A.$$

We remark that the backchain rule *BC* of [44], considered in the introduction, can be derived by applying the focusing rules until the head of a clause is deemed to unify the atom on the right and then recursively applying the $\vdash$ rules.

2.2   Lolli

Based on the first-order language freely generated by multiplicative implication $\multimap$, additive unit, implication, conjunction and universal quantification, *Lolli*'s uniform proofs system [41] uses a single-conclusion sequent calculus (Fig. 2) that distinguishes two zones, Γ containing the (reusable) program together with the possible intuitionistic dynamic assumptions and Δ, containing the linear ones, seen as a multiset. Notice that while Lolli is first-order, its type-theoretic counterpart, the Linear Logical Framework LLF [23], has functional quantification; however, they have the same proof search aspects, safe from linear unification, as we detail in Section 3.2.

$$\Sigma : \Gamma; \Delta \longrightarrow G \quad \text{Clauses } \Gamma; \Delta \text{ under signature } \Sigma \text{ uniformly entails goal } G.$$

$$\Sigma : \Gamma; \Delta \xrightarrow{D} A \quad \text{Focused clause } D \text{ from } \Gamma \text{ or } \Delta \setminus D \text{ under signature } \Sigma \text{ entails atom } A.$$

We briefly examine the crucial rules, deviating from the literature by using the same notation for additive connectives as for their intuitionistic counterparts: the fcsAt rule encodes both initial rules of a linear calculus, by requiring the linear context to be empty: in fact, if the focus is on a linear A, then this must be the only assumption that can and must be consumed. If instead the focus is intuitionistic, there must be no leftover resources, lest the computation is failed. Note also the non-deterministic partitioning of the linear context in the focusing rule for $\multimap$, highlighted by the notation $\uplus$ for multiset union, to be read backwards as resource splitting. From an additive viewpoint, rule $\vdash \top$ features an implicit weakening, while $\vdash \wedge$ an implicit contraction, both w.r.t. Δ.

We now give a first linear algorithm for reversing a list.

Example 3

```
reverse(Xs, Ys) : - once(perm(Xs, Ys)).

perm([X|Xs], Ys) ∘– (elm(X) ∘– perm(Xs, Ys)).
perm([], Ys) ∘– perm(Ys).

perm([]).
perm([X|Xs]) ∘– elm(X) ∧ perm(Xs).
```

$$\frac{}{\varSigma : \varGamma;\varDelta \longrightarrow \top} \vdash \top \qquad \frac{\varSigma : \varGamma;\varDelta \longrightarrow G_1 \qquad \varSigma : \varGamma;\varDelta \longrightarrow G_2}{\varSigma : \varGamma;\varDelta \longrightarrow G_1 \wedge G_2} \vdash \wedge$$

$$\frac{\varSigma : (\varGamma, D);\varDelta \longrightarrow G}{\varSigma : \varGamma;\varDelta \longrightarrow D \supset G} \vdash \supset \qquad \frac{\varSigma : \varGamma;(\varDelta \cup \{D\}) \longrightarrow G}{\varSigma : \varGamma;\varDelta \longrightarrow D \multimap G} \vdash \multimap$$

$$\frac{\varSigma : \varGamma;\varDelta \overset{D}{\longrightarrow} A}{\varSigma : \varGamma;\varDelta \longrightarrow A} \vdash \mathrm{fcs}_\varGamma, D \in \varGamma \qquad \frac{\varSigma : \varGamma;\varDelta \overset{D}{\longrightarrow} A}{\varGamma;(\varDelta \cup \{D\}) \longrightarrow A} \vdash \mathrm{fcs}_\varDelta$$

. . . .

$$\frac{}{\varSigma : \varGamma;\cdot \overset{A}{\longrightarrow} A} \mathrm{fcsAt} \qquad \frac{\varSigma : \varGamma;\varDelta \overset{D_i}{\longrightarrow} A}{\varSigma : \varGamma;\varDelta \overset{D_1 \wedge D_2}{\longrightarrow} A} \mathrm{fcs}\wedge_i$$

$$\frac{\varSigma : \varGamma;\cdot \longrightarrow G \qquad \varSigma : \varGamma;\varDelta \overset{D}{\longrightarrow} A}{\varSigma : \varGamma;\varDelta \overset{G \supset D}{\longrightarrow} A} \mathrm{fcs} \supset \qquad \frac{\varSigma : \varGamma;\varDelta_1 \longrightarrow G \qquad \varSigma : \varGamma;\varDelta_2 \overset{D}{\longrightarrow} A}{\varSigma : \varGamma;(\varDelta_1 \cup \varDelta_2) \overset{G \multimap D}{\longrightarrow} A} \mathrm{fcs} \multimap$$

Fig. 2. Main rules of a focused calculus for Lolli

The `perm/2` predicate simply loads (in reversed order) the elements of the input list in the linear context in the form of $\mathrm{elm}(\cdot)$ assumptions; then calls `perm/1`, which consumes those assumptions. Because of the non-deterministic splitting induced by focusing on the second clause of `perm/1`, we generate, upon backtracking, all permutations of the given list. Hence the main `reverse` predicate selects the first solution with the meta-predicate `once/1`.   ◇

2.3   LO

Linear Objects [4, 5] was the first proposal for a linear logic programming language. It extends Horn logic by generalizing clause heads to multisets of atoms connected by multiplicative disjunction ($\mathfrak{P}$), i.e. clauses have the form

$$G \multimap A_1 \mathfrak{P}, \ldots, \mathfrak{P} A_n$$

The starting point was the family of concurrent LP languages (see Chapter [35] in this volume) as a way to provide a logical account of object-oriented computations: objects are viewed as AND-concurrent, stream-communicating via shared variables (proof) processes, where the arguments in a goal are the slots and communication streams of an object. State transitions are realized with inference steps. For a canonical example, the goal

```
point(InStrm,5,7,OutStrm)
```

encodes a point with the given coordinates and communication streams `InStrm` and `OutStrm`, where a method (clause) such as

```
point([proj-x|InStrm],X,Y,OutStrm) :- point(InStrm,X,0,OutStrm)
```

specifies the transition resetting `Y` to `0` upon reception of the `proj-x` message.

In this sense, LO inherits an effect-free view of objects and does not exploit the linear logic context for state manipulation as in Lolli, since it lacks any form of scoping constructs. On the other hand, when seen as OR-concurrency ⅋ directly supports a view of objects as multiset of independent units. The above object becomes

$$\texttt{point} ⅋ \texttt{in(InStrm)} ⅋ \texttt{x(5)} ⅋ \texttt{y(7)} ⅋ \texttt{out(OutStrm)},$$

where different atoms encode a point, its coordinates and communication mediums. In this way objects are amenable of inheritance, since a more specialized objects such as

$$\texttt{point} ⅋ \texttt{in(InStrm)} ⅋ \texttt{x(5)} ⅋ \texttt{y(7)} ⅋ \texttt{out(OutStrm)} ⅋ \texttt{colour(red)}$$

can call a method (a clause with multiple heads) such as

$$\texttt{point} ⅋ \texttt{in([proj} - \texttt{x|InStrm])} ⅋ \texttt{y(Y)} \multimap \texttt{point} ⅋ \texttt{in(InStrm)} ⅋ \texttt{y(0)}$$

by matching only a sub-multiset of the atoms encoding an object. Synchronizations of this kind can be managed using multiset rewriting techniques, but, as we will see in Section 3.2, such synchronization is expensive.[2]

LO's original proof theory [4] did not make focusing explicit, but the crucial rules can be reconstructed as in Figure 3, where we use as a one-sided multi-succedent calculus; since proof search has no dynamics, we can fix the program $\mathcal{P}$ and dispose of the signature.

$\longrightarrow \mathcal{G}$ Program $\mathcal{P}$ uniformly entails multiset of goals $\mathcal{G}$.

$\xrightarrow{D} \mathcal{A}$ Focused clause D from $\mathcal{P}$ entails multiset of atoms $\mathcal{A}$.

$$\frac{\longrightarrow \{G_1, G_2\} \cup \mathcal{G}}{\longrightarrow \{G_1 ⅋ G_2\} \cup \mathcal{G}} \vdash ⅋$$

. . . .

$$\frac{\longrightarrow \{G\} \cup \mathcal{A}_1 \qquad \xrightarrow{A_1 ⅋,...,⅋ A_n} \mathcal{A}_2}{\xrightarrow{G \multimap A_1 ⅋,...,⅋ A_n} \mathcal{A}_1 \cup \mathcal{A}_2} \text{fcs} \multimap \qquad \frac{\longrightarrow G \qquad \xrightarrow{A_1 ⅋,...,⅋ A_n} \mathcal{A}}{\xrightarrow{G \supset A_1 ⅋,...,⅋ A_n} \mathcal{A}} \text{fcs} \supset$$

$$\frac{\xrightarrow{D_1} \mathcal{A}_1 \qquad \xrightarrow{D_2} \mathcal{A}_2}{\xrightarrow{D_1 ⅋ D_2} \mathcal{A}_1 \cup \mathcal{A}_2} \text{fcs}⅋$$

Fig. 3. ⅋-related rules in LO

To better illustrate the operational semantics of LO, we revisit once more the simple reverse algorithm, where we manage to attain the same behavior of Example 1 even in the absence of scoping constructs: in fact, we exploit OR-concurrency to capture the final result and pass it to the main predicate.

---

[2] Historically, this is the first observation that the operational semantics of linear LP brings into intuitionistic proof search an additional "don't know" non-determinism.

Example 4

$$\mathtt{dr} \; : \; \mathtt{reverse(Xs, Ys)} \; \mathtt{:\text{-}} \; \mathtt{rev(Xs, [])} \, \mathbin{⅋} \, \mathtt{result(Ys)}.$$

$$\mathtt{dc} \; : \; \mathtt{rev([X|Xs], Ys)} \; \mathtt{\circ\!\!-} \; \mathtt{rev(Xs, [X|Ys])}.$$

$$\mathtt{dn} \; : \; \mathtt{rev([], V)} \, \mathbin{⅋} \, \mathtt{result(V)}.$$

Intuitively, we backchain on the method dc until the input list is exhausted. Then we awaken the result(V) object by matching it with the dn method and return the instantiation for V.   ◇

This corresponds to this proof-tree, where again we informally use a *BC* rule, decorated with the label of the clause on which we focus.

$$
\dfrac{
  \dfrac{
    \dfrac{
      \dfrac{
        \dfrac{
          \dfrac{L \overset{\cdot}{=} [2,1]}{\overset{\mathtt{rev([],L)}}{\longrightarrow} rev([], [2,1])} \qquad
          \dfrac{L \overset{\cdot}{=} V}{\overset{\mathtt{result(L)}}{\longrightarrow} result(V)}
        }{\longrightarrow rev([], [2,1]), \, result(V)} \; BC, \mathtt{dn}
      }{\longrightarrow rev([], [2,1]) \mathbin{⅋} result(V)} \; \vdash \mathbin{⅋}
    }{\longrightarrow rev([1,2], []) \mathbin{⅋} result(V)} \; BC, \mathtt{dc}
  }{\longrightarrow reverse([1,2], V)} \; BC, \mathtt{dr}
}{}
$$

Note that it is crucial that dc uses linear implication, allowing one to split resources as required.

We conclude this Section noting that LO's $⅋$ can also be seen as a form of *constructive disjunction* yielding indefinite answers; we will touch upon this links between linear and disjunctive logic programming in Section 3.3.

2.4   Forum

Forum [49] can be seen as the fusion of Lolli and LO and allows one to view entire linear logic as an abstract LP language. Indeed, simply adding multiplicative falsity $\bot$ to Lolli yields a "goal-oriented" presentation of linear logic. Thus linear negation $B^{\bot}$ can be defined as expected ($B \multimap \bot$) and hence the other connectives by de Morgan dualities. In particular we can also view $B \mathbin{⅋} C$ as $(B \multimap \bot) \multimap C$. Note that while these encodings do not interfere with the soundness and completeness of focused uniform proofs, they do not yield a predictable operational semantics such as the one a programmer would expect. In fact, focusing on $\bot$ is rather non-informative, leading a naive interpreter into a tight and endless loop. Thus, the view of Forum as a *specification* language [26] and efforts, some of which we mention in Section 3.4, to find a meaningful sub-language amenable to a programming language interpretation.

The relevant judgments comprise two-sided multi-succedent sequents where Γ, Φ have intuitionistic maintenance, and $\varDelta, \mathcal{G}$ have a linear one.

$\Sigma : \Gamma; \varDelta \longrightarrow \mathcal{G}; \Phi$ Clauses $\Gamma; \varDelta$ under signature Σ uniformly entails multisets of goals $\mathcal{G}; \Phi$.

$\Sigma : \Gamma; \varDelta \overset{D}{\longrightarrow} \mathcal{A}; \Phi$ Focused clause D from Γ or $\varDelta \setminus D$ under signature Σ entails multisets of atoms $\mathcal{A}$ and goals Φ.

We refer to [50] for the twenty proof rules.

3   The Italian Contribution

The origin of the Italian interest in proof-theoretic extensions of LP can be traced back to Gabbay and Reily's N-Prolog [34,33], which featured embedded implication in goals, but no universal quantification: free variables can be shared in an implicational goal, creating certain difficulties especially when coupled with negation-as-failure. This language sparked a lot of interest, especially in Torino: A. Martelli, Giordano and others extensively researched applications w.r.t. modules and scoping constructs and extension to modal analysis, see e.g. [9]. We will not analyze this further as already well detailed in [17]. We will, however, briefly mention [37] that fixes some of the problems raised in [33]. The authors propose an operational semantics extending Stärk's ESLDNF, establishing a soundness and completeness for non-floundering queries is with respect to a completion theory interpreted in a three-valued modal logic.

3.1 λProlog

The second "wave" was initiated by Miller's sabbatical in Edinburgh, where he supervised Pareschi's thesis [58]; the latter exploited hypothetical reasoning and λ-terms to encode in a computational environment the features of certain linguistic theories, e.g. the rendering of *filler-gap dependencies*. Pareschi then hooked up with Andreoli to develop LO as we have mentioned in Section 2.3. Miller also supervised Arcelli's thesis [6] in Milano, where she related second-order λProlog to Reflective Prolog [27]. She and coauthors went on investigating applications of the language for example to program transformations [7]. Independently, Momigliano [52] extended Miller's [46], providing a way of encoding via the double negation translation of all classical logic into a focused uniform system. The language was FOHF, but the approach would apply to H^{f}OHF as well.

In [53] the issue of endowing a logical framework (namely H^{f}OHF) with a logically justified notion of negation is re-addressed, adapting the idea of *elimination* of negation [10] to the higher-order setting. This includes two separate phases. *Complementing terms*, i.e. in this case higher-order patterns: due the presence of *partially applied λ-terms*, intuitionistic λ-calculi are not closed under complementation, thus requiring one to develop a *strict*, i.e. relevant, λ-calculus, where we can directly express whether a function (here typically a higher-order logic variable) ought or not depend on its arguments. *Complementing clauses*, which can be seen as a negation normal form procedure which is consistent with intuitionistic provability. It entails finding a middle ground between the CWA usually associated with negation and the OWA typical of logical frameworks. This has come to be known as the *Regular World Assumption* that has shown to be a central notion in inductive meta-theorem proving [63,40] in systems such as *Twelf* [62].

3.2   Lolli

A problem specific to proof search in linear logic is how to effectively split resources when dealing with multiplicative connectives, without trying exponentially many partitions of the linear context. Hodas and Miller developed a lazy splitting approach for the operational semantics of Lolli, called the input-output model of resource consumption [41]. This turns out to be just an instance of a more general resource management problem in linear logic programming (and, with a somewhat different emphasis, in linear theorem proving). As pointed out and addressed in [21], a properly understood operational semantics has to deal with two additional features. First, the $\top$ connective is allowed to consume any resource, a feature which is handy to wind up with success certain computations without burdening the user with tracking and consuming any remaining assumption. Secondly, additive conjunction requires *strict* resources, i.e. those which *can* be duplicated but *must* be used during the solution of a given goal. A final contribution of this paper is the *residuation* calculus, a form of resolution for sequent calculi that pushes all non-determinism out of focusing and into the introduction rules. This has also applications in proof-theoretic compilation [19].

The (linear) spine calculus [24] is an answer to a related issue: devising an efficient representation of the (linear) λ-calculus, tailored to make building blocks of LP such as unification efficient even in the higher-order case. In fact, and differently from the first-order case, even restricting to terms of atomic type, in a token such as

$$(\ldots (h \, M_1) \ldots M_n) \tag{1}$$

the head is deeply buried and hence not immediately accessible. This is further complicated in the linear case, where destructors can be arbitrarily interleaved. In the spine calculus every atomic term has the form $H \cdot S$, where H is the *root* and S the *spine*: a term such as (1) translates into $h \cdot (U_1; \ldots; U_n; \text{NIL})$, where ';' associates to the *right*, U_i translates M_i and NIL represents the end of the spine.

The relevance of this contribution is twofold:

1. The restriction of this calculus to the intuitionistic case is the internal representation adopted in Twelf and it is also at the basis of the *Tejus* compiler for λProlog [57].
2. Exploiting the Curry-Howard correspondence, spines can be seen as a term assignment language for uniform provability, in particular for Lolli, LLF [23], and for any subsystem thereof, as we exemplify in Figure 4.

We modify the main provability judgments to account for proof-terms, unifying Σ and Γ as usual in type theory:

$\Gamma \longrightarrow U : G$ U is a term (proof) of type (goal) G given assumptions Γ

$\Gamma \xrightarrow{D} S : A$ S is a spine (proof) consisting of heads of type (clause) D to terms S of type (goal) A given assumptions Γ.

Of course, once the spine representation was in place, there was still the need to provide an unification algorithm for this language. In [22] the authors fill this gap, providing a judgmental view of a linear pre-unification procedure in the style of Huet. Being a conservative extension of ordinary higher-order unification, it may not terminate and if it does, it returns a system of equations between flexible atoms, possibly yielding infinite numbers of incomparable unifiers. The paper shows also that it is not possible to simulate higher-order linear unification by generating standard higher-order solutions and promoting those which satisfy the linearity constraints. Even more noteworthy, an analogous notion to Miller's intuitionistic higher-order patterns [45], for which mgu's can be effectively found, does not seem to exist in the linear setting.

$$\frac{\Gamma, x : D \longrightarrow U : G}{\Gamma \longrightarrow (\lambda x{:}D.\, U) : D \supset G} \vdash \supset \qquad \frac{\Gamma \xrightarrow{h:D} S : A}{\Gamma \longrightarrow (h \cdot S) : A} \vdash \mathrm{fcs}, D \in \Gamma$$

$$\frac{}{\Gamma \xrightarrow{NIL:A} A} \mathrm{fcsAt} \qquad \frac{\Gamma \longrightarrow U : G \qquad \Gamma \xrightarrow{S:D} A}{\Gamma \xrightarrow{(U;S):G \supset D} A} \mathrm{fcs} \supset$$

Fig. 4. Proof terms for focused uniform proofs

3.3   LO

Most of the research about linear logic programming as far as LO and Forum are concerned was spearheaded by Giorgio Levi and his school, in their research aiming to integrate (linear) logic programming with other paradigm such as concurrency and object-orientation, beginning with Guglielmi and Delzanno's thesis [38, 28]. The latter then moved to Genoa, where he collaborated with M. Martelli, Bozzano and others.

The relationship between linear and disjunctive LP mentioned in [4] is taken up in [12], where the authors show that LO can be seen as a sub-structural fragments of DLP, where contraction on the right is disallowed. More extensive connections between a fragment of LO and DLP are further established using abstract interpretation methods [13]. A propositional bottom-up semantics for LO (and its extension with multiplicative unit LO_1) is proposed via a fixed point operator operating on (ideals of) multisets. Note that the semantics is effective for LO, but not for LO_1; the former, in fact, lacks the expressivity of *counting* resources, while in the latter it is possible to encode formalisms such as Petri nets with transfer arcs. Emphasis on the propositional side was also motivated by earlier work on partial evaluation of LO programs [3]. This yielded an approach to model-checking where verifying a safety problem encoded in temporal logic is akin to computing the fixed point of a linear logic program. This is further studied in [14], where bottom-up evaluation is extended to first order LO programs with universally quantified goals and possibly empty heads. See for more details the Chapter [29] in this volume.

We remark that bottom-up evaluation has now gained an important role in general sequent-based automated theorem proving [42, 25], as well as in the operational semantics of LolliMon [43], the first-order logic programming language underlying the Concurrent Logical Framework [66]. The latter integrates Lolli with a *monadic* modality encapsulating synchronous connectives.

3.4   Forum

Some early work exploited the connection between linear logic and multiset rewriting to encode aspects of planning and concurrency [15, 18]. More developed research was concerned with finding a logical counterpart of object-based languages such as the Object Calculus; [16] introduced **Ob$_{\multimap}$**, an object language where methods are represented as logical formulae and whose operational semantics is realized via proof search. The language is then encoded in a *linear* extension of second-order N-Prolog, with a limited form of predicate quantification In [30] the authors present a restriction of Forum with the aim of integrating logic programming with the rewrite-based specification languages; intended applications are modelling of concurrent systems and meta-programming. Clauses have the form $G_1 \supset \dots G_n \supset (\mathcal{B}\mathcal{A} \circ\!\!- G)$ and may again incorporate a form of predicate quantification, provided the underlying term language is basically first-order. State-based computations are specified similarly as in LO, i.e. storing resources on the right-hand side of the sequent and matching them with multi-headed clauses.

4 Conclusions

We have tried to show how the proof-theoretic approach to LP has led to a series of logically motivated logic programming languages of increasing power, supporting modern abstraction mechanisms via higher-order extensions and imperative features via resource-consciousness. The Italian contribution has been both foundational and applicative, in terms of language extensions, implementation techniques and usage of the new features to capture various computation models. We cannot leave out, however, that the original emphasis on endowing logic programming with some of the more successful features of functional programming has died down or, better, it has changed emphasis. Indeed, the design of LolliMon is heavily influenced by Moggi's computational monads, which are omnipresent in functional languages such as Haskell. What has thrived, beyond a better understanding of the foundations of proof search that is showing promising fruits in general theorem proving, is the theory and practice of logical frameworks. We argue that this development from logical representation to meta-reasoning over the latter is a natural and welcomed one, which could not have happened without the proof-theoretical standpoint. We can isolate two trends in which Italian researchers have an active role:

1. The development of more expressive type-theoretic frameworks, from linear [23] to concurrent ones [65, 66].
2. The integration of HOAS and principle of (co)induction, both in standard systems [54] and in ones directly derived from logic programming such as the *Bedwyr* model-checker [8] and the *Abella* interactive theorem prover [36], see [55] for work on their logical foundations.

Acknowledgments. This survey owes to many of Miller's papers, especially "An Overview of Linear Logic Programming" [50]. We thank Iliano Cervesato and Laura Giordano for bibliographic suggestions and the anonymous referees for many useful remarks.

References

1. Alpuente, M., Sessa, M.I. (eds.): 1995 Joint Conference on Declarative Programming, GULP-PRODE 1995, Marina di Vietri, Italy (1995)
2. Andreoli, J.-M.: Logic programming with focusing proofs in linear logic. J. Log. Comput. 2(3), 297–347 (1992)

3. Andreoli, J.-M., Castagnetti, T., Pareschi, R.: Abstract interpretation of linear logic programming. In: Miller, D. (ed.) Proceedings of the International Logic Programming Symposium, Vancouver, Canada, pp. 295–314. MIT Press, Cambridge (1993)

4. Andreoli, J.-M., Pareschi, R.: LO and behold! Concurrent structured processes. In: Proceedings of OOPSLA 1990, Ottawa, Canada, October 1990, vol. 25(10), pp. 44–56. Published as ACM SIGPLAN Notices (1990)

5. Andreoli, J.-M., Pareschi, R.: Linear objects: Logical processes with built-in inheritance. New Generation Computing 9, 445–473 (1991)

6. Arcelli, F.: Aspetti di ordine superiore e di metalivello della programmazione logica. PhD thesis, DSI, Universitá di Milano (1991)

7. Arcelli, F., Formato, F.: Implementing higher-order term-rewriting for program transformation in λProlog. In: Alpuente, Sessa [1], pp. 245–256

8. Baelde, D., Gacek, A., Miller, D., Nadathur, G., Tiu, A.: The Bedwyr system for model checking over syntactic expressions. In: Pfenning, F. (ed.) CADE 2007. LNCS (LNAI), vol. 4603, pp. 391–397. Springer, Heidelberg (2007)

9. Baldoni, M., Giordano, L., Martelli, A.: A modal extension of logic programming: Modularity, beliefs and hypothetical reasoning. J. Log. Comput. 8(5), 597–635 (1998)

10. Barbuti, R., Mancarella, P., Pedreschi, D., Turini, F.: A transformational approach to negation in logic programming. Journal of Logic Programming 8, 201–228 (1990)

11. Bossi, A., Meo, M.C.: Theoretical Foundations and Semantics. In: Dovier, A., Pontelli, E. (eds.) 25 Years of Logic Programming in Italy. LNCS, pp. 15–36. Springer, Heidelberg (2010)

12. Bozzano, M., Delzanno, G., Martelli, M.: On the relations between disjunctive and linear logic programming. Electr. Notes Theor. Comput. Sci. 48 (2001)

13. Bozzano, M., Delzanno, G., Martelli, M.: An effective fixpoint semantics for linear logic programs. Theory Pract. Log. Program. 2(1), 85–122 (2002)

14. Bozzano, M., Delzanno, G., Martelli, M.: Model checking linear logic specifications. TPLP 4(5-6), 573–619 (2004)

15. Bruscoli, P., Guglielmi, A.: Expressiveness of the abstract logic programming language Forum in planning and concurrency. In: Alpuente, M., Barbuti, R., Ramos, I. (eds.) GULP-PRODE (2), pp. 221–237 (1994)

16. Bugliesi, M., Delzanno, G., Liquori, L., Martelli, M.: Object calculi in linear logic. J. Log. Comput. 10(1), 75–104 (2000)

17. Bugliesi, M., Lamma, E., Mello, P.: Modularity in logic programming. J. Log. Program. 19/20, 443–502 (1994)

18. Cervesato, I.: Petri nets and linear logic: a case study for logic programming. In: Alpuente, Sessa [1], pp. 313–320

19. Cervesato, I.: Proof-theoretic foundation of compilation in logic programming languages. In: Jaffar, J. (ed.) Proceedings of the 1998 Joint International Conference and Symposium on Logic Programming (JICSLP 1998), Manchester, UK, pp. 115–129. MIT Press, Cambridge (1998)

20. Cervesato, I., Hodas, J.S., Pfenning, F.: Efficient resource management for linear logic proof search. In: Herre, H., Dyckhoff, R., Schroeder-Heister, P. (eds.) ELP 1996. LNCS (LNAI), vol. 1050, pp. 67–81. Springer, Heidelberg (1996)

21. Cervesato, I., Hodas, J.S., Pfenning, F.: Efficient resource management for linear logic proof search. Theoretical Computer Science 232(1-2), 133–163 (2000); Extended version of [20]

22. Cervesato, I., Pfenning, F.: Linear higher-order pre-unification. In: Winskel, G. (ed.) Proceedings of the Twelfth Annual Sumposium on Logic in Computer Science (LICS 1997), Warsaw, Poland, pp. 422–433. IEEE Computer Society Press, Los Alamitos (1997)

23. Cervesato, I., Pfenning, F.: A linear logical framework. Information and Computation (1998); Special issue with invited papers from LICS 1996, Clarke, E. (ed.)

24. Cervesato, I., Pfenning, F.: A linear spine calculus. J. Log. Comput. 13(5), 639–688 (2003)
25. Chaudhuri, K., Pfenning, F., Price, G.: A logical characterization of forward and backward chaining in the inverse method. J. Autom. Reasoning 40(2-3), 133–177 (2008)
26. Chirimar, J.L.: Proof Theoretic Approach to Specification Languages. PhD thesis, University of Pennsylvania (May 1995)
27. Costantini, S., Lanzarone, G.A.: A metalogic programming language. In: ICLP, pp. 218–233 (1989)
28. Delzanno, G.: Logic and Object-Oriented Programming in Linear Logic. PhD thesis, Università di Pisa (February 1997)
29. Delzanno, G., Giacobazzi, R., Ranzato, F.: Static Analysis, Abstract Interpretation and Verification in (Constraint Logic) Programming. In: Dovier, A., Pontelli, E. (eds.) 25 Years of Logic Programming in Italy. LNCS, vol. 6125, pp. 136–158. Springer, Heidelberg (2010)
30. Delzanno, G., Martelli, M.: Proofs as computations in linear logic. Theoretical Computer Science 258(1-2), 269–297 (2001)
31. Dovier, A., Pontelli, E. (eds.): 25 Years of Logic Programming in Italy. LNCS, vol. 6125. Springer, Heidelberg (2010)
32. Felty, A.P.: Implementing tactics and tacticals in a higher-order logic programming language. J. Autom. Reasoning 11(1), 41–81 (1993)
33. Gabbay, D.M.: N-Prolog: An extension of Prolog with hypothetical implication II - logical foundations, and negation as failure. J. Log. Program. 2(4), 251–283 (1985)
34. Gabbay, D.M., Reyle, U.: N-Prolog: An extension of Prolog with hypothetical implications I. J. Log. Program. 1(4), 319–355 (1984)
35. Gabbrielli, M., Palamidessi, C., Valencia, F.D.: Concurrent and Reactive Constraint Programming. In: Dovier, A., Pontelli, E. (eds.) 25 Years of Logic Programming in Italy. LNCS, vol. 6125, pp. 231–253. Springer, Heidelberg (2010)
36. Gacek, A.: The Abella interactive theorem prover (system description). In: Armando, A., Baumgartner, P., Dowek, G. (eds.) IJCAR 2008. LNCS (LNAI), vol. 5195, pp. 154–161. Springer, Heidelberg (2008)
37. Giordano, L., Olivetti, N.: Combining negation as failure and embedded implications in logic programs. J. Log. Program. 36(2), 91–147 (1998)
38. Guglielmi, A.: Abstract Logic Programming in Linear Logic Independence and Causality in a First Order Calculus. PhD thesis, Università di Pisa (April 1996)
39. Harper, R., Honsell, F., Plotkin, G.: A framework for defining logics. Journal of the Association for Computing Machinery 40(1), 143–184 (1993)
40. Harper, R., Licata, D.R.: Mechanizing metatheory in a logical framework. J. Funct. Program. 17(4-5), 613–673 (2007)
41. Hodas, J., Miller, D.: Logic programming in a fragment of intuitionistic linear logic. Information and Computation 110(2), 327–365 (1994); A preliminary version appeared in the Proceedings of the Sixth Annual IEEE Symposium on Logic in Computer Science, pp. 32–42, Amsterdam, The Netherlands (July 1991)
42. Liang, C., Miller, D.: Focusing and polarization in linear, intuitionistic, and classical logics. Theoretical Computer Science 410(46) (2009)
43. López, P., Pfenning, F., Polakow, J., Watkins, K.: Monadic concurrent linear logic programming. In: Barahona, P., Felty, A.P. (eds.) PPDP, pp. 35–46. ACM, New York (2005)
44. Miller, D.: Lexical scoping as universal quantification. In: Levi, G., Martelli, M. (eds.) Proceedings of the Sixth International Conference on Logic Programming, Lisbon, Portugal, pp. 268–283. MIT Press, Cambridge (1989)
45. Miller, D.: A logic programming language with lambda-abstraction, function variables, and simple unification. In: Schroeder-Heister, P. (ed.) ELP 1989. LNCS (LNAI), vol. 475, pp. 253–281. Springer, Heidelberg (1991)

46. Miller, D.: A logical analysis of modules in logic programming. Journal of Logic Programming 6(1-2), 79–108 (1989)
47. Miller, D.: Abstractions in logic programming. In: Odifreddi, P. (ed.) Logic and Computer Science, pp. 329–359. Academic Press, London (1990)
48. Miller, D.: A proposal for modules in λProlog. In: Dyckhoff, R. (ed.) ELP 1993. LNCS (LNAI), vol. 798. Springer, Heidelberg (1994)
49. Miller, D.: Forum: A multiple-conclusion specification logic. Theoretical Computer Science 165(1), 201–232 (1996)
50. Miller, D.: Overview of linear logic programming. In: Ehrhard, T., Girard, J.-Y., Ruet, P., Scott, P. (eds.) Linear Logic in Computer Science. London Mathematical Society Lecture Note, vol. 316, pp. 119–150. Cambridge University Press, Cambridge (2004)
51. Miller, D., Nadathur, G., Pfenning, F., Scedrov, A.: Uniform proofs as a foundation for logic programming. Annals of Pure and Applied Logic 51, 125–157 (1991)
52. Momigliano, A.: Minimal negation and Hereditary Harrop Formulae. In: Nerode, A., Taitslin, M.A. (eds.) LFCS 1992. LNCS, vol. 620, pp. 326–335. Springer, Heidelberg (1992)
53. Momigliano, A.: Elimination of negation in a logical framework. In: Clote, P.G., Schwichtenberg, H. (eds.) CSL 2000. LNCS, vol. 1862, pp. 411–426. Springer, Heidelberg (2000)
54. Momigliano, A., Ambler, S.: Multi-level meta-reasoning with higher-order abstract syntax. In: Gordon, A.D. (ed.) FOSSACS 2003. LNCS, vol. 2620, pp. 375–391. Springer, Heidelberg (2003)
55. Momigliano, A., Tiu, A.F.: Induction and co-induction in sequent calculus. In: Berardi, S., Coppo, M., Damiani, F. (eds.) TYPES 2003. LNCS, vol. 3085, pp. 293–308. Springer, Heidelberg (2004)
56. Nadathur, G.: Correspondences between classical, intuitionistic and uniform provability. Theoretical Computer Science 232, 273–298 (2000)
57. Nadathur, G.: The metalanguage λProlog and its implementation. In: Kuchen, H., Ueda, K. (eds.) FLOPS 2001. LNCS, vol. 2024, pp. 1–20. Springer, Heidelberg (2001)
58. Pareschi, R.: Type-Driven Natural Language Analysis. PhD thesis, University of Edinburgh. University of Pennsylvania, Department of Computer and Information Science, Technical Report No. MS-CIS-89-45 (July 1989)
59. Pfenning, F.: Computation and deduction. Unpublished lecture notes, p. 217 (Revised March 2001) (May 1992)
60. Pfenning, F.: Logical frameworks. In: Robinson, A., Voronkov, A. (eds.) Handbook of Automated Reasoning. Elsevier Science Publishers, Amsterdam (1999)
61. Pfenning, F., Elliott, C.: Higher-order abstract syntax. In: Proceedings of the ACM SIGPLAN 1988 Symposium on Language Design and Implementation, Atlanta, Georgia, June 1988, pp. 199–208 (1988)
62. Pfenning, F., Schürmann, C.: System description: Twelf — A meta-logical framework for deductive systems. In: Ganzinger, H. (ed.) CADE 1999. LNCS (LNAI), vol. 1632, pp. 202–206. Springer, Heidelberg (1999)
63. Schürmann, C.: Automating the Meta-Theory of Deductive Systems. PhD thesis, Carnegie-Mellon University, CMU-CS-00-146 (2000)
64. Warren, O.H.D.: Higher-order extensions to Prolog: Are they needed? In: Hayes, J.E., Michie, D., Pao, Y.-H. (eds.) Machine Intelligence, vol. 10, pp. 441–454. Halsted Press (1982)
65. Watkins, K., Cervesato, I., Pfenning, F., Walker, D.: A concurrent logical framework: The propositional fragment. In: Berardi, S., Coppo, M., Damiani, F. (eds.) TYPES 2003. LNCS, vol. 3085, pp. 355–377. Springer, Heidelberg (2004)
66. Watkins, K., Cervesato, I., Pfenning, F., Walker, D.: Specifying properties of concurrent computations in CLF. Electr. Notes Theor. Comput. Sci. 199, 67–87 (2008)

Transformation and Debugging of Functional Logic Programs[*]

Maria Alpuente[1], Demis Ballis[2], and Moreno Falaschi[3]

[1] DSIC, Universidad Politécnica de Valencia
Camino de Vera s/n, Apdo. 22012, 46071 Valencia, Spain
alpuente@dsic.upv.es
[2] Dip. Matematica e Informatica
Via delle Scienze 206, 33100 Udine, Italy
demis@dimi.uniud.it
[3] Dip. di Scienze Matematiche e Informatiche
Pian dei Mantellini 44, 53100 Siena, Italy
moreno.falaschi@unisi.it

Abstract. The Italian contribution to functional-logic programming has been significant and influential in a number of areas of semantics, and semantics-based program manipulation techniques. We survey selected topics, with a particular regard to debugging and transformation techniques. These results as usual depend on the narrowing strategy which is adopted and on the properties satisfied by the considered programs. In this paper, we restrict ourselves to first-order functional-logic languages without non-deterministic functions. We start by describing some basic classical transformation techniques, namely folding and unfolding. Then, we recall the narrowing-driven partial evaluation, which is the first generic algorithm for the specialization of functional logic programs. Regarding debugging, we describe a goal-independent approach to automatic diagnosis and correction which applies the immediate consequence operator modeling computed answers to the diagnosis of bugs in functional logic programs. A companion bug-correction program synthesis methodology is described that attempts to correct the erroneous components of the wrong code.

1 Introduction

Functional logic languages combine the most important features of functional programming (expressivity of functions and types, higher-order functions, nested expressions, efficient reduction strategies, sophisticated abstraction facilities) and logic programming (unification, logical variables, partial data-structures, built-in search). The operational principle of integrated languages with a complete semantics is usually based on *narrowing* [37], which consists of the instan-

---

[*] This work has been partially supported by the Italian MUR under grant RBIN04M8S8, FIRB project, Internationalization 2004, and the EU (FEDER) and Spanish MEC project TIN2007-68093-C02-02.

A. Dovier, E. Pontelli (Eds.): 25 Years of Logic Programming, LNCS 6125, pp. 271–299, 2010.

tiation of variables in expressions, followed by a reduction step on the instantiated function call. *Narrowing* is complete in the sense of functional programming
(computation of normal forms) as well as logic programming (computation of
answers). Due to the huge search space of unrestricted narrowing, steadily improved strategies have been proposed, with innermost narrowing and needed
narrowing being of main interest (see [41,43] for a survey.)

Functional logic programming is an area which was pioneered by Italian researchers. For instance, the first published survey paper on this subject was [22].
Since then, the Italian contribution has been significant and influential in a number of semantics-based program manipulation techniques. The main purpose of
this work is to outline a selection of these techniques, with a particular regard to
debugging and transformation. Actually, good programs have to be both correct
(w.r.t. a given specification) and efficient, but these two aspects are often in
delicate balance.

Program transformations provide a methodology for deriving correct and possibly efficient programs. We recall first a simple transformation methodology
based on fold/unfold techniques [12,13]. Then we recall the narrowing-driven
partial evaluation, which was first proposed in [16] and is the first generic algorithm for the specialization of functional logic programs. Regarding program
debugging, we offer an up-to-date, comprehensive, and uniform presentation of
the declarative debugging of functional logic programs as developed in [6,7]. Our
method is based on a fixpoint semantics for functional logic programs that models the set of computed answers in a bottom-up manner and is parametric w.r.t.
the considered narrowing strategy, which can be either eager or lazy. The proposed methodology does not require the user to provide a symptom (a known
bug in the program) to start. Rather, our diagnoser discovers whether there is
one such bug and then tries to correct it automatically by means of inductive
learning, without asking the user to answer difficult questions about program
semantics as typically happens in algorithmic debugging. A further important
advantage of our method is the fact that we develop a finite methodology which
is also goal-independent and allows us to perform diagnosis statically. We additionally address the problem of modifying incorrect components of the initial
program in order to form an integrated debugging framework in which it is possible to detect program bugs and correct them automatically, which we first
outlined in [4]. The correction technique is driven by a set of evidence examples
that are automatically produced as an outcome by the diagnoser, and infers the
program corrections by combining top-down (unfolding-based) transformations
with a bottom-up (induction-based) program synthesis methodology. Due to the
strong relation between program transformation and program synthesis, the cooperation between these two methodologies within the debugging framework is
fruitful and extremely smooth.

We do not consider in this paper programs containing non-strict, non-deterministic functions with call-time choice semantics [58,59], as adopted by some
modern functional logic languages like Curry [42,46] or Toy [56]. This is because
there does not exist a simple and adequate notion of narrowing for call-time

choice that can replace existing efficient versions of narrowing like the strategies discussed in this paper, which are well established and appropriate operational procedures for functional logic languages [58].

All the proposed transformation and verification frameworks have been implemented into prototypical systems which have been thoroughly evaluated using large suites of benchmarks in order to assess their usefulness experimentally. Tools and experiments are freely available at the URL:

http://users.dsic.upv.es/grupos/elp/soft.html

Plan of the paper. The rest of the paper is organized as follows. Section 2 presents some preliminary basic definitions. In Section 3, we formalize narrowing along with two well-know narrowing strategies: the leftmost-innermost (*inn*) and the leftmost-outermost (*out*) narrowing strategy. We then formulate both an operational semantics and a fixpoint semantics for functional logic programs which are parametric w.r.t. the chosen narrowing strategy. We also show the correspondence between the two program denotations. Section 4 outlines the rudiments of functional logic program transformation, while Section 5 focuses on the narrowing-driven approach to functional logic program specialization. Section 6 formalizes the diagnosis framework by providing the necessary notions of rule incorrectness and uncoveredness, and describes an effective methodology based on abstract interpretation that can be used to implement declarative debuggers. Moreover, we present a bug-correction program synthesis methodology which, after diagnosing the buggy program, tries to correct the erroneous components of the wrong code automatically. Finally, in Section 7 we discuss some related work.

2   Preliminaries

Let us briefly recall some known results about rewrite systems [51] and functional logic programming (see [41,47] for extensive surveys). For simplicity, definitions are given in the one-sorted case. The extension to many-sorted signatures is straightforward, see [66].

Throughout this paper, V denotes a countably infinite set of variables and Σ denotes a non-empty, finite set of function symbols, or signature, each of which has a fixed associated arity. Throughout the paper, we will use the following notation: lowercase letters from the end of the alphabet x, y, z, possibly with subindices, denote variables, and we often write $f/n \in \Sigma$ to denote that f is a function symbol of arity n. $\tau(\Sigma \cup V)$ and $\tau(\Sigma)$ denote the non-ground term algebra and the ground term algebra built on $\Sigma \cup V$ and Σ, respectively. An *equation* is a syntactic expression of the form $t = t'$, where $t, t' \in \tau(\Sigma \cup V)$.

Terms are viewed as labelled trees in the usual way. Positions are represented by sequences of natural numbers denoting an access path in a term, where Λ denotes the empty sequence. $O(t)$ (resp. $\overline{O}(t)$) denotes the set of positions (resp. non-variable positions) of a term t. $t|_u$ is the subterm at the position u of t. $t[r]_u$ is the term t with the subterm at the position u replaced with r. These notions extend to sequences of equations in a natural way. For instance, the non-variable position set of a sequence of equations $g = (t_1 = t'_1, \ldots, t_n = t'_n)$ can be defined as follows: $\overline{O}(g) = \{i.1.u \mid i \in \{1, \ldots, n\}, u \in \overline{O}(t_i)\} \bigcup \{i.2.u \mid i \in \{1, \ldots, n\}, u \in \overline{O}(t'_i)\}$.

By $Var(s)$, we denote the set of variables occurring in the syntactic object s, while $[s]$ denotes the set of ground instances of s. Syntactic equality is denoted by $=$.

A *substitution* is a mapping from the set of variables V into the set of terms $\tau(\Sigma \cup V)$. We write $\theta_{\upharpoonright s}$ to denote the restriction of the substitution θ to the set of variables in the syntactic object s. The *empty substitution* is denoted by *id*. Composition of substitutions is denoted by juxtaposition, with identity element *id*. A substitution θ is more general than σ, denoted by $\theta \leq \sigma$, if $\sigma = \theta\gamma$ for some substitution γ. We say that a substitution σ is a *unifier* of two terms t and t' if $t\sigma = t'\sigma$. We let $mgu(t, t')$ denote a *most general unifier* of t and t'.

A *conditional term rewriting system* (CTRS for short) is a pair $(\Sigma, \mathcal{R})$, where $\mathcal{R}$ is a finite set of reduction (or rewrite) rule schemes of the form $(\lambda \rightarrow \rho \Leftarrow C)$, $\lambda, \rho \in \tau(\Sigma \cup V)$ and $\lambda \notin V$. The condition C is a (possibly empty) sequence $e_1, \ldots, e_n$, $n \geq 0$ of equations. Variables in C or ρ that do not occur in λ are called *extra-variables*. We will often write just $\mathcal{R}$ instead of $(\Sigma, \mathcal{R})$. If a rewrite rule has an empty condition, we write $\lambda \rightarrow \rho$. A TRS is a CTRS whose rules have no conditions. A *goal* g is a non-empty sequence of equations $\Leftarrow C$, i.e., a rule with no head (consequent). Sometimes we leave out the $\Leftarrow$ symbol when we write goals.

For CTRS $\mathcal{R}$, $r \ll \mathcal{R}$ denotes that r is a new variant of a rule in $\mathcal{R}$ such that r contains only *fresh* variables, i.e. contains no variable previously met during computation (standardized apart). Given a CTRS $(\Sigma, \mathcal{R})$, we assume that the signature Σ is partitioned into two disjoint sets $\Sigma = \mathcal{C} \uplus \mathcal{D}$, where $\mathcal{D} = \{f \mid (f(t_1, \ldots, t_n) \rightarrow r \Leftarrow C) \in \mathcal{R}\}$ and $\mathcal{C} = \Sigma \setminus \mathcal{D}$. Symbols in $\mathcal{C}$ are called *constructors* and symbols in $\mathcal{D}$ are called *defined functions*. The elements of $\tau(\mathcal{C} \cup V)$ are called *constructor terms*. A *constructor* substitution $\sigma = \{x_1 \mapsto t_1, \ldots, x_n \mapsto t_n\}$ is a substitution such that each t_i, $i = 1, \ldots, n$ is a constructor term. A term is linear if it does not contain multiple occurrences of the same variable. A *pattern* is a term of the form $f(\bar{d})$ where $f/n \in \mathcal{D}$ and $\bar{d}$ are constructor terms. We say that a CTRS is *constructor-based* (CB) if the left-hand sides of $\mathcal{R}$ are patterns.

A rewrite step is the application of a rewrite rule to an expression. A term s *conditionally rewrites* to a term t, $s \rightarrow_{\mathcal{R}} t$, if there exist $u \in \overline{O}(s)$, $(\lambda \rightarrow \rho \Leftarrow s_1 = t_1, \ldots, s_n = t_n) \in \mathcal{R}$, and substitution σ such that $s|_u = \lambda\sigma$, $t = s[\rho\sigma]_u$, and for all $i \in \{1, \ldots, n\}$ there exists a term w_i such that $s_i\sigma \rightarrow^*_{\mathcal{R}} w_i$ and $t_i\sigma \rightarrow^*_{\mathcal{R}} w_i$, where $\rightarrow^*_{\mathcal{R}}$ is the transitive and reflexive closure of $\rightarrow_{\mathcal{R}}$. The term $s|_u$ is said to be a *redex* of s. When no confusion can arise, we omit the subscript $\mathcal{R}$. A term s is a *normal form*, if there is no term t with $s \rightarrow_{\mathcal{R}} t$. A CTRS $\mathcal{R}$ is *strongly terminating* if there are no infinite sequences of the form $t_0 \rightarrow_{\mathcal{R}} t_1 \rightarrow_{\mathcal{R}} t_2 \rightarrow_{\mathcal{R}} \cdots$. A CTRS $\mathcal{R}$ is *confluent* if, whenever a term s reduces to two terms t_1 and t_2, both t_1 and t_2 reduce to the same common term. The program $\mathcal{R}$ is said to be canonical if the binary one-step rewrite relation $\to_\mathcal{R}$ defined by $\mathcal{R}$ is strongly terminating and confluent [51].

3   Evaluating Functional Logic Programs by Narrowing

Functional logic languages are extensions of functional languages with principles derived from logic programming [53,68]. The computation mechanism of functional logic languages is based on *narrowing* [37], a generalization of term rewriting where unification replaces matching: both the rewrite rule and the term to be rewritten can be instantiated. Under the narrowing mechanism, functional programs behave like logic programs in the sense that narrowing solves equations by computing solutions with respect to a given CTRS, which is henceforth called the "program".

Definition 1 (Narrowing). *Let $\mathcal{R}$ be a program and g be a goal. We say that g conditionally narrows into g' in $\mathcal{R}$ if there exist a position $u \in \overline{O}(g)$, $r = (\lambda \to \rho \Leftarrow C) \ll \mathcal{R}$, and a substitution σ such that: $\sigma = mgu(g|_u, \lambda)$, and g' is the sequence $C\sigma, g[\rho]_u\sigma$.*

*We write $g \overset{u,r,\sigma}{\rightsquigarrow} g'$ or simply $g \overset{\sigma}{\rightsquigarrow} g'$. The relation $\rightsquigarrow$ is called (*unrestricted or ordinary*) conditional narrowing.*

Basically, narrowing steps involve unification while functional reduction employs pattern matching. The condition that the binding substitution σ is a mgu can be relaxed to accomplish with certain narrowing strategies like needed narrowing [20], which use unifiers but not necessarily most general ones.

By using Definition 1, we can define (*successful*) narrowing derivations as follows. We use the symbol $\top$ to denote sequences of the form $true, \ldots, true$, and $\mathcal{R}_+$ denotes $\mathcal{R} \cup \{x = x \to true\}$, $x \in V$. Using this rule allows us to treat syntactical unification as a narrowing step, i.e., we use the rule $r = (x = x \to true)$ to compute mgu's: $s = t \overset{\Lambda,r,\sigma}{\rightsquigarrow} true$ holds iff $\sigma = mgu(\{s = t\})$.

Definition 2 (Narrowing derivation). *A narrowing derivation for g in $\mathcal{R}$ is defined by $g \overset{\theta}{\rightsquigarrow}{}^* g'$ iff $\exists \theta_1, \ldots, \exists \theta_n$. $g \overset{\theta_1}{\rightsquigarrow} \ldots \overset{\theta_n}{\rightsquigarrow} g'$ and $\theta = \theta_1 \ldots \theta_n$, $n > 0$. A successful derivation for g in $\mathcal{R}$ is a narrowing derivation $g \overset{\theta}{\rightsquigarrow}{}^* \top$ in $\mathcal{R}_+$, and $\theta_{\restriction Var(g)}$ is called a computed answer substitution (cas) for g in $\mathcal{R}$.*

The *narrowing* mechanism is a powerful tool for constructing complete equational unification algorithms for useful classes of CTRSs, including canonical CTRSs [48]. Similarly to logic programming, completeness means the ability to compute representatives of all solutions for one or more equations.

Example 1. Consider the following program $\mathcal{R}$ which defines the **last** element of a list in a logic programming style, by using the list concatenation function **append** (list constructors are **nil** (empty list) and [_|_] (cons constructor)):

$$\begin{aligned} &R1: &&\text{last(xs)} &&\to \text{y} \Leftarrow \text{append(zs, [y])} = \text{xs.} \\ &R2: &&\text{append(nil, xs)} &&\to \text{xs.} \\ &R3: &&\text{append([x|xs], ys)} &&\to \text{[x|append(ys, ys)].} \end{aligned}$$

Given the input goal last(ys) = 0, narrowing is able to compute in $\mathcal{R}$ infinitely many answers of the form $\{ys \mapsto [0]\}, \{ys \mapsto [z|0]\}, \ldots$ For instance, the first answer is computed by the following narrowing derivation (at each step, the narrowing relation $\rightsquigarrow$ is labelled with the applied substitution and rule[1], and the reduced subterm is underlined):

$$\underline{\text{last}(\text{ys})} = 0 \rightsquigarrow_{\{ys \mapsto xs\},R1} \underline{\text{append}(\text{ws}, [y])} = xs, y = 0$$
$$\rightsquigarrow_{\{ws \mapsto nil\},R2} (\underline{[y]} = xs, y = 0)$$
$$\rightsquigarrow_{\{y \mapsto 0\},(x=x \rightarrow true)} (true, \underline{[0]} = xs)$$
$$\rightsquigarrow_{\{xs \mapsto [0]\},(x=x \rightarrow true)} \top$$

Moreover, without assuming canonicity, Meseguer and Thati showed that narrowing is still complete as a procedure to solve reachability problems [62] (that is, to find "more general" solutions σ for the variables of s and t such that $s\sigma$ rewrites to $t\sigma$ in a number of steps). Reachability problems extend narrowing capabilities to a wider spectrum that includes the analysis of concurrent systems. Narrowing has also received much attention due to the many other important applications, such as automated proofs of termination [21], verification of cryptographic protocols [33], equational constraint solving [10], partial evaluation [16], program transformation [14] and model checking [34], among others.

Narrowing Strategies. Since unrestricted narrowing has quite a large search[2] space, several strategies to control the selection of redexes have been developed. A *narrowing strategy* (or *position constraint*) is any well-defined criterion which obtains a smaller search space by permitting narrowing to reduce only some chosen positions. A narrowing strategy φ can be formalized as a mapping that assigns a subset $\varphi(g)$ of $\overline{O}(g)$ to every input expression g (e.g. a goal different from $\top$) such that, for all $u \in \varphi(g)$, the goal g is narrowable at position u. An important property of a narrowing strategy φ is completeness, meaning that the narrowing constrained by φ is still complete. There is an inherited tradeoff coming from functional programming, between the benefits of outer evaluation of orthogonal (i.e. left-linear and overlap-free [73]), nonterminating rules and those of inner or eager evaluation with terminating, non-orthogonal rules. Also, under the eager strategy, programs are required not to contain extra-variables, that is, each program rule $\lambda \rightarrow \rho \Leftarrow C$ satisfies $Var(\rho) \cup Var(C) \subset Var(\lambda)$, whereas the weaker condition $Var(\rho) \subset Var(\lambda) \cup Var(C)$ is commonly demanded in lazy programs. A survey of results about the completeness of narrowing strategies can be found in [19]. To simplify our notation, we let $\mathbb{R}_\varphi$ denote the class of programs that satisfy the conditions for the completeness of the strategy φ.

Throughout this paper, we focus our attention on two very common narrowing strategies: the *leftmost-innermost* and the *leftmost-outermost* narrowing strategies. More specifically, we let $inn(g)$ (resp. $out(g)$) denote the narrowing strategy

---

[1] Substitutions are restricted to the input variables.

[2] Actually, there are three sources of non-determinism in narrowing: the choice of the equation within the goal, the choice of the redex within the equation, and the choice of the rewrite rule.

which selects the position p of the leftmost-innermost (resp. leftmost-outermost) narrowing redex of g. [3]

We formulate a conditional narrower with strategy φ, $\varphi \in \{inn, out\}$, as the smallest relation $\rightsquigarrow_\varphi$ satisfying

$$\frac{u = \varphi(g) \,\wedge\, (\lambda \to \rho \Leftarrow C) \ll \mathcal{R}_+^\varphi \,\wedge\, \sigma = mgu(\{g_{|u} = \lambda\})}{g \overset{\sigma}{\rightsquigarrow}_\varphi (C, g[\rho]_u)\sigma}.$$

For $\varphi \in \{inn, out\}$, $\mathcal{R}_+^\varphi = \mathcal{R} \cup Eq^\varphi$, where the set of rules Eq^φ models the equality on terms.

Namely, Eq^{out} is the set of rules that define the validity of equations as a *strict equality* between terms which is appropriate when computations may not terminate [63]:

$$c \approx c \to true \qquad\qquad\quad \% \ c/0 \in \mathcal{C}$$
$$c(x_1, \ldots, x_n) \approx c(y_1, \ldots, y_n) \to (x_1 \approx y_1) \wedge \ldots \wedge (x_n \approx y_n) \quad \% \ c/n \in \mathcal{C}$$

whereas Eq^{inn} is the standard equality defined by:

$$x = x \to true \qquad\qquad\qquad \% \ x \in \mathcal{V}$$

We also assume that equations in g and C have the form $s = t$ whenever we consider $\varphi = inn$, whereas the equations have the form $s \approx t$ when we consider $\varphi = out$. Note that an input equation like $f(a) = g(a)$ is not an acceptable goal when $\varphi = out$. In the following, this difference will be made explicit by using $=_\varphi$ to denote the standard equality $=$ of terms whenever $\varphi = inn$, whereas $=_\varphi$ is $\approx$ for the case when φ is out.

It is known that neither inn nor out are generally complete. For instance, consider $\mathcal{R} = \{f(y, a) \to true, f(c, b) \to true, g(b) \to c\}$ with input goal $f(g(x), x) =_\varphi true$. Then innermost narrowing only computes the answer $\{x \mapsto b\}$ for $f(g(x), x) = true$ whereas outermost narrowing only computes $\{x \mapsto a\}$ for the considered goal $f(g(x), x) \approx true$. For the completeness of a narrowing strategy, the following *uniformity* condition is required [66]: a confluent program is uniform iff the position selected by φ is a valid narrowing position for φ for all normalized substitutions (i.e. substitutions that only contain terms in normal form) applied to it. Note that the program $\mathcal{R}$ above does not satisfy the uniformity principle since the top position of the term $f(g(x), x)$ is not a valid narrowing position if we apply the substitution $\{x \mapsto b\}$ to this term. A sufficient condition for uniformity in constructor-based, canonical programs can be found in [32]. Moreover, there are methodologies which allow one to transform non-uniform programs into programs fulfilling the uniformity condition (e.g., see [9]).

Innermost narrowing is the foundation of several functional logic programming languages like SLOG [40], LPG [23] and (a subset of) ALF [41]. Also, the multi-paradigm language Maude [29] is equipped with a (kind of) innermost

---

[3] The leftmost-innermost position of g is the leftmost position of g that points to a pattern. A position p is leftmost-outermost in a set of positions O if there is no $p' \in O$ with either p' prefix of p, or $p' = q.i.q'$ and $p = q.j.q''$ and $i < j$, where i, j are natural numbers and q, q' sequences of natural numbers.

narrowing strategy (called *variant narrowing* [29]) that is part of an equational unification procedure. Moreover, reachability analyses for programs written in Maude rely on the so-called *topmost theories* [62], where the innermost strategy is often advantageous. Recently, the notion of *strategic narrowing* has been proposed as the main mechanism for the analysis of security policies in the strategy language Elan, relying on the confluence, termination and sufficient completeness of the underlying rewrite system [26]. In this context, innermost narrowing, innermost priority narrowing (i.e., innermost narrowing with a partial ordering on the program rules) and outermost narrowing have proven to be of prime interest [26].

Modern functional logic languages like Curry [44] and Toy [56] are based on lazy evaluation principles instead, which delay the evaluation of function arguments until their values are needed to compute a result. This allows one to deal with infinite data structures and avoids some unnecessary computations [43,41]. Needed narrowing [20] is a complete lazy narrowing strategy that is optimal w.r.t. the length of the derivations and the number of computed solutions in inductively sequential (IS) programs, Needed narrowing [20] can be easily and efficiently implemented by means of a transformation proposed in [45], which permits leftmost outermost narrowing to be used on the transformed program while preserving the answers computed by needed narrowing in the original program. Thanks to the possibility to use this transformation, we do not lose (much) generality by developing our methodology for the simpler leftmost outermost narrowing; this simplifies reasoning about computations, and consequently proving semantic properties, e.g. completeness.

Similarly to the other strategies discussed in this paper, needed narrowing adopts the classical theory of rewriting (that corresponds to run-time choice [49]) as underlying theory. However, in a run-time choice semantics, the values of the arguments are fixed as they are used, and the copies of the arguments created by parameter-passing may evolve independently afterwards [57]. Hence, classical rewriting is not valid for call-time choice evaluation, which is the operational semantics commonly adopted in functional logic languages dealing with non-strict, non-deterministic functions, and is related, at the operational level, to the sharing mechanism of lazy evaluation in functional languages [57,58]. Nevertheless, by adding a sharing mechanism to their encoding, needed narrowing implementations are sound for the call-time choice semantics of functional logic programs (for a discussion, see [57,58]). Moreover, for the deterministic programs considered in this paper, run-time and call-time are able to produce the same outcomes [58,69].

3.1   Two Functional Logic Program Denotations

The operational semantics $\mathcal{O}_{\varphi}^{ca}(\mathcal{R})$ of a functional logic program $\mathcal{R}$ w.r.t. the narrowing strategy $\varphi \in \{inn, out\}$ can be defined by considering all the possible successful narrowing derivations which can be obtained by applying the narrowing strategy φ to "most general calls". We denote by $\leadsto_{\varphi}$ the restriction of the narrowing relation that is obtained when the narrowing strategy φ is used.

Definition 3. *Let $\mathcal{R}$ be a program, $\varphi \in \{inn, out\}$. Then,*

$$\mathcal{O}_\varphi^{ca}(\mathcal{R}) = \Im_\mathcal{R}^\varphi \cup \{(f(x_1, \ldots, x_n) = x_{n+1})\theta \mid (f(x_1, \ldots, x_n) =_\varphi x_{n+1}) \rightsquigarrow_\varphi^{\theta\,*} \top$$
$$\textit{where } f/n \in \mathcal{D}, \quad x_{n+1} \textit{ and } x_i \textit{ are distinct variables,}$$
$$\textit{for } i = 1, \ldots, n \,\}$$

where $\Im_\mathcal{R}$ denotes the set of the identical equations $c(x_1, \ldots, x_n) =_\varphi c(x_1, \ldots, x_n)$ for all the constructor symbols c/n occurring in $\mathcal{R}$.

It is known that the considered operational semantics can be derived by a fixpoint computation which allows for the (bottom-up) construction of a model that is completely goal-independent. To this respect, in [6,7] we formalized a fixpoint semantics $\mathcal{F}_\varphi(\mathcal{R})$ —parametric w.r.t. the narrowing strategy φ— that can be calculated as the least fixpoint of a generalized version of the usual immediate consequence operator [47] $T_\mathcal{R}^\varphi$. Since the operator $T_\mathcal{R}^\varphi$ is continuous over the complete lattice of the Herbrand interpretations [7], the least fixpoint of $T_\mathcal{R}^\varphi$ (and hence the semantics) is generated by computing at most ω iterations of the operator $T_\mathcal{R}^\varphi$, that is $lfp(T_\mathcal{R}^\varphi) = T_\mathcal{R}^\varphi \uparrow \omega$. Therefore, the fixpoint semantics of a functional logic program can be defined as follows.

Definition 4. *The least fixpoint semantics of a program $\mathcal{R}$ in $\mathbb{R}_\varphi$ is defined as*

$$\mathcal{F}_\varphi(\mathcal{R}) = lfp(T_\mathcal{R}^\varphi) = T_\mathcal{R}^\varphi \uparrow \omega$$

where $\varphi \in \{inn, out\}$.

The fixpoint semantics $\mathcal{F}_\varphi(\mathcal{R})$ is more general than the operational semantics $\mathcal{O}_\varphi^{ca}(\mathcal{R})$ in the sense that it models both successful and partial (i.e. intermediate as well as non-terminating) computations, while $\mathcal{O}_\varphi^{ca}(\mathcal{R})$ catches only successful narrowing derivations. Therefore, a fixpoint characterization of the operational semantics can be derived from $\mathcal{F}_\varphi(\mathcal{R})$ by removing all those equations representing computations which are still incomplete or not terminating.

Given a set of equations S, let $partial(S)$ be an operator that selects those equations of S that do not model successful computations, i.e., computations that are still incomplete or do not terminate. In other words, we select all equations whose right-hand side is not a constructor term. More formally, $partial(S) = \{l = r \in S \mid r \notin \tau(\mathcal{C} \cup \mathcal{V})\}$.

Theorem 1. *[6] The following relation holds:*

$$\mathcal{O}_\varphi^{ca}(\mathcal{R}) = \mathcal{F}_\varphi(\mathcal{R}) - partial(\mathcal{F}_\varphi(\mathcal{R}))$$

4   Narrowing-Based Program Transformation

The folding and unfolding transformations, that were first introduced by Burstall and Darlington in [25] for functional programs, are the most basic and powerful techniques for a framework to transform programs. Unfolding is essentially the replacement of a call by its body, with appropriate substitutions. Folding is the inverse transformation, that is, the replacement of some piece of code by an equivalent function call. For functional programs, folding and unfolding steps involve only pattern matching. The fold/unfold transformation approach was first adapted to logic programs by Tamaki and Sato [72] by replacing matching with unification in the transformation rules. A lot of literature has been devoted to proving the correctness of fold/unfold systems w.r.t. the various semantics proposed for functional programs [25], logic programs [72], and constraint logic programs [35]. However, there are several other applications for fold/unfold rules besides providing a general theoretical basis for program transformation. For instance, such transformations have been used to formalize inductive programming frameworks for program synthesis as well as theory revision [24,4]. To this respect, an example of an unfolding-based theory revision technique for the automated repair of functional logic programs is described in Section 6.2.

Another important application is program analysis. Program analyses can be improved by iterating the unfolding of a program a finite number of times. In fact, an analysis is in general more accurate on the unfolded program than on the original program [11].

Example 2. Consider the following program $\mathcal{R}$ for addition and doubling of natural numbers in Peano's notation.

$$\begin{aligned}
\text{double}(x) &\rightarrow \text{add}(x, x). \\
\text{add}(0, x) &\rightarrow x. \\
\text{add}(s(x), y) &\rightarrow s(\text{add}(x, y)).
\end{aligned}$$

Now, given an equational unification problem $s = t$ in $\mathcal{R}$, consider the unsatisfiability analysis which is based on the idea on *non-joinability* of the root symbols of the normal forms of s and t. Namely, consider the abstract TRS $\mathcal{R}^{\alpha}$ that is obtained by abstracting the lhs's and rhs's of the rules in $\mathcal{R}$ using the abstraction function $\alpha(t) = f$ for $t = f(t_1, \ldots, t_n)$, whereas $\alpha(x) = c$, with $c \in \mathcal{C}$, for $x \in \mathcal{V}$:

$$\begin{aligned}
\text{double} &\rightarrow \text{add}. \\
\text{add} &\rightarrow 0. \\
\text{add} &\rightarrow s.
\end{aligned}$$

Then, the analysis consists in proving that $(s \downarrow)^{\alpha}$ and $(t \downarrow)^{\alpha}$ are not joinable in $\mathcal{R}^{\alpha}$, where $(u \downarrow)$ denotes the normal form of u in $\mathcal{R}$. Unfortunately, this analysis is too naïve (imprecise) to conclude the unsatisfiability of the equation $\text{double}(s(x)) = 0$, since the the normal form of $\text{double}(s(x))$ is $\text{add}(s(x), s(x))$, whose root symbol add can be reduced to 0 in $\mathcal{R}^{\alpha}$. However, by unfolding the first rule of $\mathcal{R}$ w.r.t. the rules for addition, we get the unfolded program $Unf(\mathcal{R})$ (see Definition 6 below):

$$\begin{aligned}
\text{double}(0) &\rightarrow 0. \\
\text{double}(s(x)) &\rightarrow s(\text{add}(x, s(x))). \\
\text{add}(0, x) &\rightarrow x. \\
\text{add}(s(x), y) &\rightarrow s(\text{add}(x, y)).
\end{aligned}$$

Now, by running the analysis in the unfolded program $Unf(\mathcal{R})$ instead of $\mathcal{R}$, the unsatisfiability of the considered equation $\texttt{double}(\texttt{s}(\texttt{x})) = 0$ follows.

In the functional logic setting, a natural way to program transformation is to use a form of narrowing-driven unfolding/folding, i.e., the expansion and the contraction, by means of narrowing, of program subexpressions using the corresponding definitions. A complete characterization of fold/unfold transformations w.r.t. computed answers in functional logic languages with eager/lazy semantics can be found in [12,14].

The use of narrowing empowers the fold/unfold system by implicitly embedding the instantiation rule (the operation of the Burstall and Darlington framework [25] which introduces an instance of an existing equation) into the fold/unfold operators by means of unification.

4.1   Unfolding Functional Logic Programs

Roughly speaking, *unfolding* a program $\mathcal{R}$ w.r.t. a rule r yields a new specialized version of $\mathcal{R}$ in which the rule r is replaced by new rules obtained from r by performing a narrowing step on the right-hand side of r. Typically, unfolding is non-deterministic, since several subterms in the right-hand side of a rule may be narrowable.

Definition 5 (Unfolding operators). *Let $\mathcal{R}$ be a program, and $\varphi \in \{inn, out\}$ be a narrowing strategy.*

(i) *Let $r_1, r_2 \ll \mathcal{R}$ such that $r_1 = (\lambda_1 \to \rho_1 \Leftarrow C_1)$ and $r_2 = (\lambda_2 \to \rho_2 \Leftarrow C_2)$. The* rule unfolding via φ of r_1 w.r.t. r_2 *is defined as follows*

$$\mathsf{U}^{\varphi}_{r_2}(r_1) = \{\lambda_1 \sigma \to \rho' \Leftarrow C' \mid (\rho_1 = y, C_1) \overset{\sigma, r_2, u}{\rightsquigarrow}_{\varphi} (\rho' = y, C'), u \in \overline{O}(\rho_1) \cup \overline{O}(C_1)\},$$

where y is a fresh variable.

(ii) *Let $r \ll \mathcal{R}$. The* rule unfolding of r w.r.t. $\mathcal{R}$ via φ *is as follows*

$$Unf^{\varphi}(\mathcal{R}, r) = \begin{cases} r & \text{if } \mathsf{U}^{\varphi}_{r'}(r) = \emptyset \text{ for each } r' \in \mathcal{R} \\ \bigcup_{r' \in \mathcal{R}} \mathsf{U}^{\varphi}_{r'}(r) & \text{otherwise} \end{cases}$$

Under a theoretical viewpoint, given a functional logic program, it is possible to define a semantics based on unfolding which is equivalent to its operational and fixpoint ones. This unfolding semantics helps to prove the equivalence between the operational and the fixpoint semantics of the language.

The formalization of such a semantics is as follows.

Definition 6 (Program Unfolding). *Let $\mathcal{R}$ be a program, and $\varphi \in \{inn, out\}$ be a narrowing strategy. The unfolding of a program $\mathcal{R}$ via φ is the program obtained by unfolding via φ the rules of $\mathcal{R}$ w.r.t. $\mathcal{R}$. Formally,*

$$Unf^{\varphi}(\mathcal{R}) = \bigcup_{r \in \mathcal{R}} \{Unf^{\varphi}(\mathcal{R}, r)\}.$$

The repeated application of the program unfolding operator leads to a sequence of equivalent programs which is inductively defined as follows.

Definition 7. *Let $\mathcal{R}$ be a program, and $\varphi \in \{inn, out\}$ be a narrowing strategy. The sequence:*

$$\mathcal{R}^0 \ \ = \mathcal{R}$$
$$\mathcal{R}^{i+1} = Unf^\varphi(\mathcal{R}^i), i \geq 0$$

is called the unfolding sequence starting from $\mathcal{R}$ via φ.

The unfolding semantics of a program is defined as the limit of the unfolding process described in Definition 7. Let us now formally define the *unfolding semantics* $\mathcal{U}_\varphi^{ca}(\mathcal{R})$ of a program $\mathcal{R}$. The main point of this definition is in compelling the right-hand sides of the equations in the denotation to be constructor terms. Recall that $\Im_\mathcal{R}$ be the set of identical equations $c(x_1, \ldots, x_n) =_\varphi c(x_1, \ldots, x_n)$, for each $c/n \in \mathcal{C}$.

Definition 8. *Let $\mathcal{R}$ be a program, and $\varphi \in \{inn, out\}$ be a narrowing strategy. Then,*

$$\mathcal{U}_\varphi^{ca}(\mathcal{R}) = \Im_\mathcal{R} \ \cup \ \bigcup_{i \in \omega} \{(s = d) \mid (s \to d) \in \mathcal{R}^i \ \ and \ d \in \tau(\mathcal{C} \cup \mathcal{V})\}$$

where $\mathcal{R}^0, \mathcal{R}^1, \ldots$ is the unfolding sequence starting from $\mathcal{R}$ via φ.

Finally, the following theorem formalizes a useful alternative characterization of the computed answers semantics $\mathcal{O}_\varphi^{ca}(\mathcal{R})$ in terms of unfolding.

Theorem 2. *Let $\mathcal{R} \in \mathbb{R}_\varphi$. Then, $\mathcal{U}_\varphi^{ca}(\mathcal{R}) = \mathcal{O}_\varphi^{ca}(\mathcal{R})$.*

4.2   Folding Functional Logic Programs

In the following, we introduce a folding transformation for the *inn* narrowing strategy that can be seen as an extension to functional logic programs of the reversible folding of [67] for logic programs. We have chosen this form of folding since it exhibits the useful, pursued property that the answer substitutions computed by innermost narrowing are preserved through the transformation. Actually, such a result does not hold for the *out* narrowing strategy.

Let us introduce the innermost folding operation. We use the following auxiliary notation. Let $\{r_1, \ldots r_n\}$ be a set of program rules and $\mathcal{R}$ be a program, then by $\{r_1, \ldots r_n\} \ll \mathcal{R}$, we denote the fact that $r_i \ll \mathcal{R}$, for each $i = 1, \ldots, n$.

Definition 9 (Innermost fold). *Let $\mathcal{R}$ be a program. Let $\{r_1, \ldots, r_n\} \ll \mathcal{R}$ (the "folded rules") and $R_{def} = \{r'_1, \ldots, r'_n\} \ll \mathcal{R}$ (the "folding rules") be two disjoint subsets of program rules (modulo renaming), with $r'_i = (\lambda'_i \to \rho'_i \Leftarrow C'_i)$, $i = 1, \ldots, n$. Let r be a rule[4], $u \in O(r)$ be a position of the rule r, and t be a pattern such that, for all $i = 1, \ldots, n$:*

---

[4] Roughly speaking, r is the "common skeleton" of the rules that are folded in the folding step. The occurrence u in r acts as the pointer to the "hole" where the folding call is let fall.

1. $\theta_i = mgu(\{\lambda_i' = t\})$,
2. $r_i = (\lambda \to \rho_i \Leftarrow C_i', C_i)\theta_i$ and $r[\rho_i']_u = (\lambda \to \rho_i \Leftarrow C_i)$, and
3. for any rule $r' = (\lambda' \to \rho' \Leftarrow C') \ll \mathcal{R}$ not in R_{def}, λ' does not unify with t.

Then, we define the folding of $\{r_1, \ldots, r_n\}$ in $\mathcal{R}$ using R_{def} as follows:

$$Fold(\mathcal{R}, \{r_1, \ldots, r_n\}, R_{def}) = (\mathcal{R} - \{r_1, \ldots, r_n\}) \cup \{r_{fold}\}$$

where $r_{fold} = r[t]_u$.

Intuitively, the folding operation proceeds in a contrary direction to the narrowing steps. In narrowing steps, for a given unifier of the redex and the left-hand side of the applied rule, a reduction step is performed on the instantiated redex, then the conditions of the unfolding rule are added to the unfolded one, and finally the narrowing substitution is applied. Here, first of all, folded rules are "deinstantiated" (generalized). Next, one gets rid of the conditions of the applied folding rules, and, finally, a reduction step is performed against the reversed heads of the folding rules.

Note that the folding operation has two sources of non-determinism. The first is in the choice of the folded calls; the second is in the choice of a generalization (folding call) of the heads of the instantiated function definitions which are used to substitute the folded calls.

Example 3. Let us consider the following program $\mathcal{R}$:

$$
\begin{array}{llll}
\mathtt{f(x)} & \to \mathtt{s(x)} & \Leftarrow \mathtt{h(s(x)) = 0} & (r_1) \\
\mathtt{f(s(z))} & \to \mathtt{s(s(0)))} & \Leftarrow \mathtt{z = 0} & (r_2) \\
\mathtt{num(y)} & \to \mathtt{y} & \Leftarrow \mathtt{h(y) = 0} & (r_3) \\
\mathtt{num(s(s(z)))} & \to \mathtt{s(s(0))} & \Leftarrow \mathtt{z = 0} & (r_4)
\end{array}
$$

Now, we can fold the rules $\{r_1, r_2\}$ of $\mathcal{R}$ w.r.t. $R_{def} = \{r_3, r_4\}$ using $r = (\mathtt{f(x)} \to \square)$ and $t = \mathtt{num(s(x))}$, obtaining the resulting program $\mathcal{R}'$:

$$
\begin{array}{llll}
\mathtt{f(x)} & \to \mathtt{num(s(x))} & & (r_{fold}) \\
\mathtt{num(y)} & \to \mathtt{y} & \Leftarrow \mathtt{h(y) = 0} & (r_3) \\
\mathtt{num(s(s(z)))} & \to \mathtt{s(s(0))} & \Leftarrow \mathtt{z = 0} & (r_4)
\end{array}
$$

In [12], it has been shown that the proposed fold transformation preserves the operational semantics $\mathcal{O}_{inn}^{ca}(\mathcal{R})$ of computed answer substitutions of functional logic programs under the usual conditions for the completeness of the *inn* strategy.

Theorem 3 (Strong correctness). *[12] Let $\mathcal{R} \in \mathbb{R}_{inn}$ be a program and $\mathcal{R}' = Fold(\mathcal{R}, \{r_1, \ldots, r_n\}, R_{def})$ be a folding of $\{r_1, \ldots, r_n\}$ in $\mathcal{R}$ using R_{def}. Then, we have that $\mathcal{O}_{inn}^{ca}(\mathcal{R}) = \mathcal{O}_{inn}^{ca}(\mathcal{R}')$.*

An extension of the narrowing-based fold/unfold transformation framework of [12,13] to rewriting logic theories as implemented in the functional programming language Maude [29] can be found in [3]. It allows one to deal with (non-deterministic) rules, equations, sorts and algebraic laws (like commutativity and associativity). This program transformation framework is also applied to to the problem of securing the transfer of code from a code producer to a code consumer by implement a Code Carrying Theory (CCT) system based on folding/unfolding transformations. CCT is an approach for securing delivery of code from a producer to a consumer where only a certificate (usually in the form of assertions and proofs) is transmitted from the producer to the consumer who can check its validity and then extract executable code from it. In the approach of [3], the certificate consists of a sequence of transformation steps which can be applied to a given consumer specification in order to automatically synthesize safe code in agreement with the original requirements. The key idea behind our CCT methodology is as follows. Assuming the code consumer provides the requirements in the form of a rewrite theory, the code producer can (semi-) automatically obtain an efficient implementation of the specified functions by applying a sequence of transformation rules. Moreover, having proved the correctness of the transformation system, the code producer can transmit as the required certificate just a compact representation of the sequence of transformation rules to the consumer so he does not need to manually construct any other correctness proof. By applying the transformation rules to the initial requirements, the code consumer can inexpensively obtain the executable code that can be eventually compiled to a different target language if needed.

5   Functional Logic Program Specialization

The aim of *partial evaluation* (PE) is to specialize a given program w.r.t. part of its input data (hence also called *program specialization*). PE has been widely applied in the field of functional programming (FP) [50] and logic programming (LP) [55]. Although the objectives are similar, the general methods are often different due to the distinct underlying computation models. This separation has the negative consequence of duplicated work since developments are not shared and many similarities are overlooked.

Narrowing-driven PE (NPE) [15] is the first generic algorithm for the specialization of functional logic programs. The method is formalized within the theoretical framework established in [55,61] for the partial evaluation of logic programs (also known as *partial deduction*, PD), although a number of concepts have been generalized to deal with nested function calls. The NPE approach has better opportunities for optimization thanks to the functional dimension (e.g. by the inclusion of deterministic simplification steps). Also, since unification is embedded into narrowing, it is able to automatically propagate syntactic information on the partial input (term structure) and not only constant values. The different instances of the framework which can be obtained by considering different narrowing strategies preserve some logical, strong (computed answers) program semantics under conditions easily ascertained by reusing methods and results developed for narrowing.

Given a program P and a set S of atoms, the aim of PD [55] is to derive a new program P' which computes the same answers for any input goal which is an instance of an atom in S. The program P' is obtained by gathering together the set of *resultants*, which are constructed as follows: for each atom A of S, i) first construct a finite SLD-tree, $T(A)$, for $P \cup \{\Leftarrow A\}$, then ii) consider the leaves of the non-failing branches of $T(A)$, say $G_1, \ldots, G_r$, and the computed substitutions along these branches, say $\theta_1, \ldots, \theta_r$, and finally iii) construct the clauses: $A\theta_1 \Leftarrow G_1, \ldots, A\theta_r \Leftarrow G_r$. The basic correctness of the transformation is ensured whenever P' is *S-closed*, i.e. every atom in P' is an instance of an atom in S. An *independence* condition, which holds if no two atoms in S have a common instance, is needed to guarantee that P' does not produce additional answers. The constructed SLD-trees can be viewed as (i) *symbolic computations* for the atoms in S; the S-closedness of P' illustrates the idea of (ii) *regularity* of a symbolic computation; and finally, (iii) *program extraction* from a set of SLD-trees consists basically in building up the associated set of resultant rules.

We now identify these three categories for narrowing-driven PE [15,17].

Symbolic Execution. It is similar to PD, but we use narrowing in the place of SLD-resolution. For a set S of terms (possibly with nested function calls) and a functional logic program $\{\lambda_i \to \rho_i \Leftarrow C_i\}_{i=1}^n$, a partial (finite) narrowing tree is constructed for each term in S. The inclusion of a deterministic, normalization process between narrowing steps improves the elimination of intermediate data structures and reduces the size of the specialized program since less choices are unfolded [8]. By exploiting the results on *normalizing narrowing* [41], this is achieved in a principled way which does not compromise termination. Control issues are managed by using standard techniques as in [61].

Search for Regularities. Our notion of regularity is similar to the PD closedness condition, which we have generalized to recurse over the terms in order to handle nested function calls. Informally, a term t is considered S-closed iff it only contains constructors and variables, or i) there exists a substitution θ such that $t\theta \in S$, and ii) the terms in θ are recursively S-closed. For instance, the term $f(g(0))$ is closed w.r.t. the set of calls $\{f(x), g(x)\}$.

Program Extraction. In order to extend the notion of resultant to our setting, we specialize single terms s, and consider derivations for initial goals $s = y$, where y is a fresh variable not occurring in s, that we extend down to the leaves $(C, t = y)$ (where C are the equations brought by the conditions of the applied program rules), and we extract the resultant as $(s\theta \to t \Leftarrow C)$.

There are two issues of correctness for a PE procedure: termination, i.e., given any input goal, execution should always reach a stage for which there is no way to continue; and (partial) correctness, i.e., (if execution terminates, then) the operational semantics of the goal with respect to the residual program and with respect to the original program should coincide.

As for termination, NPE involves two classical termination problems: the so-called *local* termination problem (the termination of unfolding, or how to control and keep the expansion of the narrowing trees which provide partial evaluations for individual calls finite), and the *global* termination (which concerns termination of recursive unfolding, or how to stop recursively constructing narrowing

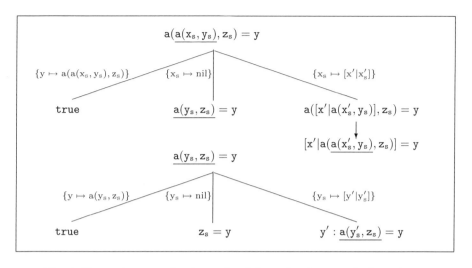

Fig. 1. Narrowing trees for the goals $a(a(x_s, y_s), z_s) = y$ and $a(x_s, y_s) = y$

trees while still guaranteeing that the desired amount of specialization is retained and that the closedness condition is reached). Actually, the set of terms S appearing in the goals with which the specialization is performed usually needs to be augmented in order to fulfill the closedness condition. This brings up the problem of how to keep this set finite throughout the PE process by means of some appropriate abstraction operator which guarantees termination. Control issues in narrowing-driven partial evaluation can be controlled by using standard techniques as in [61]. A detailed algorithm for the partial evaluation of functional logic programs can be found in [15], which is able to guarantee the termination of the specialization process.

A partial evaluation is defined as the set of resultants extracted from the derivations of the constructed partial narrowing trees, as illustrated in the following example.

Example 4. Consider again the function append of Example 1 with initial goal $\mathsf{append}(\mathsf{append}(x_s, y_s), z_s) = y$. This goal appends three lists by appending the first two, yielding an intermediate list, and then appending the last one to that. We evaluate the goal by using *normalizing* conditional narrowing (that is, each narrowing step is followed by the normalization of the narrowed goal w.r.t. the given CTRS). Starting with the sequence $q = \mathsf{append}(\mathsf{append}(x_s, y_s), z_s)$, we compute the trees depicted in Figure 1 for the sequence of terms

$$q' = (\mathsf{append}(\mathsf{append}(x_s, y_s), z_s), \mathsf{append}(x_s, y_s)).$$

Note that append has been abbreviated to a in the picture. Then we get the following residual program $\mathcal{R}'$:

$$\text{append}(\text{append}(\text{nil}, y_s), z_s) \rightarrow \text{append}(y_s, z_s)$$
$$\text{append}(\text{append}([x|x_s], y_s), z_s) \rightarrow [x|\text{append}(\text{append}(x_s, y_s), z_s)]$$
$$\text{append}(\text{nil}, z_s) \rightarrow z_s$$
$$\text{append}([y|y_s], z_s) \rightarrow [y|\text{append}(y_s, z_s)]$$

which is able to append the three lists by passing over its input only once. This result has been obtained thanks to using normalization. Note that no specific strategy has been employed for executing the goal, while the intended specialization has been achieved.

The use of efficient forms of narrowing can significantly improve the accuracy of the specialization method and increase the efficiency of the resulting program, because runtime optimizations are also performed at specialization time.

The behavior of a concrete narrowing-driven partial evaluator greatly depends on the narrowing strategy being used, since different strategies have quite different semantic properties. It is accepted that the use of an eager narrowing strategy is less convenient than a lazy one regarding the elimination of intermediate data structures, although the use of normalization may alleviate the problem, as said before. Generally speaking, if the operational semantics $\mathcal{O}_\varphi^{ca}(\mathcal{R})$ is to be preserved by program transformations, then the only reasonable class (for eager as well as for lazy PE) is that of left-linear, constructor-based (CB) programs, which are known to produce only constructor answers. These programs are generalized to the more general class of left-linear *rnf-based* programs, where all arguments of the left-hand sides of the rules are rigid normal forms, i.e. unnarrowable. Unfortunately, the construction of resultants may produce rewrite rules whose left-hand side contain nested function symbols, if the terms to be partially evaluated contain nested function symbols. If this kind of programs are allowed by the narrowing strategy being considered (e.g. unrestricted narrowing), there is no problem at all. However, when dealing with narrowing strategies which require constructor-based programs, a post-processing renaming transformation is mandatory in order to have an executable residual program [8]. Complex terms are 'folded' recursively, by replacing them by calls to new functions which satisfy the CB constraint. Furthermore, it can automatically guarantee that no additional answer is computed in the specialized program, which is otherwise ensured by an *independence condition* on the set of partially evaluated terms (as explained in [15]) guaranteeing that no *overlaps* exist between the specialized function calls.

Example 5. In Example 4, the resulting set of terms

$$\{\text{append}(\text{append}(x_s, y_s), z_s), \text{append}(x_s, y_s)\}$$

in q' is not independent. This example illustrates the need for an extra renaming phase able to produce an independent set of terms such as $\{\text{app\_3}(x_s, y_s, z_s), \text{app\_2}(x_s, y_s)\}$ and associated specialized program

$$\text{app\_3}(\text{nil}, y_s, z_s) \rightarrow \text{app\_2}(y_s, z_s)$$
$$\text{app\_3}([x|x_s], y_s, z_s) \rightarrow [x|\text{app\_3}(x_s, y_s, z_s)]$$
$$\text{app\_2}(\text{nil}, z_s) \rightarrow z_s$$
$$\text{app\_2}([y|y_s], z_s) \rightarrow [y|\text{app\_2}(y_s, z_s)]$$

which has the same computed answers as the original program append for the query $\mathrm{app\_3(x_s, y_s, z_s)}$ (modulo the renaming transformation).

The use of lazy narrowing during partial evaluation gives a better overall behavior regarding both the elimination of intermediate data structures and the propagation of information. Unfortunately, this approach introduces new drawbacks into the partial evaluation process. Firstly, the class of programs is not preserved by the transformation; for instance, orthogonality may be destroyed. On the other hand, the quality of the partially evaluated program may be degraded by introducing e.g. infinite computations which could not be proven in the original program [18]. The use of needed narrowing during partial evaluation overcomes both problems, since the structure of programs is preserved and no redundant or undesirable derivations are encoded in the residual program. Nevertheless, a new difficulty arises when the operation principle of *residuation* is integrated within the NPE framework. Namely, the difficulty lies in preserving the *floundering* behaviour of the original program, ensuring that there is a precise correspondence between the computations that suspend in the original and the specialized programs [2].

6   Declarative Debugging

Debugging programs with the combination of user-defined functions and logic variables is a difficult but important task which has deserved some interest in recent years, and different debugging techniques have been proposed. The idea behind declarative error diagnosis is to collect information about what the program is intended to do and compare this with what it actually does. Starting from these premises, a diagnoser can find errors. The information needed can be found in many different ways. It can be built by asking the user (as an oracle), or by means of a formal specification (or an older, correct, version of the program), or some combination of both.

Abstract diagnosis [30] is a declarative debugging framework that extends the methodology in [38,70] (which is based on using the immediate consequence operator to identify bugs in logic programs) to diagnoses w.r.t. computed answers. An important advantage of this framework is that it is goal-independent and does not require the determination of symptoms in advance. In [6,7], we generalized the declarative diagnosis methodology of [30] to the debugging of wrong as well as missing answers of functional logic programs.

In our setting, correctness as well as completeness of a program $\mathcal{R}$ are established by comparing the operational semantics $\mathcal{O}_\varphi^{ca}(\mathcal{R})$ to an intended semantics $\mathcal{I}_{ca}$ modeling the successful narrowing derivations that a programmer has in mind. More formally,

Definition 10. *Let $\mathcal{I}_{ca}$ be the intended success set semantics for program $\mathcal{R}$.*

1. *$\mathcal{R}$ is partially correct w.r.t. $\mathcal{I}_{ca}$, if $\mathcal{O}_\varphi^{ca}(\mathcal{R}) \subseteq \mathcal{I}_{ca}$.*
2. *$\mathcal{R}$ is complete w.r.t. $\mathcal{I}_{ca}$, if $\mathcal{I}_{ca} \subseteq \mathcal{O}_\varphi^{ca}(\mathcal{R})$.*
3. *$\mathcal{R}$ is totally correct w.r.t. $\mathcal{I}_{ca}$, if $\mathcal{O}_\varphi^{ca}(\mathcal{R}) = \mathcal{I}_{ca}$.*

If a program contains errors, these are signalled by corresponding *symptoms*. The "intended success set semantics" allows us to establish the validity of an atomic equation by a simple "membership" test, in the style of the *s*-semantics [36].

Definition 11. *Let $\mathcal{I}_{ca}$ be the intended success set semantics for $\mathcal{R}$. An incorrectness symptom is an equation e such that $e \in \mathcal{O}_{\varphi}^{ca}(\mathcal{R})$ and $e \notin \mathcal{I}_{ca}$. An incompleteness symptom is an equation e such that $e \in \mathcal{I}_{ca}$ and $e \notin \mathcal{O}_{\varphi}^{ca}(\mathcal{R})$.*

For the detection of buggy rules, however, we need to consider a "well-furnished" intended fixpoint semantics $\mathcal{I}_{\mathcal{F}}$ (such that $\mathcal{I}_{ca} \subseteq \mathcal{I}_{\mathcal{F}}$), which models successful as well as "in progress" (i.e., partial) computations, and enjoys the semantic properties of the denotation formalized in Definition 4, that is, $\mathcal{I}_{\mathcal{F}}$ should correspond to the fixpoint semantics of the correct program and $\mathcal{I}_{ca} = \mathcal{I}_{\mathcal{F}} - partial(\mathcal{I}_{\mathcal{F}})$.

An equation e is *uncovered* if it cannot be derived by any program rule using the intended fixpoint semantics, in symbols $e \in \mathcal{I}_{\mathcal{F}}$ and $e \notin T_{\mathcal{R}}^{\varphi}(\mathcal{I}_{\mathcal{F}})$. Having such a semantics, the diagnosis of buggy rules as well as the detection of uncovered equations can be performed by exploiting the following definitions.

Definition 12. *Let $\mathcal{I}_{\mathcal{F}}$ be the intended fixpoint semantics for $\mathcal{R}$. If there exists an equation $e \in T_{\{r\}}^{\varphi}(\mathcal{I}_{\mathcal{F}})$ s.t. e is not covered by $\mathcal{I}_{\mathcal{F}}$, then the rule $r \in \mathcal{R}$ is incorrect on e.*

Therefore, the incorrectness of rule r is signalled by a simple transformation of the intended semantics $\mathcal{I}_{\mathcal{F}}$.

Definition 13. *Let $\mathcal{I}_{\mathcal{F}}$ be the intended fixpoint semantics for $\mathcal{R}$. An equation e is uncovered in $\mathcal{R}$ if $e \in \mathcal{I}_{\mathcal{F}}$ and e is not covered by $T_{\mathcal{R}}^{\varphi}(\mathcal{I}_{\mathcal{F}})$.*

By the above definition, an equation e is uncovered if it cannot be derived by any program rule using the intended fixpoint semantics. In particular, we are interested in the equations of $\mathcal{I}_{ca} \subseteq \mathcal{I}_{\mathcal{F}}$ that are uncovered, i.e., $e \in \mathcal{I}_{ca}$ and e is not covered by $T_{\mathcal{R}}^{\varphi}(\mathcal{I}_{\mathcal{F}})$, since such equations represent missing computed answers.

Partial correctness of a program is established by the following proposition.

Proposition 1. *[7] If there are no incorrect rules in $\mathcal{R}$ w.r.t. the intended fixpoint semantics $\mathcal{I}_{\mathcal{F}}$, then $\mathcal{R}$ is partially correct w.r.t. the intended success set semantics $\mathcal{I}_{ca}$.*

Assuming that $\mathcal{I}_{\mathcal{F}}$ is finite, Proposition 1 shows a simple methodology to prove partial correctness. In the case when $\mathcal{I}_{\mathcal{F}}$ is not finite, this methodology can be still applied by considering finite approximations of the program semantics, as explained in Section 6.1 below. Completeness is harder, since it not possible to detect all possible uncovered equations by comparing the specification of the intended fixpoint semantics $\mathcal{I}_{\mathcal{F}}$ to $T_{\mathcal{R}}^{\varphi}(\mathcal{I}_{\mathcal{F}})$. In other words, the absence of uncovered equations does not allow us to derive that the program under examination is complete.

It is worth noting that checking the conditions of Definitions 12 and 13 requires just one application of $T_{\mathcal{R}}^{\varphi}$ to $\mathcal{I}_F$, while the standard detection based on symptoms [70] would require either an external oracle or the construction of the semantics, and therefore a fixpoint computation.

6.1   Abstract Diagnosis

In general, the diagnosis methodology we presented in Section 6 cannot be used to directly derive practical debuggers, since the correctness as well as completeness tests of Definitions 12–13 cannot be implemented in an effective way when the intended semantics $\mathcal{I}_{\mathcal{F}}$ is infinite, which is a very common case.

Following an idea inspired by [30], we defined an effective diagnosis methodology in [6,7], which is based on abstract interpretation [31].

Abstract interpretation formalizes the idea of "approximate computation" in which computation is performed with descriptions of data rather than with the data themselves. In particular, the semantics operators are replaced by abstract operators that are shown to 'safely' approximate the standard ones. In this context, our abstract diagnosis framework allows one to work on finite representations of the intended semantics $\mathcal{I}_F$ giving support to the implementation of finite diagnosis procedures.

More specifically, the basic idea is to consider two *finite* sets: $\mathcal{I}^+$ which over-approximates the intended fixpoint semantics $\mathcal{I}_{\mathcal{F}}$ and $\mathcal{I}^-$ which under-approximates $\mathcal{I}_{\mathcal{F}}$. In our methodology, an executable specification R_{spec} is given in order to effectively compute over- and under-approximations of the intended fixpoint semantics. Basically, we take the set which results from a finite number of iterations of the concrete immediate consequence operator $T_{\mathcal{R}_{Spec}}^{\varphi}$ as under-approximation $\mathcal{I}^-$, while $\mathcal{I}^+$ corresponds to the abstract fixpoint semantics of the abstract specification $R_{spec}^{\sharp}$, which is obtained from R_{spec} by replacing recursive function calls appearing in the specification's rules with occurrences of the special symbol $\sharp$. Such an abstraction allows us to avoid non-termination of the fixpoint computation, and provides a simple methodology for computing $\mathcal{I}^+$ which is satisfactory in practice.

We then use these sets $\mathcal{I}^+$ and $\mathcal{I}^-$ as shown in Theorems 4–5 in order to implement the abstract effective versions of the correctness/completeness tests of Definitions 12–13. Basically, the immediate consequence operator, $T_{\mathcal{R}}^{\varphi}$, (w.r.t. the program $\mathcal{R}$) is applied to $\mathcal{I}^-$ to check incorrectness w.r.t. $(\mathcal{I}^+, \mathcal{I}^-)$ and the abstract version of the immediate consequence operator, $T_{\mathcal{R}}^{\sharp \varphi}$ is applied to $\mathcal{I}^+$ to check incompleteness w.r.t. $(\mathcal{I}^+, \mathcal{I}^-)$.

Theorem 4. *Let $(\mathcal{I}^+, \mathcal{I}^-)$ be a correct approximation of the intended semantics $\mathcal{I}_{\mathcal{F}}$. If r is abstractly incorrect w.r.t. $(\mathcal{I}^+, \mathcal{I}^-)$ on e, then r is incorrect on e.*

Theorem 5. *Let $(\mathcal{I}^+, \mathcal{I}^-)$ be a correct approximation of the intended semantics $\mathcal{I}_{\mathcal{F}}$. If $\mathcal{R}$ is abstractly incomplete w.r.t. $(\mathcal{I}^+, \mathcal{I}^-)$ on e, then e is uncovered in $\mathcal{R}$.*

The previous theorems provide a compact description of the results proved in [6,7] and are the basis of the correctness of our abstract diagnosis framework.

The diagnosis w.r.t. approximate properties is always effective because the abstract specifications are finite. If no error is found, we say that $\mathcal{R}$ is *abstractly correct and complete w.r.t.* $(\mathcal{I}^+, \mathcal{I}^-)$. As one can expect, the results may be weaker than those that can be achieved on the concrete domain just because of the approximation: the fact that $\mathcal{R}$ is abstractly correct and complete w.r.t. $(\mathcal{I}^+, \mathcal{I}^-)$ does not generally imply the total correctness of $\mathcal{R}$ w.r.t. $\mathcal{I}$. The method is sound[5] in the sense that each error which is found by using $\mathcal{I}^+, \mathcal{I}^-$ is really a bug w.r.t. $\mathcal{I}$.

Example 6. Let us consider the following (wrong) Fibonacci program $\mathcal{R}$.

$$\begin{aligned}
\text{fib}(0) &\to 0. & \text{add}(0, x) &\to x. \\
\text{fib}(x) &\to \text{fibaux}(0, 0, x). & \text{add}(s(x), y) &\to s(\text{add}(x, y)). \\
\text{fibaux}(x, y, 0) &\to x. \\
\text{fibaux}(x, y, s(z)) &\to \text{fibaux}(y, \text{add}(x, y), z).
\end{aligned}$$

The specification is given by the following program $\mathcal{R}_{Spec}$:

$$\begin{aligned}
\text{fib}(0) &\to s(0). & \text{add}(0, x) &\to x. \\
\text{fib}(s(0)) &\to s(0). & \text{add}(s(x), y) &\to s(\text{add}(x, y)). \\
\text{fib}(s(s(x))) &\to \text{add}(\text{fib}(s(x)), \text{fib}(x)).
\end{aligned}$$

The abstract specification $\mathcal{R}^{\sharp}_{Spec}$ is

$$\begin{aligned}
\text{fib}(0) &\to s(0). & \text{add}(0, x) &\to x. \\
\text{fib}(s(0)) &\to s(0). & \text{add}(s(x), y) &\to s(\sharp). \\
\text{fib}(s(s(x))) &\to \text{add}(\sharp, \sharp).
\end{aligned}$$

Let $\varphi = inn$; then. After 2 iterations of the $T^{inn}_{\mathcal{R}_{Spec}}$ operator, we get the following under-approximation.

$$\begin{aligned}
\mathcal{I}^- = \{& 0 = 0, s(x) = s(x), \text{add}(x, y) = \text{add}(x, y), \text{fib}(x) = \text{fib}(x), \text{add}(0, x) = x, \\
& \text{add}(s(x), y) = s(\text{add}(x, y)), \text{add}(s(0), y) = s(y), \text{fib}(0) = s(0), \\
& \text{fib}(s(0)) = s(0), \text{fib}(s(s(x))) = \text{add}(\text{fib}(s(x)), \text{fib}(x)), \\
& \text{add}(s^2(x), y) = s^2(\text{add}(x, y)), \text{fib}(s(s(0))) = \text{add}(s(0), \text{fib}(0)), \\
& \text{fib}(s(s(0))) = \text{add}(\text{fib}(s(0)), s(0)) \\
& \text{fib}(s^3(x)))) = \text{add}(\text{add}(\text{fib}(s(x)), \text{fib}(x)), \text{fib}(s(x)))\}
\end{aligned}$$

The over-approximation $\mathcal{I}^+$ is given by the following set of equations (after three iterations of the $T^{\sharp inn}_{\mathcal{R}^{\sharp}_{Spec}}$ operator, we get the fixpoint):

$$\begin{aligned}
\mathcal{I}^+ = \{& 0 = 0, s(x) = s(x), \text{add}(x, y) = \text{add}(x, y), \\
& \text{fib}(x) = \text{fib}(x), \text{add}(0, x) = x, \text{add}(s(x), y) = s(\sharp), \\
& \text{fib}(0) = s(0), \text{fib}(s(0)) = s(0), \text{fib}(s(s(x))) = \text{add}(\sharp, \sharp), \\
& \text{fib}(s(s(x))) = \sharp, \text{fib}(s(s(x))) = s(\sharp)\}
\end{aligned}$$

---

[5] This is in contrast with the abstract diagnosis methodologies of [5,30], which work as follows: when the diagnoser finds that the program is correct, then it is certainly free of errors, whereas if an (abstract) error is reported, then it can be either a (concrete) error or not.

Now, consider the equation $\mathtt{fib(x)} = \mathtt{fib(x)}$ of $\mathcal{I}^-$. By applying $T^{inn}_{\{fib(0)\rightarrow 0\}}$ to this equation, we get the equation $e = \mathtt{fib(0)} = 0$, which is not covered by $\mathcal{I}^+$, i.e., it is not subsumed by any abstract equation of $\mathcal{I}^+$. This proves that r is incorrect on e.

6.2 Automated Program Correction

Inductive Logic Programming (ILP) is the field of Machine Learning concerned with learning logic programs from positive and negative examples, generally in the form of ground literals [64]. A challenging subfield of ILP is known as *inductive theory revision,* which is close to program debugging under the *competent programmer* assumption of [70]. In other words, the initial program $\mathcal{R}$ is assumed to be written with the intention of being correct and, if it is not, then a close variant $\mathcal{R}^c$ of it is. Automatic program correction attempts to find such a variant.

In this context, the correction problem can be stated as follows. Let $\mathcal{R}$ be a wrong program such that $\mathcal{R}' \subseteq \mathcal{R}$ is a set of wrong rules w.r.t. an intended semantics $\mathcal{I}_{\mathcal{F}}$. Let E^p and E^n be two disjoint sets of ground equations witnessing the correct as well as the wrong computational behaviour of $\mathcal{R}$. Equations in E^p (respectively, E^n) are called *positive examples* (respectively, *negative examples*).

The correction problem amounts to synthesizing a set of rules $\mathcal{X}$ such that

$$\mathcal{R}^c = (\mathcal{R} \setminus \mathcal{R}') \cup \mathcal{X}, \ \mathcal{R}^c \vdash_\varphi E^p \text{ and } \mathcal{R}^c \nvdash_\varphi E^n.$$

where $\mathcal{R}$ *entails* E using the narrowing strategy $\varphi \in \{inn, out\}$ (in symbols, $\mathcal{R} \vdash_\varphi E$) iff each $e \in E$ is successfully derived in $\mathcal{R}$ using the narrowing strategy φ (that is, $e \leadsto^*_\varphi \top$ in $\mathcal{R}$), and $\mathcal{R}$ *disproves* E using the narrowing strategy (in symbols, $\mathcal{R} \nvdash_\varphi E$) iff no $e \in E$ can be successfully derived in $\mathcal{R}$ using the narrowing strategy φ.

Program $\mathcal{R}^c$ is called *corrected* program (w.r.t. E^p and E^n). Roughly speaking, a corrected program $\mathcal{R}^c$ is a program that entails all the positive examples and disproves all the negative examples.

In [4], we developed an automated procedure for program correction which mainly follows the top-down, inductive learning approach known as example-guided unfolding [24], which uses unfolding as specialization operator to discriminate positive from negative examples. The basic idea of the method is to first specialize the program $\mathcal{R}$ by unfolding function calls in the right-hand sides of the rules yielding a close variant $\mathcal{R}'$ of $\mathcal{R}$. Then, we obtain $\mathcal{R}^c$ by removing those rules of $\mathcal{R}'$ which are responsible for the derivation of the negative examples.

For example, consider the following program $\mathcal{R}$

$$\mathtt{even(0)} \rightarrow \mathtt{true} \qquad\qquad (r_1)$$
$$\mathtt{even(s(x))} \rightarrow \mathtt{even(x)} \qquad\qquad (r_2)$$

which is wrong w.r.t. the usual intended semantics of the **even** function. Moreover, let E^p be $\{\mathtt{even(0)} = \mathtt{true},\ \mathtt{even(s^2(0))} = \mathtt{true},\ \mathtt{even(s^4(0))} = \mathtt{true}\}$ and E^n be $\{\mathtt{even(s(0))} = \mathtt{true},\ \mathtt{even(s^3(0))} = \mathtt{true},\ \mathtt{even(s^5(0))} = \mathtt{true}\}$.

The wrong program $\mathcal{R}$ can be first transformed into an equivalent program $\mathcal{R}'$ by unfolding rule (r_2) as follows:

$$\text{even}(0) \rightarrow \text{true} \qquad\qquad (r_1)$$
$$\text{even}(\text{s}(0) \rightarrow \text{true} \qquad\qquad (r_2{}')$$
$$\text{even}(\text{s}(\text{s}(\text{x}))) \rightarrow \text{even}(\text{x}) \qquad\qquad (r_2{}'')$$

Then, note that we can obtain the desired corrected program by simply removing rule r_2' from $\mathcal{R}'$.

Soundness of this approach has been proven in [4].

The unfolding-based correction procedure presented above is known to produce a correction when the initial program is *overly general* (with some extra outfit which is needed to specialize recursive definitions [24]); that is, it allows us to prove all positive examples and some incorrect ones. Unfortunately, most of the programs to be debugged are not overly general, and hence our correction methodology cannot be directly applied. Therefore, we coupled the example-guided unfolding approach with a generalization technique in order to correct programs that do not fulfill the applicability condition (*over-generality*). The methodology consists in applying a bottom-up pre-processing to "generalize" the initial wrong program, before proceeding to the usual top-down correction. Roughly speaking, we extend the original erroneous program with new synthesized rules so that the entire example set E^p succeeds w.r.t. the generalized program, and hence the top-down corrector can be effectively applied.

The generalization method exploits the bottom-up technique for the inductive learning of functional logic programs developed by Ferri, Hernández and Ramírez [39] which automatically infers new program rules from sets of ground examples. The induction process is based on *inverse narrowing* — a variant of Muggleton's inverse resolution operator [64]— which essentially reverses the classical deductive inference process in order to generate valid premises (typically, in the form of logic programs) from known consequences (i.e. examples).

The resulting blend of top-down and bottom-up synthesis is conceptually cleaner than more sophisticated, purely top-down or bottom-up ones and combines the advantages of both techniques.

7   Related Work and Concluding Remarks

Finding program bugs is a long-standing problem in software construction. Unfortunately, the debugging support is rather poor for functional languages (see [60] and references therein), and there are no good general-purpose semantics-based debuggers available.

In the field of multi-paradigm declarative languages, standard trace debuggers are based on suitably extended box models which help to display the execution. Due to the complexity of the operational semantics of (functional) logic programs, the information obtained by tracing the execution is difficult to understand. Several authors follow the idea of algorithmic declarative debugging in the style proposed by Shapiro [70]: an oracle (typically the user) is supposed to endow the debugger with error symptoms, as well as to correctly answer oracle questions driven by proof trees aimed at locating the actual source of errors. A debugger for the functional logic language Escher based on this methodology is proposed in [54]. Unfortunately, when debugging real code, the questions are often textually large and may be difficult to answer. Following the generic scheme which is based on proof trees of [65], a procedure for the declarative debugging of wrong answers in higher-order functional logic programs is proposed in [28]. This is a semi-automatic debugging technique where the debugger tries to locate the node in an execution tree which is ultimately responsible for a visible bug symptom. A declarative debugger (for wrong answers) based on this methodology was developed for the lazy functional logic language TOY and adapted to Curry. The methodology in [28] includes a formalization of computation trees which is precise enough to prove the logical correctness of the debugger and also helps to simplify oracle questions. Missing answers are debugged in [27].

As far as we know, none of the above-mentioned debuggers integrates both diagnosis and correction capabilities in a uniform and seamless way. As a matter of fact, program correction has scarcely been studied in the context of declarative programming. In [70], a theory revision framework for correction purposes has been proposed; however, it requires the user either to strongly interact with the debugger or to manually correct the code. Automated correction of faulty codes has been investigated in concurrent logic programming. In [1], a framework for the diagnosis and the correction of Moded flat GHC programs has been developed. This framework exploits strong mode/typing and constraint analysis in order to locate bugs; then, symbols which are likely sources of error are syntactically replaced by other program symbols so that new slightly different programs (mutations) are produced. Finally, mutations are newly checked for correctness. This approach is essentially able to correct *near misses* (i.e., wrong variable/constant occurrences), but no mistakes involving predicates or function symbols can be repaired. Moreover, only modes and types are employed to come up with a corrected program; no finer semantic information is taken into consideration which might improve the quality of the repair.

We are not aware of any formal antecedent of the narrowing-driven approach in the PE literature. A closer, automatic approach is that of positive supercompilation [71], whose basic operation is *driving* [74], a unification-based transformation mechanism which is somewhat similar to (lazy) narrowing. Another related work is the framework of *conjunctive partial deduction* (CPD), which aims at achieving unfold/fold-like transformations within fully automated PD [52]. Similarly to conjunctive partial deduction [52] and supercompilation [74], NPE combines some good features of deforestation [75], partial evaluation [50], and PD [55,61].

All the narrowing–based techniques for program transformation and debugging that are overviewed in this paper have been implemented in a collection of tools that are publicly available at www.dsic.upv.es/users/elp/soft.html

Acknowledgements

This paper is a modest attempt to summarize twenty years of research on narrowing-based program manipulation in Italy by reviewing the main lines of research and contributions of the authors in the following fields: program transformation, partial evaluation and program debugging of functional logic programs. Many thanks are due to Francisco Correa, Ginés Moreno, and Germán Vidal, large part of the material here reported was developed with their collaboration. Finally, we are very thankful to the anonymous referees for their remarks that allowed us to improve our paper.

References

1. Ajiro, Y., Ueda, K.: Kima — an Automated Error Correction System for Concurrent Logic Programs. Automated Software Engineering 19, 67–94 (2002)
2. Albert, E., Alpuente, M., Hanus, M., Vidal, G.: A Partial Evaluation Framework for Curry Programs. In: Ganzinger, H., McAllester, D., Voronkov, A. (eds.) LPAR 1999. LNCS (LNAI), vol. 1705, pp. 376–395. Springer, Heidelberg (1999)
3. Alpuente, M., Baggi, M., Ballis, D., Falaschi, M.: A Fold/Unfold Transformation Framework for Rewrite Theories and its Application to CCT. In: Proc. 2010 ACM SIGPLAN Symp. on Partial Evaluation and Semantics-based Program Manipulation (PEPM), pp. 43–52. ACM, New York (2010)
4. Alpuente, M., Ballis, D., Correa, F.J., Falaschi, M.: Automated Correction of Functional Logic Programs. In: Degano, P. (ed.) ESOP 2003. LNCS, vol. 2618, pp. 54–68. Springer, Heidelberg (2003)
5. Alpuente, M., Comini, M., Escobar, S., Falaschi, M., Lucas, S.: Abstract Diagnosis of Functional Programs. In: Leuschel, M., Bueno, F. (eds.) LOPSTR 2002. LNCS, vol. 2664, pp. 1–16. Springer, Heidelberg (2003)
6. Alpuente, M., Correa, F., Falaschi, M.: Declarative Debugging of Functional Logic Programs. In: Gramlich, B., Lucas, S. (eds.) Proc. Int'l Workshop on Reduction Strategies in Rewriting and Programming, WRS 2001. ENTCS, vol. 57. Elsevier, Amsterdam (2001)
7. Alpuente, M., Correa, F., Falaschi, M.: A Debugging Scheme for Functional Logic Programs. In: Hanus, M. (ed.) Proc. 10th Int'l Workshop on Functional and (Constraint) Logic Programming, WFLP 2001. ENTCS, vol. 64. Elsevier, Amsterdam (2002)
8. Alpuente, M., Falaschi, M., Julián, P., Vidal, G.: Specialization of Lazy Functional Logic Programs. In: Proc. ACM SIGPLAN Conf. on Partial Evaluation and Semantics-Based Program Manipulation, PEPM 1997. Sigplan Notices, vol. 32(12), pp. 151–162. ACM Press, New York (1997)
9. Alpuente, M., Falaschi, M., Julián, P., Vidal, G.: Uniform Lazy Narrowing. Journal of Logic and Computation 13(2), 287–312 (2003)
10. Alpuente, M., Falaschi, M., Levi, G.: Incremental Constraint Satisfaction for Equational Logic Programming. Theoretical Computer Science 142(1), 27–57 (1995)
11. Alpuente, M., Falaschi, M., Manzo, F.: Analyses of Unsatisfiability for Equational Logic Programming. Journal of Logic Programming 22(3), 221–252 (1995)
12. Alpuente, M., Falaschi, M., Moreno, G., Vidal, G.: Safe folding/unfolding with conditional narrowing. In: Hanus, M., Heering, J., Meinke, K. (eds.) ALP 1997 and HOA 1997. LNCS, vol. 1298, pp. 1–15. Springer, Heidelberg (1997)

13. Alpuente, M., Falaschi, M., Moreno, G., Vidal, G.: An Automatic Composition Algorithm for Functional Logic Programs. In: Jeffery, K., Hlaváč, V., Wiedermann, J. (eds.) SOFSEM 2000. LNCS, vol. 1963, pp. 289–297. Springer, Heidelberg (2000)
14. Alpuente, M., Falaschi, M., Moreno, G., Vidal, G.: Rules + Strategies for Transforming Lazy Functional Logic Programs. Theoretical Computer Science 311(1-3), 479–525 (2004)
15. Alpuente, M., Falaschi, M., Vidal, G.: Narrowing-driven Partial Evaluation of Functional Logic Programs. In: Riis Nielson, H. (ed.) ESOP 1996. LNCS, vol. 1058, pp. 45–61. Springer, Heidelberg (1996)
16. Alpuente, M., Falaschi, M., Vidal, G.: Partial Evaluation of Functional Logic Programs. ACM Transactions on Programming Languages and Systems 20(4), 768–844 (1998)
17. Alpuente, M., Falaschi, M., Vidal, G.: A Unifying View of Functional and Logic Program Specialization. ACM Computing Surveys 30(3es), 9es (1998)
18. Alpuente, M., Hanus, M., Lucas, S., Vidal, G.: Specialization of functional logic programs based on needed narrowing. TPLP 5(3), 273–303 (2005)
19. Antoy, S.: Evaluation strategies for functional logic programming. J. Symb. Comput. 40(1), 875–903 (2005)
20. Antoy, S., Echahed, R., Hanus, M.: A Needed Narrowing Strategy. Journal of the ACM 47(4), 776–822 (2000)
21. Arts, T., Giesl, J.: Termination of Term Rewriting using Dependency Pairs. TCS 236(1-2), 133–178 (2000)
22. Bellia, M., Levi, G.: The relation between logic and functional languages. Journal of Logic Programming 3, 217–236 (1986)
23. Bert, D., Echahed, R.: On the Operational Semantics of the Algebraic and Logic Programming Language LPG. In: Reggio, G., Astesiano, E., Tarlecki, A. (eds.) Abstract Data Types 1994 and COMPASS 1994. LNCS, vol. 906, pp. 132–152. Springer, Heidelberg (1995)
24. Bostrom, H., Idestam-Alquist, P.: Induction of Logic Programs by Example–guided Unfolding. Journal of Logic Programming 40, 159–183 (1999)
25. Burstall, R.M., Darlington, J.: A Transformation System for Developing Recursive Programs. Journal of the ACM 24(1), 44–67 (1977)
26. Santana de Oliveira, A., Kirchner, C., Kirchner, H.: Analysis of Rewrite-Based Access Control Policies. In: Proc. 3rd Int'l Workshop on Security and Rewriting Techniques, SecreT 2008. ENTCS. Elsevier, Amsterdam (2008)
27. Caballero, R., Rodríguez Artalejo, M., del Vado Vírseda, R.: Declarative Diagnosis of Missing Answers in Constraint Functional-Logic Programming. In: Garrigue, J., Hermenegildo, M.V. (eds.) FLOPS 2008. LNCS, vol. 4989, pp. 305–321. Springer, Heidelberg (2008)
28. Caballero-Roldán, R., López-Fraguas, F.J., Rodríquez Artalejo, M.: Theoretical Foundations for the Declarative Debugging of Lazy Functional Logic Programs. In: Kuchen, H., Ueda, K. (eds.) FLOPS 2001. LNCS, vol. 2024, pp. 170–184. Springer, Heidelberg (2001)
29. Clavel, M., Durán, F., Eker, S., Lincoln, P., Martí-Oliet, N., Meseguer, J., Talcott, C.: All About Maude - A High-Performance Logical Framework. Springer, New York (2007)
30. Comini, M., Levi, G., Meo, M.C., Vitiello, G.: Abstract diagnosis. Journal of Logic Programming 39(1-3), 43–93 (1999)
31. Cousot, P., Cousot, R.: Abstract Interpretation: A Unified Lattice Model for Static Analysis of Programs by Construction or Approximation of Fixpoints. In: Proc. Fourth ACM Symp. on Principles of Programming Languages, pp. 238–252 (1977)

32. Echahed, R.: On completeness of narrowing strategies. In: Dauchet, M., Nivat, M. (eds.) CAAP 1988. LNCS, vol. 299, pp. 89–101. Springer, Heidelberg (1988)
33. Escobar, S., Meadows, C., Meseguer, J.: A Rewriting-Based Inference System for the NRL Protocol Analyzer and its Meta-Logical Properties. TCS 367(1-2), 162–202 (2006)
34. Escobar, S., Meseguer, J.: Symbolic model checking of infinite-state systems using narrowing. In: Baader, F. (ed.) RTA 2007. LNCS, vol. 4533, pp. 153–168. Springer, Heidelberg (2007)
35. Etalle, S., Gabbrielli, M.: Modular Transformations of CLP Programs. In: Sterling, L. (ed.) Proc. 12th Int'l Conf. on Logic Programming. The MIT Press, Cambridge (1995)
36. Falaschi, M., Levi, G., Martelli, M., Palamidessi, C.: A new Declarative Semantics for Logic Languages. In: Kowalski, R., Bowen, K. (eds.) Proc. Fifth Int'l Conf. on Logic Programming, pp. 993–1005. The MIT Press, Cambridge (1988)
37. Fay, M.: First Order Unification in an Equational Theory. In: Proc. of 4th Int'l Conf. on Automated Deduction, CADE 1979, pp. 161–167 (1979)
38. Ferrand, G.: Error Diagnosis in Logic Programming, and Adaptation of E.Y.Shapiro's Method. Journal of Logic Programming 4(3), 177–198 (1987)
39. Ferri, C., Hernández, J., Ramírez, M.J.: Incremental Learning of Functional Logic Programs. In: Kuchen, H., Ueda, K. (eds.) FLOPS 2001. LNCS, vol. 2024, pp. 233–247. Springer, Heidelberg (2001)
40. Fribourg, L.: SLOG: a logic programming language interpreter based on clausal superposition and rewriting. In: Proc. Second IEEE Int'l Symp. on Logic Programming, pp. 172–185. IEEE, New York (1985)
41. Hanus, M.: The Integration of Functions into Logic Programming: From Theory to Practice. Journal of Logic Programming 19&20, 583–628 (1994)
42. Hanus, M.: A unified computation model for functional and logic programming. In: Proc. 24th ACM Symp. on Principles of Programming Languages, Paris, pp. 80–93. ACM, New York (1997)
43. Hanus, M.: Multi-paradigm Declarative Languages (invited tutorial). In: Dahl, V., Niemelä, I. (eds.) ICLP 2007. LNCS, vol. 4670, pp. 45–75. Springer, Heidelberg (2007)
44. Hanus, M., Kuchen, H., Moreno-Navarro, J.J.: Curry: A Truly Functional Logic Language. In: Proc. ILPS 1995 Workshop on Visions for the Future of Logic Programming, pp. 95–107 (1995)
45. Hanus, M., Prehofer, C.: Higher-Order Narrowing with Definitional Trees. Journal of Functional Programming 9(1), 33–75 (1999)
46. Hanus, M. (ed.): Curry: An Integrated Functional Logic Language (ver. 0.8.2) (2006), http://www.informatik.uni-kiel.de/~curry
47. Hölldobler, S.: Foundations of Equational Logic Programming. LNCS (LNAI), vol. 353. Springer, Heidelberg (1989)
48. Hullot, J.M.: Canonical Forms and Unification. In: Bibel, W. (ed.) CADE 1980. LNCS, vol. 87, pp. 318–334. Springer, Heidelberg (1980)
49. Hussman, H.: Unification in Conditional-Equational Theories. In: Caviness, B.F. (ed.) EUROCAL 1985. LNCS, vol. 204, pp. 543–553. Springer, Heidelberg (1985)
50. Jones, N.D., Gomard, C.K., Sestoft, P.: Partial Evaluation and Automatic Program Generation. Prentice-Hall, Englewood Cliffs (1993)
51. Klop, J.W.: Term Rewriting Systems. In: Abramsky, S., Gabbay, D., Maibaum, T. (eds.) Handbook of Logic in Computer Science, vol. I, pp. 1–112. Oxford University Press, Oxford (1992)

52. Leuschel, M., De Schreye, D., de Waal, A.: A Conceptual Embedding of Folding into Partial Deduction: Towards a Maximal Integration. In: Maher, M. (ed.) Proc. the Joint Int'l Conf. and Symp. on Logic Programming, JICSLP 1996, pp. 319–332. The MIT Press, Cambridge (1996)
53. Levi, G., Palamidessi, C., Bosco, P.G., Giovannetti, E., Moiso, C.: A complete semantics caracterization of K-LEAF, a logic language with partial functions. In: Proc. Second IEEE Symp. on Logic In Computer Science, pp. 318–327. IEEE, New York (1987)
54. Lloyd, J.W.: Debugging for a declarative programming language. Machine Intelligence 15 (1998)
55. Lloyd, J.W., Shepherdson, J.C.: Partial Evaluation in Logic Programming. Journal of Logic Programming 11, 217–242 (1991)
56. López-Fraguas, F.J., Sánchez Hernández, J.: Toy: A multiparadigm declarative system. In: Narendran, P., Rusinowitch, M. (eds.) RTA 1999. LNCS, vol. 1631, pp. 244–247. Springer, Heidelberg (1999)
57. López-Fraguas, F.J., Rodríguez-Hortalá, J., Sánchez-Hernández, J.: A simple rwwrite notion for call-time choice semantics. In: Proc. 9th International ACM SIGPLAN Conference on Principles and Practice of Declarative Programming (PPDP 2007), pp. 197–208. ACM, New York (2007)
58. López-Fraguas, F.J., Rodríguez-Hortalá, J., Sánchez-Hernández, J.: A flexible framework for programming with non-deterministicfunctions. In: Proc. 2009 ACM SIGPLAN Symp. on Partial Evaluation and Semantics-based Program Manipulation (PEPM), pp. 91–100. ACM, New York (2009)
59. López-Fraguas, F.J., Rodríguez-Hortalá, J., Sánchez-Hernández, J.: Narrowing for first order functional logic programs with call-time choice semantics. In: Seipel, D., Hanus, M., Wolf, A. (eds.) INAP 2007. LNCS, vol. 5437, pp. 206–222. Springer, Heidelberg (2009)
60. Marlow, S., Iborra, J., Pope, B., Gill, A.: A Lightweight Interactive Debugger for Haskell. In: Keller, G. (ed.) Proceedings of the ACM SIGPLAN Workshop on Haskell, Haskell 2007, Freiburg, Germany, September 30, pp. 13–24. ACM, New York (2007)
61. Martens, B., Gallagher, J.: Ensuring Global Termination of Partial Deduction while Allowing Flexible Polyvariance. In: Sterling, L. (ed.) Proc. ICLP 1995, pp. 597–611. MIT Press, Cambridge (1995)
62. Meseguer, J., Thati, P.: Symbolic reachability analysis using narrowing and its application to verification of cryptographic protocols. Higher-Order and Symbolic Computation 20(1-2), 123–160 (2007)
63. Moreno-Navarro, J.J., Rodríguez-Artalejo, M.: Logic Programming with Functions and Predicates: The language Babel. Journal of Logic Programming 12(3), 191–224 (1992)
64. Muggleton, S.: Inductive Logic Programming. New Generation Computing 8(3), 295–318 (1991)
65. Naish, L.: A declarative debugging scheme. Journal of Functional and Logic Programming 1997(3) (April 1997)
66. Padawitz, P.: Computing in Horn Clause Theories. EATCS Monographs on Theoretical Computer Science, vol. 16. Springer, Berlin (1988)
67. Pettorossi, A., Proietti, M.: Transformation of Logic Programs: Foundations and Techniques. Journal of Logic Programming 19,20, 261–320 (1994)
68. Reddy, U.S.: Narrowing as the Operational Semantics of Functional Languages. In: Proc. Second IEEE Int'l Symp. on Logic Programming, pp. 138–151. IEEE, New York (1985)

69. Riesco, A., Rodríguez-Hortalá, J.: Programming with Singular and Plural Non-deterministic Functions. In: Proc. 2010 ACM SIGPLAN Symp. on Partial Evaluation and Semantics-based Program Manipulation (PEPM), pp. 83–92. ACM, New York (2010)

70. Shaphiro, E.Y.: Algorithmic Program Debugging. The MIT Press, Cambridge (1982)

71. Sørensen, M.H., Glück, R., Jones, N.D.: A Positive Supercompiler. Journal of Functional Programming 6(6), 811–838 (1996)

72. Tamaki, H., Sato, T.: Unfold/Fold Transformations of Logic Programs. In: Tärnlund, S. (ed.) Proc. Second Int'l Conf. on Logic Programming, Uppsala, Sweden, pp. 127–139 (1984)

73. TeReSe (ed.): Term Rewriting Systems. Cambridge University Press, Cambridge (2003)

74. Turchin, V.F.: The Concept of a Supercompiler. ACM Transactions on Programming Languages and Systems 8(3), 292–325 (1986)

75. Wadler, P.L.: Deforestation: Transforming programs to eliminate trees. Theoretical Computer Science 73, 231–248 (1990)

25 Years of Applications of Logic Programming in Italy

Alessandro Dal Palù[1] and Paolo Torroni[2]

[1] Dip. di Matematica, Università di Parma
alessandro.dalpalu@unipr.it
[2] DEIS, Università di Bologna
paolo.torroni@unibo.it

Abstract. We present a review of practical applications of Logic Programming appeared in Italy since 1985. We classify them according to their area of application and discuss some trends emerged in the latest developments. Notwithstanding this survey is far to be comprehensive, it shows that Logic Programming successfully evolved and quickly adapted to new challenges offered by a notable variety of application areas.

1 Introduction

The beginning of Logic Programming (LP) applications was driven by the investments and interest by private industries, attracted by the novelty and potentialities of this technology. The enthusiasm of those years was great. Around the time GULP was founded in 1987, the main Italian ICT and consumer electronics event, SMAU, was discovering Artificial Intelligence (AI) [16].

AI made its debut in the Italian market from a variety of stands. Around the same time, IJCAI was being held in Milan, witnessing an already intense research activity in the academic world. The heterogeneous mix of AI promoters included small enterprises of academic roots, such as Delphi, a University of Pisa's spin-off then located in Viareggio, and big actors such as IBM, and included many more in between. Back then AI mainly meant Expert Systems, and the use of Prolog inference engines and the adoption of declarative technologies in general was considered a very promising approach. Nixdorf Italia, involved in Esprit-2 research projects and in the development of air fleet optimization tools for Alitalia, was using a development environment written in Prolog, called Twaice. IBM, Unisys, Pirelli Informatica and Datitalia Processing, among others, were all promoting expert systems for configuration and diagnosis which made use of knowledge bases and declarative rules. IBM was pushing expert systems technologies by announcing a series of AI courses.

The mid of 90s witnessed a change in the impact of LP into industries [130]. Even if, e.g., Constraint Logic Programming started to be employed as an effective tool to solve complex and industrial problems, applications were being developed as stand-alone projects rather than integrated in the production line mainstream. Programmers were not familiar with Prolog and thus it was difficult

A. Dovier, E. Pontelli (Eds.): 25 Years of Logic Programming, LNCS 6125, pp. 300–328, 2010.

to estimate costs and performances of projects that included LP critical components. Often the choice was then to move to traditional technologies. Finally, LP
was a tool developed and used by researchers and little low-level programming
skills were transferred to students [88].

In the latest 15 years, the scenario evolved and expanded widely. We are witnessing the birth and rise of new application domains, alongside the evolution of
older application domains in which LP had been present since the 80s, addressed
with renewed vigor motivated by more modern LP-based solutions. Probably,
LP technologies are now ripe and stakeholders ready for their adoption. We can
rely on more efficient algorithms, on the achievements of CLP, on more powerful
machines, and on a better understanding of the their theoretical underpinnings.
Another reason why there are so many academic LP applications in Italy nowadays may be that, recently, funding agencies privilege applied research projects
rather than basic research. This happens both at the National level, and at the
European level, where most of Italian research is seeking funding.

On the downside, comparatively little Prolog is used for applications - despite
the availability of many implementations of it. We still note, as 15 years ago,
that only some companies use Prolog. However, from the LP education point
of view, the situation has improved since 1995. In 2007, GULP ran a survey
to evaluate the extent of computational logic teaching at Italian universities. It
turns out that nowadays declarative programming is being taught in 20 out of 94
Italian universities in around 50 courses, at various levels in computer science
and engineering curricula. Some of these courses have been running for as long
as 20 years. In 80% of the cases, the syllabus includes practical lab sessions
that teach students how to use SWI Prolog, SICStus, ECLiPSe or other Prolog
engines, Answer Set Programming (ASP) solvers such as DLV, SAT solvers and
model checkers.

This is the first work that surveys LP applications in Italy in 25 years. Due
to abundance of material we came across, we had to leave out many interesting
applications. Exhaustiveness was not our aim. In particular, we restricted this
survey to applications developed in Italy. Even if these are only a fraction of the
international panorama, they represent a significant body of work. Besides, our
resources did not permit us to run a more accurate investigation of industrial
applications. However, the managed to put together a very rich selection of
experiences, and at the same time to keep the scope of this overview as broad
as possible.

Our presentation is mainly thematic, and only marginally historical. The
reader interested in the historical developments of LP research and applications
in Italy will find more information in [121]. However, it is interesting to note
the extent to which applications domains have increased in the last 25 years. In
the early days, applications of LP especially focussed on a little number of domains, such as robotics. During two decades instead many other new areas were
explored and, more importantly, in very recent years each application domain,
new or old, was addressed by at least one application.

2   Methodology and Organization of Contributions

The applications reported in this chapter, and the information we gathered around them, is mainly the result of bibliography search and polls submitted to mailing lists. In particular, we used the GULP mailing list and similar means in order to contact colleagues that are or were involved in projects dealing with LP. The answers to our poll gave us a non exhaustive, but significant collection of projects and experiences, often described with passion and enthusiasm.

There are two drawbacks of our investigation that we could not handle with our resources. The first one is that older applications are more difficult to dig and for this reason our report appears to be biased towards more recent ones. Some information about the early applications of LP in Italy can be found in [130]. The other drawback is that potentially many industrial applications are left out, since we had no means to reach projects that are completely independent from academia. Considering also the relevant number of students graduated with (at least) the basics of LP, it is reasonable to imagine a number of independent solutions that we are not aware of. Even if strongly biased by our poll method, it is our belief that in the years the number of industrial applications has not dramatically increased. We can also observe the lack of an organized and international developer's network that is often a necessary condition for technology transfer and dissemination in the industrial and commercial entourage.

In the presentation we arbitrarily divided the various applications into 14 categories, which correspond to individual subsections (see below):

1. industrial and commercial applications;
2. knowledge and information extraction, management and integration;
3. time tabling and rostering;
4. robotics;
5. graphics and design;
6. agent systems;
7. education, learning and cultural heritage;
8. software engineering;
9. verification;
10. natural language;
11. health care;
12. reasoning;
13. bioinformatics;
14. decision support, risk analysis and alarms.

It is a complex task to create independent partitions, thus some applications may fall into more than one category. We tried to cluster them according to their main application area.

Figure 1 shows the number of mentioned applications, as we categorized them. The figure does not reflect any statistically meaningful information, but it serves to show that LP has been broadly applied to very diverse areas.

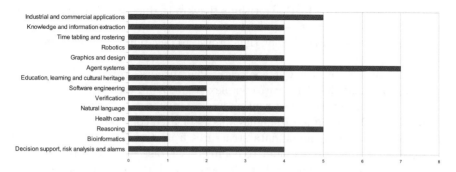

Fig. 1. The distribution of applications reported in this chapter

3   Applications

We present here the list of applications, divided according to the main application areas. We wish to clarify beforehands the intended meaning for the concept of "application", since this could be interpreted in different manners. As commonly used, the term "application" refers to some kind of project, developed outside of the academia, in which a certain technology plays a central role. In our survey, this kind of application goes under the category of *industrial and commercial applications*. However these application are the minority, if compared to the number of projects in which, for example, LP is used to solve problems arising from other academic disciplines. Therefore, in this manuscript we also consider to be an *application*, in an extended meaning, any project that was inspired by any context (including academic and public administration), that involved some LP technology and that developed at least to the stage of a working prototype. Nevertheless, it is worthwhile noticing that many of the "non-industrial/non-commercial" LP applications we reviewed have evolved into prototypes and systems that have been, or are being further developed by industrial or commercial actors.

3.1   Industrial and Commercial Applications

One of the first commercial LP applications was OMAR (late 80s): an interactive system, developed by Momigliano and colleagues, for predictive and reactive routing of the Alitalia fleet(s) [134]. Its kernel was developed entirely in Quintus Prolog and it consisted of 20000 lines of code. It was the largest system of its kind. OMAR was not an expert system, but a scheduler in which routing was modeled as a Constraint Satisfaction Problem (CSP), and constraints were about maintenances, schedule optimization/flexibility, geopolitical issues and so on. It had a GUI written in C that made use of the Quintus interface, whereas the interface with the Alitalia flight information system was handled separately in SQL.

Since at that time there were no dedicated CLP languages yet, Prolog proved to be an ideal modeling and solving tool. Performance was improved by OMAR relying on approximated CSP algorithms (backward/forward checking, network consistency) rather than by standard backtracking, and by making non-determinism as local as possible. Prolog thus proved to be a powerful prototyping and delivery language.

Scientifically, the project was a success. The authors proved and experimentally confirmed that the worst-case complexity was n^2 in the number of aircrafts and tasks. That meant that fleet routing was completed, e.g., in half a minute for the DC-9 fleet (26 aircrafts, 170 flights plus associated constraints) including the time to compute the dynamic data structures from the Alitalia database. Human experts took half to one hour to complete the task. Quality-wise, OMAR's routing solutions were comparable with those of a (human) senior scheduler.

Unfortunately, in spite of the technical quality and academic recognition of the OMAR project, because of mere data ownership issues, OMAR could never be used in practice.

In the early 90s we find two notable applications which also made it to the market. IDEA [124] was an intelligent data retrieval system designed by Sancassani and colleagues at DS Logics srl in Bologna for the Epidemiological Observatory of Emilia Romagna. IDEA attempted to formalize the interaction between the data logical view and its real physical organization in databases. SICStus-Objects (a Prolog distribution extended with object-oriented features) was used to achieve an efficient implementation of an intelligent inference agent, whose task was to translate an epidemiologist's request for data into appropriate database queries and vice versa. Logic played a central role. The use of Prolog, with its well-understood semantics, made the semantics of IDEA reasoning rules clear and easy to debug. Prolog was used both to implement logical inference but also other non-logical components.

At the same time as IDEA, the SECReTS expert system [32] was being developed, in Prolog, by Chiopris and colleagues at ICON srl in collaboration with an Italian credit institute (BPS). SECReTS was sold to and used by several Italian banks to support the analysis of client-specific data. The application was successful in dealing with large amounts of data coming from the Italian Central Bank's Risk Center, whereas other traditional tools used earlier often failed to identify meaningful events and to set a clear boundary between monitoring and diagnosis. LP could address situations that required non-trivial reasoning, which languages such as SQL could not, and it was suitable for implementing classification and diagnosis procedures that help providing the user with an intuitive general view of the framework, instead of a sequence of raw alarms.

In recent times, Mascardi and colleagues (Genova) implemented a MAS prototype [89,21] for Ansaldo STS, that monitors processes running in a railway signalling plant, detects functioning anomalies, and provides support to the early notification of problems to the Command and Control System Assistance. The MAS has been implemented using DCaseLP, a multi-language prototyping environment that provides libraries for integrating tuProlog[1] agents into Jade

---

[1] tuProlog [54,116] is a Prolog engine written in Java. See
http://www.alice.unibo.it/xwiki/bin/view/Tuprolog/

(see [13] for more information on these tools). Prolog has proven to be extremely well suited to the implementation of monitoring agents because of their intrinsic rule-based nature. The developed prototype was essential to demonstrate the functioning of the railway signalling system, which is now being developed using procedural languages.

Another optimization problem coming from a famous shoe industry was solved by Meneghetti (Udine) [92], with the application of Constraint Logic Programming (CLP). Floor storage system is used in the shoe industry to store fashion products of seasonal collections with low quantity and high variety. Since space is costly and order picking must be rapid, stacking of shoe boxes should be optimized. The problem is modeled by assigning an integer code to each box basing on shoe characteristics (model, type, color, and size) and trying to force similar boxes into near locations to improve workers ability of fast order retrieval. The model is encoded in CLP and solved comparing different strategies, also using Large Neighborhood Search. Mixing CLP and LNS revealed a powerful methodology for solving allocation problems in floor storage systems for the shoe industry. Furthermore, the declarative nature of CLP allows the programmer to easily describe what properties are required to the desired solution. Requirements can be modified, added or deleted to adhere to a dynamic industrial environment without changing the basic model, but only declaring new constraints, making it adaptable and transferrable to different industrial realities.

3.2 Knowledge and Information Extraction, Management and Integration

LP technologies have been used in the late 90s until 2004 by Milanese and colleagues (Udine) in the realization of MIRAGGIO: a prototypical design support system oriented to the dynamic generation and modification of design data and process models [38,37,96]. The MIRAGGIO prototype was implemented in LPA[2] Prolog++ [136]. It was based on a mixed Object-Oriented and frame-based approach, with the aim of defining a flexible environment for design process modeling, for the reuse of previous project results, the dynamic reconfiguration and consistency check of data structures and the reuse of previously developed design policies. To this end MIRAGGIO was using a two-level KDB. At the base level lied a knowledge database managing rule-based engineering and design knowledge, whereas at the meta-level lied the design system management rules, aimed to implement autonomous design strategy planning and decision support capabilities based on historical records and to specify its interactive behaviour with the designer.

More recently information integration has largely been oriented towards ontologies. The use of LP languages such as Prolog to implement taxonomic reasoning is not new. We mention the Omega system for taxonomies by Attardi (DELPHI s.p.a., Viareggio) and Simi (Pisa) dating back to 1986 [8]. However, it is with the recent integration of ontological reasoners into high-performance LP

---

[2] Logic Programming Associated ltd, London.

reasoners such as SWI-Prolog and DLV that we start to have the first applications with concrete potential for real-world exploitation. In the Italian landscape, Leone and colleagues (Calabria) have been developing ontology-based knowledge management solutions for several application domains based on the DLV system. We mention some of them, and forward the interest reader to [18,75].

The industrial exploitation of DLV [82] in the area of Knowledge Management is explored by two University of Calabria's spin-off companies: Exeura s.r.l. and DLVSYSTEM s.r.l. The latter licenses and maintains DLV. The former maintains three industrial products: OntoDLV (ontology management) [117,118], OLEX (document classification) [125,46], and HiLeX (information extraction) [123,122], incorporating the DLV system as the computational core. OntoDLV is an ontology management and reasoning system, OLEX is a document classification system and HiLeX is an information extraction system.

DLV has also been applied in the context of knowledge management at CERN (the European Laboratory for Particle Physics) for knowledge manipulation on large databases; in the context of E-Tourism and Automatic Itinerary Search (in projects funded by the Government of the Calabria Region), and in the Team Building application, developed for the port authority of the Gioia-Tauro Seaport.

Finally, INFOMIX [81] is an information integration application capable of dealing with incomplete and inconsistent data, built in cooperation with RODAN systems, a commercial DBMS developer.

In all cases, according to Leone [80], the key factor for the success of this sort of applications was the expressiveness of the ASP language and the ease of use of the ASP system DLV, more than its efficiency.

3.3   Time Tabling and Rostering

Time tabling and rostering are some of the oldest applications of Logic Programing. Since 1985, Monfroglio has been developing and maintaining a high school time-tabling application written in LPA Prolog [100]. The application has been successfully experimented and adopted in educational institutions with more than 50 classes. In the 80s the application would take a couple of hours to return a results (much less on today's hardware). The advantages of this application with respect to other commercially available software is that it is flexible, since the user can propose an initial personal timetable, and that its output is of higher quality, since it is possible to achieve a better distribution of the teaching load.

Since 2004, Gavanelli (Ferrara) has been applying CLP for timetabling at Ferrara University's School of Engineering [72]. The application was written in CLP(FD) ECLiPSe. This is because timetabling is a classical combinatorial optimization problem, for which CLP(FD) is particularly suited. The system accommodates non-overlapping constraints about students, teaching staff and venues, and other constraints specifying student and teaching staff requirements, such as those regarding the equipment, capacity of classrooms, distance, etc. Gavanelli's system is currently used. The application is successful. It provides timely results, months before the start of lessons, thus enabling students to plan their curriculum well in advance. Moreover, it was able to eliminate the overlapping between elective and compulsory courses, thus addressing an existing unfortunate situation which was previously unsolved. Finally, CLP made it possible to specify some constraints which were out of bound for hand-written timetables, such as constraints on the timetables of students who switch curriculum from one year to another. The application, which has produced its output as a HTML document, is now interfaced instead with Google calendar: a more flexible format which enables the integration with other applications developed at the University of Ferrara.

Outside of the educational environment, among the first timetabling applications we also mention the industrial-level one developed by Momigliano and colleagues within the OMAR project (see above).

Similar to the school timetabling problem, the crew rostering problem was also addressed using LP by Caprara, Focacci, Lamma, Mello, Milano, Toth and Vigo (Bologna) in 1998. To this end, the authors use a mixture of AI and OR techniques, namely OR efficient procedures based on a mathematical approach to the problem, and CLP for knowledge representation, achieving declarativeness, non-determinism and an incremental style of programming. This line of research would later very successfully evolve into the integration of CP, AI, and OR techniques [15].

3.4   Robotics

The very birth of LP in Italy happened in tight connection with research on Robotics. The first Prolog installation dates back in 1974, when Pagello brought from Grenoble a couple of tapes and two boxes full of punchcards with a copy of Alain Colmerauer's Prolog interpreter written in FORTRAN IV. Pagello had the interpreter running on a Univac 1108 machine at Politecnico di Milano. Later in 1975 a Prolog interpreter was also available at the University of Padova. In this way, Pagello and his colleagues could start developing Prolog applications for planning in robotic environments, addressing the Robot plan-formation problem introduced in Edinburgh by Bobrow. At that time, Bobrow's problem was being addressed by Hewitt, Sussman, McDermott and others in the US, mainly by way of procedural constructs.

The results of this line of research was tested on a physical robot: a Cartesian manipulator provided by Olivetti. The main purpose was the automatic generation of action sequences for robotic mechanical assembly. The Olivetti robot was programmed in a language developed by Pagello and colleagues at the University of Padova and Politecnico di Milano. LP would provide the tools needed to design more easily a higher level planning system to automatically generate lower level instructions for the robot. The practical impact of this application was however limited. First order predicate calculus turned out to be suitable only for a not too complex and quite stable working environment with little indeterminism, whereas more complex assembly problems requiring sophisticated sensors, and mobile robotics, would be out of reach. The dichotomy between procedural and declarative programming grew larger while the limitations of the former were emphasized by the increasing complexity of robotic devices.

The interest in robotic application did not die with this first experience. In 1993 Natali, Omicini and Zanichelli (Bologna) [109] proposed an architecture for a robot programming environment based on Prolog, extended with control capability towards program structuring and concurrence. The architecture reflected the interest, growing at that time, for the intelligent agent programming paradigms and theories. Thanks to 15 years of advances in LP, this research had overcome much of the limitations of earlier attempts, however its outcomes were not directly applied to physical robots. Ten years later, Galizia [70] uses Disjunctive Logic Programming, and answer set planning in particular, to improve the Space Shuttle's planning capabilities under system malfunction conditions. Galizia implemented a system in DLV, building on work previously conducted at Texas Tech University's KR Lab [110]. Experimental evaluation was done by comparing the outputs of the newer system with those of the older one, but not on physical robots.

Interestingly, we find that in recent times the application of LP techniques is attracting the attention of strong Robotics research groups again. In 2007 Scalzo, Nardi and colleagues [25] define a robotic architecture that enables the design of robotic systems which exploit context information to adapt the behaviour of all their subsystems. Context information is modeled explicitly, to allow for automated reasoning to operate and suitably affect the robot's behaviour. LP rules specify the robot subsystems' control. During a control cycle, the system acquires data from the sensors and from its subsystems (e.g., navigation and planning), extracts symbolic contextual information from the analysis of data, and feeds it into a knowledge base and rule-based reasoning system, which in turn feeds control parameters back to the robot's subsystems. The approach has been implemented and validated on simulated and on physical robots.

3.5   Graphics and Design

In the graphics domain, we find applications of LP research mainly on computer-aided design and manufacturing and in 3D recognition.

In the late 80s, Milanese and colleagues (Udine) investigated the integration of the Graphics Kernel System (GKS) standard and Prolog, both for graphical entities modeling and for the implementation of a distributed graphics system [94]. Prolog was considered for its descriptive programming style, enabling modeling of graphical entities via their basic elements and relations. Besides, it would accommodate simple transformation rules to modify or create objects of increasing complexity. In fact, one of the limitations of GKS, which was starting to become a wide-spread standard, was its inability to build complex objects bottom-up starting from simpler ones, to modify parts of graphical objects, and to associate them with information regarding their nature and functionality. Milanese and colleagues extended Prolog with communication and modularization constructs to propose a distributed, multi-processor GKS model, using a two-level architecture. The base level would manage general implementation schemas, while the meta-level would accommodate their instantiation in the overall graphics scheme. Meta-programming would accommodate customization and reconfiguration methodologies in a flexible way.

In the early 90s, Milanese worked with Dulli and Visentin (Padova) on the definition of the KADMOS declarative language and its KAMPE developing environment to create and manipulate hierarchical models of graphical entities for CAD/CAM applications [95,59,58,60]. KADMOS was a mixed logic programming and object-oriented language. Inference was used to create graphical entities that need to be validated with respect to design constraints, whereas object-orientation was needed for inheritance, classification, and modularization. Prolog was chosen as the implementation language for KADMOS.

In a research carried out between 1999 and 2002, Gavanelli, Lamma, Piccardi (Ferrara), Cucchiara (Modena), Mello, and Milano (Bologna) applied the CSP paradigm to 3D object recognition [44,45]. They proposed an approach for recognizing 3D CAD-made objects in complex range images containing several overlapped and different objects. The reasoning engine was based on Interactive CSP, to guide the acquisition of surfaces on-demand and focus only on significant image parts.

More recently, Farenzena and Fusiello (Verona) and Dovier (Udine) studied the use of interval analysis and CLP to obtain an accurate geometric model of a scene that rigorously takes into account the propagation of data errors and roundoff [68].

3.6 Agent Systems

LP has given a large contribution to the development of agent and Multi-Agent System (MAS) specifications and verification languages and techniques. Yet, before agents had become so popular, at the end of the 80s Terna (Torino) implemented a Prolog for microeconomic behaviour simulation which we can also include in this section [132]. The program was modeling the Bank of Italy, Industry and Unions, to build an interaction system and implement in this way a sort of economics game.

Towards the end of the 90s, Ciampolini, Mello and Torroni (Bologna) and Lamma (Ferrara) developed an architecture and language for multi-agent hypothetical reasoning based on Abductive Logic Programming (ALP). The architecture, ALIAS [33], was considering agents made of two modules: at the bottom level one for hypothetical reasoning, and at the top level one for communication. The language, LAILA [34], was offering communication primitives to accommodate distributed reasoning in collaboration, to produce globally consistent results, or in competition, to produce locally consistent results. ALIAS has been only applied to toy examples in agent negotiation, recommendation systems and judicial evaluation of criminal evidence [35]. However, the ideas proposed in ALIAS have been further developed in the DARE system developed at Imperial College London [84].

Contemporary to the development of ALIAS, we find work on logic-based agents by Torroni (Bologna) and Toni and Sadri (Imperial College London) [127,128]. The use of ALP was crucial to the definition of negotiation policies in a declarative way, with an operational execution model underneath. These ideas remained central in the later EU-funded SOCS project,[3] in which 6 universities in Europe, including Ferrara, Bologna and Pisa, collaborated in the definition of a computational logic model for the description, analysis and verification of global and open societies of heterogeneous logic based agents, named computees [133]. The SOCS models of agents and agent interaction were heavily based upon proof procedures for (various extensions of) LP. In particular, the operational model for KGP agents [79] relies upon CIFF [85], a proof procedure for ALP with constraints, and Gorgias, for LP with priorities [53]. The operational model for agent societies [1] instead relies upon SCIFF [2], a proof procedure for ALP with arbitrarily quantified variables, CLP constraints, dynamic event handling and reasoning with expectations. KGP and SCIFF have been applied to a variety of domains, including normative MAS [126], recommendation systems, ambient intelligence [131], business process interaction, medical guidelines, Web service choreographies [29,30], agent-oriented requirements engineering [24], and argumentation [135].

One of the most recent applications of SCIFF is the modeling and verification of declarative and open interaction models specified in the graphical ConDec language [115]. A mapping has been defined between ConDec and SCIFF [103]. Thanks to such a mapping, and to a number of SCIFF-based tools and extensions, it is possible to monitor and verify at run-time business process executions with respect to the model, and analyze traces after execution (process mining) [26,27]. To this end, a Pro-M[4] plug-in has been implemented based on SCIFF. The fully-fledged specification, verification and analysis framework is called CLIMB (Computational Logic for the verification and Modeling of Business processes and choreographies)[5]. CLIMB was used in national projects and on some case studies of chemical and physical analysis of waste water [83] in collaboration with HERA, a private agency, and ENEA, the Italian authority for energy and environment, and also in collaboration with other private manufacturing companies. The application of SCIFF for the static verification of business processes has been evaluated and successfully compared with state-of-the-art model checking techniques [104] to find that it offers greater flexibility and scalability in a number of realistic cases.

The SOCS project also produced the PROSOCS agent platform [20] incorporating KGP agents and SCIFF to support the social infrastructure for inter-agent interactions. PROSOCS has been extended within the more recent ARGUGRID project,[6] which involved 8 partners, including Pisa and 3 industrial European partners. ARGUGRID focused on e-business applications of LP-based

---

[3] See http://lia.deis.unibo.it/research/socs/
[4] See http://prom.win.tue.nl/tools/prom/
[5] See http://lia.deis.unibo.it/research/climb/
[6] See http://www.argugrid.eu

argumentation agents to support the decision-making of intelligent agents "representing" buyers and sellers of products in electronic marketplaces (e-marketplaces).[7] Agents in ARGUGRID are built using the GOLEM agent environment [22], which is a generalisation of the PROSOCS platform. The GOLEM platform is developed using Java and LP tools such as tuProlog (see above). GOLEM is specified in the ambient event calculus [23], a logic-based formalism that supports the representation of a distributed agent environment as a persistent composite structure evolving over time. Such a complex structure supports the interaction between agents, objects, and containers, entities that have their own external observable state and can be distributed over a network. Following the successful deployment of GOLEM in ARGUGRID, the system is being deployed in a commercial setting in collaboration with Thinking Safe Ltd, UK,[8] to provide resilience in autonomic networks and support business continuity.[9]

Since 1999, Costantini and Tocchio (L'Aquila) have been developing the DALI platform to specify agents and MAS based on computational logics [43]. DALI is interoperable with other FIPA-compliant agent platforms.[10] DALI is a general-purpose and agent-oriented logical language implemented in SICStus Prolog. DALI has been used for industrial applications and is patent pending. It is also used to teach AI and agents at L'Aquila.[11]

Since 1998, Omicini, Ricci, Denti, Viroli (Bologna, Cesena campus) Zambonelli (Modena and Reggio-Emilia) and Cremonini (Milano, Crema campus) have been working on the TuCSoN service infrastructure[12] [113,114] for the coordination and communication among independent and concurrent software components, such as agents. Interaction relies on tuple centers, i.e., programmable tuple spaces characterized by a reactive behaviour that can be programmed at run-time. Tuples are first-order Prolog terms. The behaviour of tuple spaces is expressed via the ReSpecT language [112]. TuCSoN is written in Java, while ReSpecT relies on the tuProlog library (see above). To date, TuCSoN has been used in a number of applications, including the implementation of the Agent Coordination Context [119], to manage the interaction space between agent and environment, of the minority game [111], for experimentation of new simulation models for systems biology, based on MAS, especially in relation with the Agents & Artifacts model [101], in the implementation of a workflow management system prototype for virtual organizations [120], and for pervasive smart environments.

See [13] for a more detailed account of the relations between the declarative and the MAS communities in Italy.

---

[7] For more insight on argumentation in LP, see [74], in this book.

[8] See http://www.thinkingsafe.com/

[9] See http://cacm.acm.org/news/44273

[10] FIPA is an IEEE Computer Society standards organization that promotes agent-based technology and the interoperability of its standards with other technologies. See http://www.fipa.org/

[11] See http://www.di.univaq.it/stefcost/

[12] See http://alice.unibo.it/xwiki/bin/view/TuCSoN/

3.7   Education, Learning and Cultural Heritage

Since its early days, dating back to 1985, LP has been used for education of high school teachers and in experimental projects with their students by Casadei (Bologna). Prolog was introduced and used for problem solving in diverse disciplines, to foster discussion and improve general problem solving skills, and to run problem solving competitions. From the previously mentioned enquiry made by GULP in 2007, about the teaching of LP at Italian Universities, it emerges that Prolog and other LP languages are currently being used in many curricula for teaching subjects such as AI reasoning, knowledge representation, logics and theorem proving.

Among others, Bandini, Mosca and Palmonari (Milano Bicocca) used DLV for education-related initiatives, such as archeological analysis and classification of antique ceramics [108,87,107]. This project was conducted in collaboration with Bologna's Archeology department and more recently with the Archeometry research group at the University of Barcelona (EURAB). The project delivered a system for the automatic generation of stratigraphic diagrams, and for related abductive (diagnostic) reasoning tasks [86].[13] DLV lends itself very well to the integration with external computational resources, that is components, written in other procedural languages, for the efficient computation of functions and data structures such as lists and sets. Data integration in DLV was particularly useful. Preliminary results on this application are available at the *Ipotesi di Preistoria* Web site.[14]

DLV has also been used in two other e-learning applications: MASEL and EXAM. The former [71] is an e-learning platform that features an intelligent core tha is able to build semi-automatically learning paths form a database of learning objects by exploiting the DLV system. EXAM [77] is a complete on-line exam taking portal. Teachers and students are assisted in the whole process of assessment test building, exam taking, and test correction. One of the most interesting features of the portal is the possibility to automatically generate assessment tests based on user dened constraints. The assessment test generation engine of the EXAM portal exploits DLV.

From 2005 to 2007, Costantini, Mostarda, Tocchio and Tsintza (L'Aquila) exploited DALI agents (see above) to implement two ambient-intelligence scenarios. One involves cultural assets fruition, i.e., the possibility of accessing and enjoying cultural assets. This scenario concerns the dissemination of information about cultural assets; for example, users can visit a museum or archaeological site and receive on their mobile devices appropriate, personalized information about that place. The second scenario involves cultural assets monitoring, which concerns securely transporting cultural assets from the owner organization to a renter organization and back. The DALICA system [40,41], which implements these scenarios in DALI, has been demonstrated in Villa Adriana (tivoli, Rome) to an international audience of EC officers, local institutions, CUSPIS partners,

---

[13] See [74] to know more about abduction in LP.

[14] See http://ipotesidipreistoria.cib.unibo.it/article/view/1604 (in Italian).

representatives of the Italian Ministry of Cultural Heritage, and a delegation of the Chinese Ministry of Cultural Heritage.

In 2007, Gennari and Mich (Bolzano) started developing an educational Web-based tool called LODE, for teaching learning-impaired children, especially deaf children [73,93,7]. LODE presents children with stories and exercises for reasoning, globally, on the temporal dimension of the stories. On the server side, a first prototype of LODE used the ECLiPSe CLP system to generate exercises and to check their temporal consistency. Since 2009, LODE is a project financed by CARITRO.[15]

3.8   Software Engineering

The Oikos project [4,5,6,36] was mainly carried on by Montangero (Pisa) and Ciancarini (Bologna) in the first half of the 90s. The goal of the Oikos Project was to describe, in a declarative style, the software development processes and to use LP to program and execute software processes.

Oikos is a distributed software development environment where the activities' workflow can be modeled. It is specified and implemented using Extended Shared Prolog (ESP), a parallel logic language that deals with concurrency and distribution. It provides a blackboard-based communication framework in which experiments about different architectures can be performed and evaluated. The processes modeled by Oikos are the multi-user distributed nature, a long life span, open endedness and executability of models. Oikos predefines a number of services offering basic facilities, like access to data bases, workspaces, user interfaces, etc. Services are customizable, in a declarative way that matches naturally the way ESP defines and controls the software process. ESP allows to define services, to structure them in a dynamic hierarchy, and to coordinate them according to the blackboard paradigm.

The project produced a real-case application where it was possible to enact software processes. The example considered a non trivial task of specification of a small language and the implementation of its compiler.

After this project the technology evolved towards the notion of workflow and workflow engine. Notable, in the context of workflows, is the work done by Greco, Guzzo and Saccà (Calabria) on workflow executions [76] in which a rich graph representation of workflow schemes is combined with simple (i.e., stratified), yet powerful DATALOG rules to express complex properties and constraints on executions. The high expressive power of both the graphical and rule-based formalism provides the designer with powerful mechanisms for reasoning on workflows. Another notable body of work is the CLIMB framework (see above) by Montali and colleagues (Bologna), and presented in Montali's PhD thesis [102] which received the GULP 2009 distinguished dissertation award.

3.9   Verification

An application of LP to software verification has been carried on by Bagnara and Zaffanella (Parma), in collaboration with Hill (Leeds) [12]. The project started

---

[15] See http://lode.fbk.eu

in the 90s and is still alive. In 2005 the project switched gear and currently a prototype for industrial application is being developed.

In this work, LP is the framework used for definition, analysis and automatic verification of syntactic and semantic properties of imperative languages (e.g., C, C++ and Java). In particular, the specification of concrete semantics is based on structured operational semantics and it can model runtime exceptions and non-structured flow control mechanisms. The specification, being a logic program, is executable and it is thus possible to verify the adherence of the specification against the reference standards.

Moreover, abstract semantics can be applied as well by implementing abstract interpretation techniques. This creates a general static analyzer that deals with an abstract specification.

Finally, it is possible to define some additional rules that restrict some syntactic and semantic possibilities that are a known possible source of errors, ranging from some unfortunate lexical choices in the language (identifiers containing "l" and "1") to some syntactical rules (each *switch* must have a *default* case) and some runtime errors (e.g., deallocation of null pointers). These kinds of rules are seen as important contribution to standard compilers in industrial applications.

The project produced some prototypes that are currently being merged to a tool for showing the feasibility in industrial applications. Preformance-wise, LP reveals to be successful for these kinds of applications: a prototype is able to verify the compliance of the Linux kernel to 80 coding rules in only a few minutes.

Another important body of work in this area concerns run-time interaction verification in open systems. This relates to the SOCS project (see above). The main outcomes of the project are definitions of the agent and society models and the proof-procedures implementing the operational models. Run-time interaction verification accounts to monitoring and checking whether a particular implemented, running agent does indeed operate according to its specification. The SOCS project has produced logic-based tools that reason upon the externally observable behavior of interacting agents and verify whether it complies to predefined norms or protocols. In particular, the SOCS-SI tool [3] uses the ALP SCIFF proof-procedure [2] to consider messages exchanged by agents plus other events and carries out such a run-time interaction verification task.

3.10 Natural Language Processing

The Natural Language Processing (NLP) application area has flourished in recent years. We report here four different applications that are stimulated by the exponential growth and availability of text in natural language from the Web and repositories of the last years.

Since 1996, Stefano Ferilli (Bari) coordinated a project for a general-purpose system for automatic learning of Datalog programs, starting from positive and negative concepts that are possibly correlated [65,64,63,66]. It supports multi-strategies (namely induction, abduction and abstraction) and it is inherently incremental. It is completely written in Prolog and it is currently being developed and maintained. The system is integrated in the DOMINUS system (intelligent and automatic handling of electronic documents) and included in phases of analysis and elaboration of documents.

Another project, started in 2004, by Bos (La Sapienza) [47], focuses on NLP, in particular on computational semantics, i.e. mapping syntactic structures produced by a parser to first-order languages. The system designed is based on syntactic and semantic formalisms from theoretical linguistics and the implemented prototype is able to analyze the entire Gigaword corpus (1 billion words) in less than 5 days. The system is built around a wide-coverage Combinatory Categorial Grammar (CCG) parser and connected to the Boxer module [19] to produce interpretable structures in the form of Discourse Representation Structures (DRSs).

The resulting open-domain QA system is well suited to analyzing large amounts of text containing a potential answer, because of its efficiency. The grammar is also well suited to analyzing questions, because of CCGs treatment of long-range dependencies. The system is active and available online and it has been downloaded by 800 people until the time of writing.[16]

Another recent project is a prototype for the semantic search module of the LC3 project (MIUR Public-Private Lab). In particular, the sub-goal coordinated by Di Martino (Napoli II) since 2007 is to build a natural language query parser, written in Prolog, that is able to detect conceptual patterns and to build a Query Ontology.[17] The project is still active.

Finally, Mnemosine, started in 2007, is a project run by Costantini and Paolucci (L'Aquila) [42]. They designed a semantic search engine capable of accepting natural language queries as well as answering in natural language. Results are divided in classes of pertinence. The system is based on an iteration of refinement steps, where the user specifies interactively his/her search query. Mnemosine has been implemented into a working prototype and tested on the Italian pages of Wikipedia. The main features related to LP are: the use of SE-DCG, that are an extension of Definite Clause Grammar (DCG) of Prolog; answers to queries are generated from a knowledge base where a Prolog reasoner computes the results. The prototype is stable and scalable and currently being extended.

3.11   Health Care

We have mentioned above the IDEA project on health-care support systems in the early 90s. Many years later, we again find LP laying at the core of health-care applications. We report on three applications that involve tumor prevention, assisted living and classification of clinical diagnoses. They are all quite recent and they show that LP can be effectively employed to improve the quality of life.

The first project, named SPINNER and PRITT SPRING [28], is a collaboration between academic and private parties. The project was developed between

---

[16] See: http://svn.ask.it.usyd.edu.au/trac/candc

[17] See http://lc3.spacespa.it

2004 and 2006 by Mello, Montali, Chesani (Bologna), Storari (Ferrara), in collaboration with Dianoema S.p.A (Bologna).

The goal of the project was to realize a careflow system that implements workflow concepts in the clinical domain in order to administer, support and monitor the execution of health care services performed by different health care professionals and structures. The project concentrated on the monitoring aspects and provided a solution for the conformance verification of careflow process executions.

Given a careflow model, expressed with a simple graphical language for the specification of the careflow (GOSPEL), the system translates it to a formal language based on computational logic and ALP (SCIFF). The main advantage of this formalism lies in its operational proof-theoretic counterpart, which is able to verify the conformance of a given careflow process execution (in the form of an event log) w.r.t. the model.

The feasibility of the approach has been tested on a case study related to the careflow process described in the cervical cancer screening protocol, based on the data provided by regional Health District.

The second application, started in 2007, is part of the *Secure and INDependent LIving* (SINDI) system. The system offers advanced tools for monitoring dynamical, clinical and physical parameters. SINDI caters for two kinds of people: elder people that are clinically stable and chronic patients that can stay at home. The latter type of people often needs to be educated to correct behaviors in order to limit health risks. SINDI offers medical professionals the tools to act before the verification of potential events that may limit the autonomy of the patients. The system is made of three parts: a wireless sensor network, an interface with the patient and a Reasoning Component that is in charge of understanding the context by applying inference rules for meaningful data aggregation and interpretation.

Work by Bisiani, Mileo, Merico and Pinardi (Nomadis Lab, Milano Bicocca) [17,97,98] used non monotonic reasoning as part of SINDI's reasoning component. There is a working prototype and its main reasoning features focus on situation assessment and evaluation of the patient's risk status (by comparing gathered data and clinical knowledge base). Moreover the reasoner depicts possible future scenarios, as prevention and feedback output, based on the current risk status. The latest version of the prototype is currently under experimentation at the Monza Hospital.[18]

Finally, we mention another DLV-based application. OLEX was employed for developing a system able to classify automatically case histories and documents containing clinical diagnoses. The system was commissioned, with the goal of conducting epidemiological analysis, by a local health authority in the Veneto region (ULSS of Asolo). The system classifies available case histories, in order to help the analysts while browsing and searching documents regarding specific pathologies, supplied services, or patients living in a given place etc. The application exploits an ontology of clinical case histories based on both the MESH

---

[18] See www.nomadis.unimib.it/flex/cm/pages/ServeBLOB.php/L/IT/IDPagina/20

(Medical Subject Headings) ontology and ICD9-CM, a system employed by the Italian Ministry of Heath for handling data regarding medical services (e.g. X-Rays analysis, plaster casts, etc.). The analyzed documents are stored in PDF documents and contain medical reports, hospital discharge forms, clinical analysis results etc. Classification rules take into account both the extracted linguistic information and the metadata contained in the case history forms. The system has been deployed and is currently employed by the personnel of the ULSS of Asolo.

3.12   Reasoning

Prolog is a programming language that lends itself particularly well to the implementation of other languages for reasoning. Although this way of exploiting Prolog is not motivated by the needs of industry, it is nonetheless an LP application. The list of languages and Prolog extensions would be very long. We mention only some of them, targeting different domains. Other ones are mentioned in other parts of this chapter and of this book (see, e.g., several implementations of non-monotonic reasoners discussed by Giordano and Toni [74], of agent-oriented languages surveyed by Baldoni and colleagues [13], and of higher-order LP extensions mentioned by Momigliano and Ornaghi [99]).

In the 90s Costantini (L'Aquila), Dell'Acqua (Linköping), Lanzarone (Insubria), Barklund (Uppsala) worked on a Prolog extension, named Reflective Prolog [14]. The system allows to express meta-knowledge and includes an evaluation meta-level that is invoked when needed from the base level. The language supports three different kinds of variables: object variables, predicate meta-variables and function meta-variables. The rules of substitution ensure that these may only be substituted by, respectively, an object term, a representation of a predicate, and a representation of a function. There are syntactic restrictions to keep the meta-levels distinct and prevent self reference within a single atom. A reflective Prolog program distinguishes between the meta-evaluation level and the base level. The former is at the top of the meta-level architecture and the latter, containing an amalgamated theory, comprises the remaining meta-levels below it and can not refer to any predicates in the meta-evaluation level. Procedurally, a definite Reflective Prolog program uses SLD-resolution whenever possible but automatically switches between the levels in certain circumstances. The declarative semantics for such programs, called the least reflective Herbrand Model, is an adapted form of the well-known least Herbrand model. The prototype, written in Quintus, has been later used to start new projects, e.g. DALI.

Started in 2004, Badaloni, Giacomin and Falda (Padova) realized a Temporal Reasoner capable of handling quantitative and qualitative uncertainty and vagueness [10,67]: the qualitative fuzzy temporal constraints are based on the IA^{fuz} framework formalized in [11]. Temporal uncertainty is modeled in terms of possibility distributions and fuzzy relations. A Fuzzy Temporal Constraint

Network is used to represent the knowledge about the considered scenario. Temporal reasoning inferences are performed by checking the consistency of the underlying network. The user interface is written in SWI Prolog, with less than 3K lines of code. The constraint solver is written in C++ and connects to the interface with XML files. In particular, the knowledge base manager normalizes the temporal expressions and defines a method for the consistent interpretation of expressions involving uncertainty. User scenarios are described with a simplified language and passed to the solver by XML files. Solver's output is used to generate answers in the same language. The prototype can handle fuzzy constraints (quantitative intervals and points, both precise and/or uncertain) and to generate temporal expressions similar to natural language.

Since 2007, Costantini (L'Aquila) and Formisano (Perugia) have been developing the P-RASP system (Resourced ASP with Preferences): an extension of ASP to manage reasoning with bounded resources [39]. The authors have developed a P-RASP inference engine.

Finally, there has been conspicuous research on the implementation of engines to reason about action and time. We mention recent work by Dovier (Udine), Formisano (Perugia) and Pontelli (NMSU) [56] on implementation of action languages which makes use of CLP(FD), and the Reactive Event Calculus proposed by Chesani, Mello, Montali and Torroni (Bologna) [31] based on SCIFF (see above). In both cases, the underlying CLP framework is a key factor for achieving a solution with is both declarative and efficient.

3.13   Bioinformatics

Bioinformatics, in broad terms, deals with the use of computational techniques to organize and extract knowledge from biological data. It has successfully addressed problems in areas like recognition and analysis of DNA sequences, biological systems simulations, prediction of the spatial conformation of biological polymers, and ontological analysis of biomedical knowledge. An application of LP to bioinformatics started in 2003 by Dal Palù (Parma), Dovier (Udine), Pontelli (New Mexico State University, US) and Fogolari (Udine). They address the problem of tertiary structure prediction using ab initio techniques, from the perspective of folding a protein sequence in a discretized representation of the three-dimensional space (viewed as a crystal lattice structure), optimizing an objective function which is related to the potential energy function of the resulting configuration. The problem translates into a CSP, where constraints are derived from physical properties of the molecules, and a set of heuristics that explore the search space effectively. A survey on the project is in [52] A prototype was developed using Sicstus Prolog, CLP(FD) and parallelism [50,49]. Another optimized solver was entirely rewritten in C++ [51] and extended traditional FD variables to three dimensional point variables. The work was also presented in Dal Palù's PhD thesis [48] which received the GULP 2006 distinguished dissertation award.

3.14   Decision Support, Risk Analysis and Alarms

A project carried on in the early 90s, by Sardu (System & Management), Serrecchia, Omodeo (La Sapienza), Li, Schuerman, and Véron (ECRC[19]), was an application for Decision Support System (DSS) for the environmental pollution in the Venice lagoon [129]. The project was about the specification and design of an application based on parallel CLP. The DSS includes a database describing pollution sources and a lagoon hydrodynamic model, integrated through a knowledge-based core. The prototyping of the knowledge-based core was implemented in ElipSys (developed by ECRC), a parallel CLP system derived from CHIP [55].

Another application was developed by Avanzini, Rocchesso, Belussi, Dal Palù and Dovier (Verona) [9] and aimed at creating a new auditory alert system for high tides in Venice designed to replace the existing network of electromechanical sirens. The work was developed in collaboration with the Municipality of Venice (Center for Tide Prediction and Warning) in 2003. The project is composed of different parts including the analysis of the current alert system (sound simulation); the realization of a CLP tool to determine the optimal placement of loudspeakers in Venice, a complex task with many physical, economic, and social constraints (modeled with FD variables); the creation of alert sounds for the demanding listening environment. The final phase of the project involved iteratively validating and redesigning the alert signals using human testing. After some years, the project was actually installed in Venice, in particular the location of the loudspeakers followed the results of the optimization program.

A very recent collaboration, started in 2009, between Mascardi, Martelli, Traverso (Genova), and Montolivo (Elsag-Datamat, a FinMeccanica company), focuses on risk analysis of complex infrastructures (harbours, airports, etc). Prolog was used to implement a first prototype for evaluating the feasibility of the approach. The prototype is able to computationally evaluate whether an attacker can violate the security apparatus of a given, simplified, infrastructure. A second prototype, implemented in Java extended with a Prolog-like backtracking mechanism, is much more sophisticated and might develop into a product. The project is protected by a non-disclosure agreement and the patent application has been recently filed.

Within the ARGUGRID project (see above), Mancarella (Pisa), Toni (Imperial College London) and Dung (AIT) led the development of LP-based argumentation engines, in Prolog, to identify "best" decisions in uncertain environments. The decisions may be supported by assumptions (similar to abducibles in ALP) and the rationale for decisions is presented to users in the form of a debate (arguments and counter-arguments). These engines (MARGO[20] [105], CaSAPI[21] [69],

---

[19] ECRC (European Computer-Industry Research Centre GmbH, Munich, Germany) was a a joint venture of Bull, ICL and Siemens, formed in 1984 to research new software technologies.

[20] See http://margo.sourceforge.net

[21] See http://www.doc.ic.ac.uk/~dg00/casapi.html

and MoDiSo[22] [61]) all extend, albeit in different ways and with different aims, the abductive proof procedures for LP and ALP of [62] and [78] respectively. These systems have been deployed as follows:

- MARGO for supporting the decision of the most suitable type of electronic auction to be used by a seller/buyer in an e-marketplace [106] (in collaboration with cosmoONE Hellas MarketSite S.A, Greece[23]);
- CaSAPI for selecting an e-ordering system [90](in collaboration cosmoONE Hellas MarketSite S.A, Greece) and for selecting satellites for the acquisitions fo best images [91] (in collaboration with GMV S.A., Spain[24]);
- MoDiSo for supporting the resolution of legal disputes [61].

4   Conclusions

In conclusion, LP applications are many and diverse in several, traditional and new application domains. This survey suffers from our poll methodology, by which most of the LP applications above are of academic inspiration. We believe, however, that the actual landscape is not too different from the one we depicted in this chapter.

Notwithstanding the increased education and diffusion of LP at the students level, there is still a remarkable gap between the growth of academic research and industrial applications. In our opinion, this may be due to the difficult international situation of computer and software industries, worsened by a specific Italian weakness in advanced industrial research, the crisis of AI technology and its influence on LP.

On the upside, our work emphasizes many collaborations between research groups and industrial and commercial partners, which makes us believe that now time is ripe for pushing the adoption of LP outside of academic entourage. While most of the private investors that were interested in LP 25 years ago have apparently left the stage, other new actors are coming into play. Exeura s.r.l. is a successful example of a company that is actually doing business and providing services with LP technology. There are many collaborations and projects with Public Administrations, such as municipalities and hospitals, that rely on LP and extensions. Indeed, we are now in a very different situation from that of 15 year ago. The main obstacles to LP adoptions, such as lack of LP-education and problems of efficiency and integration, seem to have been overcome in many cases. 15 years ago we were wondering why LP was not used and what was missing, while today we can get some insights from many success stories.

It is still true that average programmers and engineers are unable to write (correct and efficient) declarative programs, although we believe that the situation is better than it used to be. Programming methodologies and environments, debugging techniques, friendly interfaces did not evolve significantly compared

---

[22] See http://www.cs.ait.ac.th/~dung/modiso/About.html
[23] See http://www.cosmo-one.gr/en
[24] See http://www.gmv.com

to those of other popular imperative languages. However, these issues are confined to the production of LP-based solutions and do not affect the quality of the solutions themselves. LP technologies can now rely on efficient implementations, and offer unique degrees of flexibility. We can observe that the current trend is to develop competitive LP-based solutions for hard problems, which requires a solid background, education and high programming skills. This high quality profile, in the perspective of market globalization and considering the constant increase in the number of new and complex applications, is not necessarily a negative and penalizing aspect. We believe, instead, that competencies in declarative programming will become even more valuable in the next years.

Acknowledgements. We would like to thank the anonymous reviewers and all the colleagues who helped us by providing insight, feedback, comments, material and summary of their activities. A particular thank to (in alphabetical order): Roberto Bagnara, Johan Bos, Giorgio Casadei, Paolo Ciancarini, Marco Colombetti, Stefania Costantini, Beniamino Di Martino, Agostino Dovier, Marco Falda, Stedano Ferilli, Marco Gavanelli, Rosella Gennari, Giuseppina Gini, Maria Gini, Viviana Mascardi, Paola Mello, Vitaliano Milanese, Alessandra Mileo, Alberto Momigliano, Angelo Monfroglio, Marco Montali, Alessandro Mosca, Daniele Nardi, Andrea Omicini, Enrico Pagello, Francesco Ricca, Carlo Matteo Scalzo, Pietro Terna, and Francesca Toni.

References

1. Alberti, M., Chesani, F., Gavanelli, M., Lamma, E., Mello, P., Torroni, P.: The SOCS computational logic approach to the specification and verification of agent societies. In: Global Computing, pp. 314–339 (2004)
2. Alberti, M., Chesani, F., Gavanelli, M., Lamma, E., Mello, P., Torroni, P.: Verifiable agent interaction in abductive logic programming: The SCIFF framework. ACM Trans. Comput. Log. 9(4) (2008)
3. Alberti, M., Gavanelli, M., Lamma, E., Chesani, F., Mello, P., Torroni, P.: Compliance verification of agent interaction: a logic-based software tool. Applied Artificial Intelligence 20(2-4), 133–157 (2006)
4. Ambriola, V., Ciancarini, P., Montangero, C.: Enacting software processes in Oikos. In: Software Development Environments. SIGSOFT, vol. 15(6), pp. 12–23 (1990)
5. Ambriola, V., Ciancarini, P., Montangero, C.: The logic language ESP and its programming environment. In: Workshop on Logic Programming Environments, Technical Report IR-LP-31-25 of ECRC (June 1990)
6. Ambriola, V., Ciancarini, P., Montangero, C.: Software processes as a hierarchy of services in the Oikos meta environment. In: Soft. Proc. Workshop, pp. 57–60 (1990)
7. Arfé, B., Gennari, R., Mich, O.: Before, while and after with LODE and hearing novice readers. Tech. Rep. KRDB09-1, University of Bolzano (2009)
8. Attardi, G., Simi, M.: A description-oriented logic for building knowledge bases. IEEE 74(10) (1986)
9. Avanzini, F., Rocchesso, D., Belussi, A., Dal Palù, A., Dovier, A.: Designing an urban-scale auditory alert system. Computer 37(9), 55–61 (2004)

10. Badaloni, S., Falda, M., Giacomin, M.: Integrating quantitative and qualitative constraints in fuzzy temporal networks. AI Communications 17(4), 183–272 (2004)

11. Badaloni, S., Giacomin, M.: The algebra IA^{fuz}: a framework for qualitative fuzzy temporal reasoning. Artificial Intelligence 170(10), 872–908 (2006)

12. Bagnara, R., Hill, P.M., Pescetti, A., Zaffanella, E.: On the design of generic static analyzers for imperative languages. Quaderno 485, Dipartimento di Matematica, Università di Parma, Italy (2008)

13. Baldoni, M., Baroglio, C., Mascardi, V., Omicini, A., Torroni, P.: Agents, Multi-Agent Systems and Declarative Programming: What, When, Where, Why, Who, How? In: Dovier, A., Pontelli, E. (eds.) 25 Years of Logic Programming in Italy, ch. 10. LNCS, vol. 6125, pp. 204–230. Springer, Heidelberg (2010)

14. Barklund, J., Costantini, S., Dell'Acqua, P., Lanzarone, G.: Reflection principles in computational logic. Journal of Logic and Computation 10, 6 (December 2000)

15. Barták, R., Milano, M. (eds.): CPAIOR 2005. LNCS, vol. 3524. Springer, Heidelberg (2005)

16. Bazzocchi, L.: Lo SMAU scopre l'intelligenza artificiale. Office Automation, 86–90 (November 1988)

17. Bisiani, R., Merico, D., Mileo, A., Pinardi, S.: A logical approach to home health-care with intelligent sensor-network support. The Comp. J. Adv. Access (2009)

18. Bonatti, P., Calimeri, F., Leone, N., Ricca, F.: Answer Set Programming. In: 25 Years of Logic Programming in Italy. LNCS, vol. 6125, pp. 159–182. Springer, Heidelberg (2010)

19. Bos, J.: Towards wide-coverage semantic interpretation. In: IWCS-6, pp. 42–53 (2005)

20. Bracciali, A., Endriss, U., Demetriou, N., Kakas, A.C., Lu, W., Stathis, K.: Crafting the mind of PROSOCS agents. Appl. Artif. Intelligence 20(2-4), 105–131 (2006)

21. Briola, D., Mascardi, V., Martelli, M., Arecco, G., Caccia, R., Milani, C.: A Prolog-based MAS for railway signalling monitoring: Implementation and experiments. In: WOA 2008 (2008)

22. Bromuri, S., Stathis, K.: Situating Cognitive Agents in GOLEM. In: Weyns, D., Brueckner, S.A., Demazeau, Y. (eds.) EEMMAS 2007. LNCS (LNAI), vol. 5049, pp. 115–134. Springer, Heidelberg (2008)

23. Bromuri, S., Stathis, K.: Distributed agent environments in the ambient event calculus. In: Gokhale, A.S., Schmidt, D.C. (eds.) DEBS. ACM, New York (2009)

24. Bryl, V., Mello, P., Montali, M., Torroni, P., Zannone, N.: B-Tropos: Agent-oriented requirements engineering meets computational logic for declarative business process modelling and verification. In: Sadri, F., Satoh, K. (eds.) CLIMA VIII 2007. LNCS (LNAI), vol. 5056, pp. 157–176. Springer, Heidelberg (2008)

25. Calisi, D., Iocchi, L., Nardi, D., Scalzo, C.M., Ziparo, V.A.: Context-based design of robotic systems. Robotics and Autonomous Systems 56(11), 992–1003 (2008)

26. Chesani, F., Lamma, E., Mello, P., Montali, M., Riguzzi, F., Storari, S.: Exploiting inductive logic programming techniques for declarative process mining. In: Jensen, K., van der Aalst, W.M.P. (eds.) Transactions on Petri Nets and Other Models of Concurrency II. LNCS, vol. 5460, pp. 278–295. Springer, Heidelberg (2009)

27. Chesani, F., Mello, P., Montali, M., Riguzzi, F., Sebastianis, M., Storari, S.: Checking compliance of execution traces to business rules. In: Business Process Management Workshops, pp. 134–145 (2008)

28. Chesani, F., Mello, P., Montali, M., Storari, S.: Testing careflow process execution conformance by translating a graphical language to computational logic. In:

Bellazzi, R., Abu-Hanna, A., Hunter, J. (eds.) AIME 2007. LNCS (LNAI), vol. 4594, pp. 479–488. Springer, Heidelberg (2007)

29. Chesani, F., Mello, P., Montali, M., Storari, S., Torroni, P.: On the integration of declarative choreographies and commitment-based agent societies into the SCIFF logic programming framework. Multiagent and Grid Systems 2 (2010)

30. Chesani, F., Mello, P., Montali, M., Torroni, P.: Verification of choreographies during execution using the Reactive Event Calculus. In: Bruni, R., Wolf, K. (eds.) WS-FM 2009. LNCS, vol. 5387, pp. 55–72. Springer, Heidelberg (2009)

31. Chesani, F., Mello, P., Montali, M., Torroni, P.: Commitment tracking via the reactive event calculus. In: Boutilier, C. (ed.) IJCAI, pp. 91–96 (2009)

32. Chiopris, C.: The SECReTS banking expert system from phase 1 to phase 2. In: Comyn, G., Ratcliffe, M.J., Fuchs, N.E. (eds.) LPSS 1992. LNCS, vol. 636, pp. 91–99. Springer, Heidelberg (1992)

33. Ciampolini, A., Lamma, E., Mello, P., Toni, F., Torroni, P.: Co-operation and competition in ALIAS: a logic framework for agents that negotiate. Annals of Mathematics and Artificial Intelligence 37(1-2), 65–91 (2003)

34. Ciampolini, A., Lamma, E., Mello, P., Torroni, P.: LAILA: a language for coordinating abductive reasoning among logic agents. Comp. Lang. 27(4), 137–161 (2001)

35. Ciampolini, A., Torroni, P.: Using abductive logic agents for modeling the judicial evaluation of criminal evidence. Appl. Artif. Intelligence 18(3-4), 251–275 (2004)

36. Ciancarini, P.: Coordinating rule-based software processes with ESP. ACM Trans on Sw Engineering and Methodolgy 2(3), 203–227 (1993)

37. Concheri, G., Milanese, V.: Interaction as an issue in the development of effective tools for the management of the engineering knowledge base. In: XI ADM Conference, vol. B, pp. 101–108 (1999)

38. Concheri, G., Milanese, V.: MIRAGGIO: a system for the dynamic management of product data and design models. Advances in Engineering Software 32(7), 527–543 (2001)

39. Costantini, S., Formisano, A.: Modeling preferences and conditional preferences on resource consumption and production in ASP. Algorithms 64(1), 3–15 (2009)

40. Costantini, S., Mostarda, L., Tocchio, A., Tsintza, P.: User profile agents applied to a cultural heritage scenario. In: SEKE (2007)

41. Costantini, S., Mostarda, L., Tocchio, A., Tsintza, P.: DALICA: Agent-based ambient intelligence for cultural-heritage scenarios. IEEE Intelligent Systems 23(2), 34–41 (2008)

42. Costantini, S., Paolucci, A.: Semantically augmented DCG analysis for next-generation search engine. In: CILC (July 2008)

43. Costantini, S., Tocchio, A.: The DALI logic programming agent-oriented language. In: Alferes, J.J., Leite, J. (eds.) JELIA 2004. LNCS (LNAI), vol. 3229, pp. 685–688. Springer, Heidelberg (2004)

44. Cucchiara, R., Gavanelli, M., Lamma, E., Mello, P., Milano, M., Piccardi, M.: Constraint propagation and value acquisition: Why we should do it interactively. In: IJCAI, pp. 468–477 (1999)

45. Cucchiara, R., Gavanelli, M., Lamma, E., Mello, P., Milano, M., Piccardi, M.: From eager to lazy constrained data acquisition: A general framework. New Generation Comput. 19(4), 339–368 (2001)

46. Cumbo, C., Iiritano, S., Rullo, P.: OLEX – A reasoning-based text classifier. In: Alferes, J.J., Leite, J. (eds.) JELIA 2004. LNCS (LNAI), vol. 3229, pp. 722–725. Springer, Heidelberg (2004)

47. Curran, J.R., Clark, S., Bos, J.: Linguistically motivated large-scale NLP with C&C and Boxer. In: ACL, pp. 29–32 (2007)
48. Dal Palù, A.: Constraint Programming approaches to the Protein Structure Prediction Problem. PhD thesis, University of Udine (2006)
49. Dal Palù, A., Dovier, A., Fogolari, F.: Constraint logic programming approach to protein structure prediction. BMC Bioinformatics 5(1), 186 (2004)
50. Dal Palù, A., Dovier, A., Pontelli, E.: Heuristics, optimizations, and parallelism for protein structure prediction in CLP(FD). In: PPDP, pp. 230–241 (2005)
51. Dal Palù, A., Dovier, A., Pontelli, E.: A constraint solver for discrete lattices, its parallelization, and application to protein structure prediction. Softw. Pract. Exper. 37(13), 1405–1449 (2007)
52. Dal Palù, A., Dovier, A., Pontelli, E.: Logic programming techniques in protein structure determination: Methodologies and results. In: Erdem, E., Lin, F., Schaub, T. (eds.) LPNMR 2009. LNCS, vol. 5753, pp. 560–566. Springer, Heidelberg (2009)
53. Demetriou, N., Kakas, A.C.: Argumentation with abduction. In: Panhellenic Symposium on Logic (2003)
54. Denti, E., Omicini, A., Ricci, A.: tuProlog: A light-weight Prolog for internet applications and infrastructures. In: Ramakrishnan, I.V. (ed.) PADL 2001. LNCS, vol. 1990, pp. 184–198. Springer, Heidelberg (2001)
55. Dorochevsky, M., Li, L.-L., Reeve, M., Schuerman, K., Véron, A.: ElipSys - a parallel programming system based on logic. In: Voronkov, A. (ed.) LPAR 1992. LNCS, vol. 624, pp. 469–471. Springer, Heidelberg (1992)
56. Dovier, A., Formisano, A., Pontelli, E.: Multi-valued action languages with constraints in CLP(FD). Theory and Practice of Logic Programming 10, 167–235 (2010)
57. Dovier, A., Pontelli, E. (eds.): 25 Years of Logic Programming in Italy. LNCS, vol. 6125. Springer, Heidelberg (2010)
58. Dulli, S., Galbiati, G., Milanese, V.: Hierarchical data structures and geometric modeling: a unified approach. YUGRAPH 31(1/2), 37–42 (1990)
59. Dulli, S., Milanese, V.: A graphic programming environment based on KADMOS. Comput. Graph. Forum 11(1), 3–16 (1992)
60. Dulli, S., Milanese, V., Visentin, A.: A multiple windows user interface. In: CAD/Graphics New Advances in Computer Aided Design, pp. 186–188 (1993)
61. Dung, P.M., Thang, P.M.: Modular argumentation for modelling legal doctrines in common law of contract. Artificial Intelligence and Law 17(3) (June 2009)
62. Eshghi, K., Kowalski, R.A.: Abduction compared with negation by failure. In: ICLP, pp. 234–254 (1989)
63. Esposito, F., Fanizzi, N., Ferilli, S., Basile, T., Mauro, N.D.: Incremental multistrategy learning for document processing. Applied AI 17(8/9), 859–883 (2003)
64. Esposito, F., Fanizzi, N., Ferilli, S., Basile, T., Mauro, N.D.: Incremental learning and concept drift in INTHELEX. Intelligent Data Analysis J. 8(3), 213–237 (2004)
65. Esposito, F., Fanizzi, N., Ferilli, S., Basile, T., Mauro, N.D.: Multistrategy operators for relational learning and their cooperation. Fund. Inf. 69(4), 389–409 (2006)
66. Esposito, F., Fanizzi, N., Ferilli, S., Mauro, N.D.: Multistrategy theory revision: Induction and abduction in INTHELEX. Machine Learning Journal 38(1/2), 133–156 (2000)
67. Falda, M.: Translating fuzzy temporal constraints in more natural expressions. In: ECAI 2008 workshop on Spatial and Temporal Reasoning, pp. 11–15 (2008)

68. Farenzena, M., Fusiello, A., Dovier, A.: Reconstruction with interval constraints propagation. In: CVPR, pp. 1185–1190 (2006)
69. Gaertner, D., Toni, F.: Hybrid argumentation and its properties. In: COMMA, pp. 183–195 (2008)
70. Galizia, S.: Generazione automatica di manovre per lo space shuttle mediante la programmazione logica disgiuntiva. In: APPIA-GULP-PRODE, pp. 97–109 (2003)
71. Garro, A., Palopoli, L., Ricca, F.: Exploiting agents in e-learning and skills management context. AI Commun. 19(2), 137–154 (2006)
72. Gavanelli, M.: University timetabling in ECLiPSe. ALP Newsletter 19(3) (August 2006)
73. Gennari, R., Mich, O.: LODE: A logic-based e-learning tool for deaf children. Tech Rep. KRDB07-3, University of Bolzano (2007)
74. Giordano, L., Toni, F.: Knowledge representation and non-monotonic reasoning. In: Dovier, A., Pontelli, E. (eds.) 25 Years of Logic Programming in Italy. LNCS, vol. 6125, pp. 87–111. Springer, Heidelberg (2010)
75. Grasso, G., Iiritano, S., Leone, N., Ricca, F.: Some DLV applications for knowledge management. In: Erdem, E., Lin, F., Schaub, T. (eds.) LPNMR 2009. LNCS, vol. 5753, pp. 591–597. Springer, Heidelberg (2009)
76. Greco, G., Guzzo, A., Saccà, D.: A logic framework for reasoning on workflow executions. In: AGP 2004 (2004)
77. Ianni, G., Panetta, C., Ricca, F.: Specification of assessment-test criteria through ASP specifications. In: Answer Set Programming, CEUR Workshop 142 (2005)
78. Kakas, A.C., Mancarella, P.: Generalized stable models: a semantics for abduction. In: ECAI, pp. 385–391 (1990)
79. Kakas, A.C., Mancarella, P., Sadri, F., Stathis, K., Toni, F.: Computational logic foundations of KGP agents. J. Artif. Intell. Res. (JAIR) 33, 285–348 (2008)
80. Leone, N.: Exploiting ASP in real-world applications: Main strengths and challenges. In: Erdem, E., Lin, F., Schaub, T. (eds.) LPNMR 2009. LNCS, vol. 5753, pp. 628–630. Springer, Heidelberg (2009)
81. Leone, N., Greco, G., Ianni, G., Lio, V., Terracina, G., Eiter, T., Faber, W., Fink, M., Gottlob, G., Rosati, R., Lembo, D., Lenzerini, M., Ruzzi, M., Kalka, E., Nowicki, B., Staniszkis, W.: The INFOMIX system for advanced integration of incomplete and inconsistent data. In: ACM SIGMOD, pp. 915–917 (2005)
82. Leone, N., Pfeifer, G., Faber, W., Eiter, T., Gottlob, G., Perri, S., Scarcello, F.: The DLV system for knowledge representation and reasoning. ACM Trans. Comput. Logic 7(3), 499–562 (2006)
83. Luccarini, L., Bragadin, G.L., Mancini, M., Mello, P., Montali, M., Sottara, D.: Formal verification of wastewater treatment processes using events detected from continuous signals by means of artificial neural networks. Environmental Modelling and Software (2009) (in press)
84. Ma, J., Russo, A., Broda, K., Clark, K.: DARE: a system for distributed abductive reasoning. Autonomous Agents and Multi-Agent Systems 16(3), 271–297 (2008)
85. Mancarella, P., Terreni, G., Sadri, F., Toni, F., Endriss, U.: The CIFF proof procedure for abductive logic programming with constraints: Theory, implementation and experiments. CoRR, abs/0906.1182 (2009)
86. Mantegari, G., Mosca, A., Cattani, M.: Formal knowledge representation and automated reasoning for the study of archaeological stratigraphy. In: 12th International Congress Cultural Heritage and New Technologies (2007)

87. Mantegari, G., Mosca, A., Rondelli, B., Vizzari, G.: A semantic based approach to GIS: the PO-BASyN project. In: Computer Applications and Quantitative Methods in Archaeology (2008)

88. Martelli, M.: Constraint logic programming: Theory and applications. In: Sessa [130], pp. 137–166

89. Mascardi, V., Briola, D., Martelli, M., Caccia, R., Milani, C.: Monitoring and diagnosing railway signalling with logic-based distributed agents. In: MFCS 1977. LNCS, vol. 53, pp. 108–115. Springer, Heidelberg (2009)

90. Matt, P.-A., Toni, F., Stournaras, T., Dimitrelos, D.: Argumentation-based agents for eprocurement. In: AAMAS (Industry Track), pp. 71–74 (2008)

91. Matt, P.-A., Toni, F., Vaccari, J.: Dominant Decisions by Argumentation Agents. In: Workshop ArgMAS (2009)

92. Meneghetti, A.: Optimizing allocation in floor storage systems for the shoe industry by Constraint Logic Programming. In: ISDA, pp. 467–472. IEEE, Los Alamitos (2009)

93. Mich, O.: Constraint-based temporal reasoning and e-learning tools for deaf users. Tech Rep KRDB08-1, University of Bolzano (2008)

94. Milanese, V.: A Prolog environment for GKS-based graphics. Comput. Graph. Forum 7(1), 9–20 (1988)

95. Milanese, V.: KADMOS: A clausal language for CAD modeling systems with morphological constraints. Comput. Graph. Forum 9(1), 39–51 (1990)

96. Milanese, V.: Using semantics in engineering design. In: CIM, pp. 369–378 (2003)

97. Mileo, A., Merico, D., Bisiani, R.: A logic programming approach to home monitoring for risk prevention in assisted living. In: Garcia de la Banda, M., Pontelli, E. (eds.) ICLP 2008. LNCS, vol. 5366, pp. 145–159. Springer, Heidelberg (2008)

98. Mileo, A., Merico, D., Bisiani, R.: Wireless sensor networks supporting context-aware reasoning in assisted living. In: PETRA, p. 54 (2008)

99. Momigliano, A., Ornaghi, M.: Proof-theoretic and Higher-order Extensions of Logic Programming. In: Dovier, A., Pontelli, E. (eds.) 25 Years of Logic Programming in Italy, ch. 12. LNCS, vol. 6125, pp. 254–270. Springer, Heidelberg (2010)

100. Monfroglio, A.: Timetabling through a deductive database: a case study. Data and Knowledge Engineering 3(1), 1–27 (1988)

101. Montagna, S., Ricci, A., Omicini, A.: A&A for modelling and engineering simulations in systems biology. Int. J. Agent-Oriented Softw. Eng. 2(2), 222–245 (2008)

102. Montali, M.: Specification and Verification of Open Declarative Interaction Models: a Logic-Based Framework. PhD thesis, University of Bologna (2009)

103. Montali, M., Pesic, M., van der Aalst, W.M.P., Chesani, F., Mello, P., Storari, S.: Declarative specification and verification of service choreographies. ACM Transactions on the Web (2009)

104. Montali, M., Torroni, P., Alberti, M., Chesani, F., Gavanelli, M., Lamma, E., Mello, P.: Verification from declarative specifications using logic programming. In: Garcia de la Banda, M., Pontelli, E. (eds.) ICLP 2008. LNCS, vol. 5366, pp. 440–454. Springer, Heidelberg (2008)

105. Morge, M.: The hedgehog and the fox. In: Rahwan, I., Parsons, S., Reed, C. (eds.) Argumentation in Multi-Agent Systems. LNCS (LNAI), vol. 4946, pp. 114–131. Springer, Heidelberg (2008)

106. Morge, M., Mancarella, P., Stournaras, T.: Argumentation pour la sélection et la négociation de services. cas d'étude de télé-procédure. In: JFSMA, pp. 149–158 (2008)

107. Mosca, A., Bernini, D.: Ontology-driven geographic information system and dlvhex reasoning for material culture analysis. In: RCRA (2008)
108. Mosca, A., Rondelli, Mantegari, G.: Integrating a knowledge-based system and a geographical information system for the study of the archaeological material culture. In: Cultural Heritage Workshop, AIxIA, pp. 84–91 (2008)
109. Natali, A., Omicini, A., Zanichelli, F.: Exploiting logic programming in robot applications. In: GULP, pp. 535–548 (1993)
110. Nogueira, M., Balduccini, M., Gelfond, M., Watson, R., Barry, M.: An A-Prolog decision support system for the Space Shuttle. In: Ramakrishnan, I.V. (ed.) PADL 2001. LNCS, vol. 1990, pp. 169–183. Springer, Heidelberg (2001)
111. Oliva, E., Viroli, M., Omicini, A.: Simulation of minority game in TuCSoN. In: WOA, pp. 6–9 (2006)
112. Omicini, A., Denti, E.: From tuple spaces to tuple centres. Sci. Comput. Program. 41(3), 277–294 (2001)
113. Omicini, A., Zambonelli, F.: Coordination of mobile information agents in TuC-SoN. Internet Research: El. Networking Appl. and Policy 8(5), 400–413 (1998)
114. Omicini, A., Zambonelli, F.: Coordination for internet application development. Autonomous Agents and Multi-Agent Systems 2(3), 251–269 (1999)
115. Pesic, M., van der Aalst, W.M.P.: A declarative approach for flexible business processes management. In: Business Process Management Workshops, pp. 169–180 (2006)
116. Piancastelli, G., Benini, A., Omicini, A., Ricci, A.: The architecture and design of a malleable object-oriented Prolog engine. In: SAC, pp. 191–197 (2008)
117. Ricca, F., Gallucci, L., Schindlauer, R., Dell'Armi, T., Grasso, G., Leone, N.: OntoDLV: an ASP-based system for enterprise ontologies. J. Log. and Comput. 19(4), 643–670 (2009)
118. Ricca, F., Leone, N.: Disjunctive logic programming with types and objects: The DLV+ system. Journal of Applied Logics 5(3), 545–573 (2007)
119. Ricci, A., Omicini, A.: Agent coordination contexts: Experiments in TuCSoN. In: WOA, pp. 14–21 (2002)
120. Ricci, A., Omicini, A., Denti, E.: The TuCSoN coordination infrastructure for virtual enterprises. In: WETICE, pp. 348–353 (2001)
121. Rossi, G.: Logic Programming in Italy: A Historical Perspective. In: Dovier, A., Pontelli, E. (eds.) 25 Years of Logic Programming in Italy, ch. 1. LNCS, vol. 6125, pp. 1–14. Springer, Heidelberg (2010)
122. Ruffolo, M., Leone, N., Manna, M., Saccà, D., Zavatto, A.: Exploiting ASP for semantic information extraction. In: Answer Set Programming, pp. 248–262 (2005)
123. Ruffolo, M., Manna, M.: HiLeX: A system for semantic information extraction from Web documents. Enterprise Information Systems 3(3), 194–209 (2008)
124. Ruggieri, C., Sancassani, M., Dore, N., Russo, F., Manfredi, U.: Intelligent data retrieval in Prolog: An illuminating idea. J. Log. Program. 26(2), 169–198 (1996)
125. Rullo, P., Policicchio, V.L., Cumbo, C., Iiritano, S.: OLEX: Effective rule learning for text categorization. Knowledge and Data Engineering 21, 1118–1132 (2008)
126. Sadri, F., Stathis, K., Toni, F.: Normative KGP agents. Comput. Math. Organ. Theory 12(2-3), 101–126 (2006)
127. Sadri, F., Toni, F., Torroni, P.: Dialogues for negotiation: Agent varieties and dialogue sequences. In: Meyer, J.-J.C., Tambe, M. (eds.) ATAL 2001. LNCS (LNAI), vol. 2333, pp. 405–421. Springer, Heidelberg (2002)
128. Sadri, F., Toni, F., Torroni, P.: An abductive logic programming architecture for negotiating agents. In: Flesca, S., Greco, S., Leone, N., Ianni, G. (eds.) JELIA 2002. LNCS (LNAI), vol. 2424, pp. 419–431. Springer, Heidelberg (2002)

129. Sardu, G., Serrecchia, G., Omodeo, E., Li, L., Schuerman, K., Véron, A.: Safeguarding the Venice lagoon: Ann aplication of a knowledge-based DSS. In: GULP, pp. 519–534 (1993)

130. Sessa, M. (ed.): 1985 – 1995: Ten years of Logic Programming in Italy. Palladio, Salerno (1995)

131. Stathis, K., Toni, F.: Ambient intelligence using KGP agents. In: Markopoulos, P., Eggen, B., Aarts, E., Crowley, J.L. (eds.) EUSAI 2004. LNCS, vol. 3295, pp. 351–362. Springer, Heidelberg (2004)

132. Terna, P.: Rassegna di strumenti informatici. Giappichelli, Torino (1988)

133. Toni, F., Torroni, P. (eds.): CLIMA 2005. LNCS (LNAI), vol. 3900. Springer, Heidelberg (2006)

134. Torquati, F., Paltrinieri, M., Momigliano, A.: A constraint satisfaction approach to operative management of aircraft routing. In: IEA/AIE, vol. 2, pp. 1140–1146 (1990)

135. Torroni, P., Gavanelli, M., Chesani, F.: Argumentation in the Semantic Web. IEEE Intelligent Systems 22(6), 66–74 (2007)

136. Vasey, P.: Prolog++ 2.0–Programmer Reference. Logic Programming Associates, London (1995)

Author Index

Printing: Mercedes-Druck, Berlin
Binding: Stein+Lehmann, Berlin